THE
GEOGRAPHY
OF
STRABO

LITERALLY TRANSLATED
BY
H. C. HAMILTON
AND
W. FALCONER, M.A.,

IN THREE VOLUMES.
VOL. III..

STRABO, the author of this work, was born at Amasia, or Amasijas, a town situated in the gorge of the mountains through which passes the river Iris, now the Ieschil Irmak, in Pontus, which he has described in the 12th book.[1] He lived during the reign of Augustus, and the earlier part of the reign of Tiberius; for in the 13th book[2] he relates how Sardes and other cities, which had suffered severely from earthquakes, had been repaired by the provident care of Tiberius the present Emperor; but the exact date of his birth, as also of his death, are subjects of conjecture only. Coraÿ and Groskurd conclude, though by a somewhat different argument, that he was born in the year B. C. 66, and the latter that he died A. D. 24. The date of his birth as argued by Groskurd, proceeds on the assumption that Strabo was in his thirty-eighth year when he went from Gyaros to Corinth, at which latter place Octavianus Cæsar was then staying on his return to Rome after the battle of Actium, B. C. 31. We may, perhaps, be satisfied with following Clinton, and place it not later than B. C. 54.

In the 17th book our author speaks of the death of Juba as a recent occurrence. This event took place A. D. 21, or A. D. 18 or 19, according to other chronologists; he, therefore, outlived that king, but for how long a period we have no means of ascertaining.

The only information which we can obtain of the personal history of Strabo is to be collected from the scanty references made to himself in the course of this work;[3] for although a writer of the Augustan age, his name and his works appear [Pg vi]to have been generally unknown to his contemporaries, and to have been passed over in silence by subsequent authors who occupied themselves with the same branch of study. The work being written in Greek, and the subject itself not of a popular kind, would be hindrances to its becoming generally known; and its voluminous character would prevent many copies being made; moreover, the author himself, although for some time a resident at Rome, appears to have made Amasia his usual place of residence, and there to have composed his work. But wherever it was, he had the means of becoming acquainted with the chief public events that took place in the Roman Empire.

It is remarkable that of his father and his father's family he is totally silent, but of his mother and her connexions he has left us some notices. She was of a distinguished family who had settled at Cnossus in Crete, and her ancestors had been intimately connected with Mithridates Euergetes and Mithridates Eupator, kings of Pontus; their fortunes consequently depended on those princes.

Dorylaüs, her great grandfather, was a distinguished officer, and friend of Euergetes; but the latter being assassinated at Sinope, whilst Dorylaüs was engaged in levying troops in Crete, he determined to remain there. In that island he obtained the highest honours, having successfully, as general of the Cnossians, terminated a war between that people and the Gortynians. He married a Macedonian lady, of the name of Sterope; the issue of which marriage was Lagetas, Stratarchas, and a daughter. He died in Crete. Lagetas had a daughter, who, says Strabo, was "the mother of my mother."

Mithridates Eupator, who succeeded to the kingdom of Pontus on the death of his father, had formed from infancy a close friendship with another Dorylaüs, son of Philetærus (brother of the first-mentioned Dorylaüs), and besides conferring on him distinguished honours, appointed him high priest of Comana Pontica. The king extended also his protection to his cousins, Lagetas and Stratarchas, who were recalled from Crete. The prosperity of the family suddenly terminated by the discovery of an intrigue carried on by Dorylaüs with the Romans, for the overthrow of his benefactor. The motives assigned by Strabo for his disaffection and treachery were the declining [Pg vii]prospects of the king, and the execution of his son Theophilus and a nephew Tibius.

Dorylaüs made overtures to Lucullus for the revolt of the kingdom of Pontus to the Romans, and in return received great promises of reward, which were never fulfilled. Lucullus ceased to command in the war, and was succeeded by Pompey, who, through enmity and jealousy, prevailed on the senate not to confirm the conditions entered into by his predecessor. As before observed, there is no mention of Strabo's father in the works which have come down to us. Malte-Brun, in his Life of Strabo in the Biographie Universelle, collects several passages tending to show that he was a Roman. The name of

Strabo, or "squinting," originally Greek, was used by the Romans, and applied to the father of Pompey the Great, among others. How the geographer acquired this name is not related.

When a very young man, he received instruction in grammar and rhetoric from Aristodemus, at Nysa in Caria.[4] He afterwards studied philosophy under Xenarchus of Seleucia, the Peripatetic philosopher.[5] Strabo does not say whether he heard him at Seleucia in Cilicia, or at Rome, where he afterwards taught.

Strabo also attended the lessons of Tyrannio of Amisus,[6] the grammarian. This must have been at Rome; for Tyrannio was made prisoner by Lucullus, B. C. 71, and carried to Rome, probably not later than B. C. 66.

In book xvi.,[7] Strabo states that he studied the philosophy of Aristotle with Boethus of Sidon, who afterwards became a Stoic philosopher. Notwithstanding all these advantages, Strabo was not possessed of all the knowledge of his times, particularly in astronomy and mathematics, but he was well acquainted with history and the mythological traditions of his nation. He was a devout admirer of Homer, and acquainted with the other great poets.

The philosophical sect to which he belonged was the Stoic, as plainly appears from many passages in his Geography.

He wrote a History, which he describes (vol. i. p. 21) as composed in a lucid style; it is cited by Plutarch, and also by [Pg viii]Josephus in his Jewish Antiquities, xiv. 7. It consisted of forty-three books, which began where the history of Polybius ended, and was probably continued to the battle of Actium. This valuable History is lost.

Strabo was a great traveller, and apparently had no professional or other occupation. We may therefore conclude that his father left him a good property. Much of his geographical information is the result of personal observation. In a passage of his 2nd book[8] he thus speaks: "Our descriptions shall consist of what we ourselves have observed in our travels by land and sea, and of what we conceive to be credible in the statements and writings of others; for in a westerly direction we have travelled from Armenia to that part of Tyrrhenia which is over against Sardinia; and southward, from the Euxine to the frontiers of Ethiopia. Of all the writers on geography, not one can be mentioned who has travelled over a wider extent of the countries described than we have. Some may have gone farther to the west, but then they have never been so far east as we have; again, others may have been farther east, but not so far west; and the same with respect to north and south. However, in the main, both we and they have availed ourselves of the reports of others, from which to describe the form, size, and other peculiarities of the country." He mentions having been in Egypt, the island Gyarus, Populonium near Elba, Comana in Cappadocia, Ephesus, Mylasa, Nysa, and Hierapolis in Phrygia. He visited Corinth, Argos, Athens, and Megara; but, on the whole, he does not appear to have seen more of Greece than in passing through it on his way to Brundusium, while proceeding to Rome. Populonium and Luna in Italy were the limit of his travels northwards. It is probable he obtained his information as to Spain, France, Britain, and Germany, while staying at Rome.

The first systematic writer on geography was Eratosthenes, who died at the age of 80, about B. C. 196. His work consisted of three books.

There is no ground for considering the Geography of Strabo an improved edition of that of Eratosthenes. Strabo's work was intended for the information of persons in the higher departments of administration, and contains such geographical and historical information as those engaged in political employments[Pg ix] cannot dispense with. Consistently with this object he avoids giving minute descriptions, except where the place is of real interest, but supplies some account of the important political events that had occurred in various countries, and sketches of the great men who had flourished or laboured in them. It is a lively, well-written book, intended to be read, and forms a striking contrast to the Geography of Ptolemy. His language is simple, appropriate to the matter, without affectation, and mostly clear and intelligible, except in those passages where the text has been corrupted. Like many other Greeks, Strabo looked upon Homer as the depository of all knowledge, but he frequently labours to interpret the poet's meaning in a manner highly uncritical. What Homer only partially knew or conjectured, Strabo has made the basis of his description, when he might have given an independent description, founded on the actual knowledge of his time: these observations apply especially to his books on Greece. He does not duly appreciate Herodotus; nor does he discriminate between the stories which Herodotus tells simply as stories he had heard, and the accounts he relates as derived from personal observation. He likewise rejects the evidence of Pytheas of Marseilles as to the northern regions of Europe, and on more than one occasion calls him a liar, although it is very certain that Pytheas coasted along the whole distance from Gadeira, now Cadiz, in Spain, to the river he calls Tanaïs, but which was probably the Elbe; however, from the extracts which have been preserved it seems that he did not give simply the results of his own observations, but added reports which he collected respecting distant countries, without always drawing a distinction between what he saw himself and what was derived from the report of others.

Strabo's authorities are for the most part Greek, and he seems to have neglected the Latin memoirs and historical narratives of the campaigns of the Romans, which might have furnished him with many valuable geographical facts for the countries as well of Asia as of Europe. He made some use of Cæsar's description of France, the Alps, and Britain; he alludes to the voyage of Publius Crassus in speaking of the Cassiterides, and also the writings of Asinius Pollio, Fabius Pictor, and an anonymous writer whom he calls the Chorographer; but he might have obtained much additional information[Pg x] if he had taken pains to avail himself of the materials he could have procured during his stay at Rome.

Strabo considered that mathematical and astronomical knowledge was indispensable to the science of geography; he says in book i.,[2] that without some such assistance it would be impossible to be accurately acquainted with the configuration of the earth; and that every one who undertakes to give an accurate description of a place, should describe its astronomical and geometrical relations, and explain its extent, distance, latitude, and climate.[10] As the size of the earth, he says, has been demonstrated by other writers, we shall take for granted what they have advanced. We shall also assume that the earth is spheroidal, and that bodies have a tendency towards its centre. He likewise says, the convexity of the sea is a further proof that the earth is spheroidal to those who have sailed; for they cannot perceive lights at a distance when placed at the same level as their eyes, but if raised on high, they at once become perceptible.[11] He also observes, "our gnomons are, among other things, evidence of the revolution of the heavenly bodies, and common sense at once shows us that if the depth of the earth were infinite, such a revolution could not take place."[12] But Strabo did not consider the exact division of the earth into climates or zones, in the sense in which Hipparchus used the term, and the statement of the latitudes and longitudes of places, which in many instances were pretty well determined in his time, as essential to his geographical description.

With regard to the lost continent of Atlantis, Strabo is very cautious in criticising[13]Poseidonius; he observes, "he did well, too, in citing the opinion of Plato, that the tradition concerning the island of Atlantis might be received as something more than a mere fiction, it having been related by Solon, on the authority of the Egyptian priests, that this island, almost as large as a continent, was formerly in existence, although now it had disappeared," and remarks that Poseidonius thought it better to quote this than to say, He who brought it into existence can also cause it to disappear, as the poet did the wall of the Achivi.

The measure adopted by Strabo was the stadium. In book [Pg xi]vii. chap. vii. § 4, he says, "From Apollonia to Macedonia is the Egnatian Way; its direction is towards the east, and the distance is measured by pillars at every mile, as far as Cypsela and the river Hebrus. The whole distance is 535 miles. But reckoning, as the generality of persons do, a mile at eight stadia, there may be 4280 stadia. And, according to Polybius, who adds two plethra, which are a third of a stadium, to every eight stadia we must add 178 stadia more,—a third part of the number of miles." In book xi. chap. xi. § 5, he compares the parasang with the stadium, and states that some writers reckoned it at 60, others at 40, and others at 30 stadia.

Dr. Smith, in his Dictionary of Greek and Roman Antiquities, says, "We think that Ukert has satisfactorily shown an accurate description of a place should be particular to add its astronomical and geometrical relations, explaining carefully its extent, distance, degrees of latitude, and temperature of atmosphere. He says likewise, as the size of the earth has been demonstrated by other writers, we shall take for granted that the Greeks had not different standards of length, but always used the Olympic stadium and the foot corresponding to it. He states that the stadium was equal to 600 Greek, or 625 Roman feet, or to 125 Roman paces, and the Roman mile contained 8 stadia. Hence the stadium contained 606 feet 9 inches, English. This standard prevailed throughout Greece under the name of the Olympic stadium, because it was the exact length of the stadium or foot-race course at Olympia, measured between the pillars at the two extremities of the course." Still Dr. Smith further observes, "But although the stadium and the foot connected with it were single definite measures throughout Greece, yet we find in the eastern countries, Babylon, Syria, and Egypt, and in some neighbouring Greek states, feet longer than the Olympic, the origin of which is to be explained by the coëxistence, in the Babylonian system, of a *royal* or *sacred* and a *common* foot and cubit, which were so related to one another, that the royal cubit was three finger-breadths longer than the common."

We may conclude that Strabo's stadium varied considerably, as he sometimes received his distance from personal observation or credible report, and often quoted other writers, and reduced other standards, as the mile, the parasang, and the[Pg xii]schœnus, to the stadium. In addition to this, the most ancient mode of reckoning distances was by the number of days required to perform the journey, and this was transferred into stadia by reckoning a certain number of stadia to a day's journey.[14]

Siebenkees and Heeren (De Fontibus Geographicorum Strabonis) have examined the authorities to which Strabo had, or might have had, access, and Groskurd has availed himself of their researches.

The following is a short summary of the seventeen books from these sources, but for a more detailed account of their contents the translation itself must be referred to.

The first two books may be considered as an independent treatise, and by themselves form a remarkable contrast with the rest of the work, in the manner of treating the subjects, and in the difficulties which they present both of language and matter.

In the 1st book, the author enters into a long discussion on the merits of Homer, whom he considers to have been the earliest geographer, and defends him against the errors and misconceptions of Eratosthenes. He corrects some faults of Eratosthenes, and, in his inquiry concerning the natural changes of the earth's surface defends Eratosthenes against Hipparchus. In conclusion, he again corrects Eratosthenes as regards the magnitude and divisions of the inhabited world. The most remarkable passage in this book is that in which he conjectures the existence of the great Western Continents.[15]

The 2nd book is chiefly occupied with some accounts of mathematical geography, and the Author defends against Hipparchus the division of the inhabited world adopted by Eratosthenes into sections. Then follows a criticism of the division of the earth into six zones, as taught by Poseidonius and Polybius. The pretended circumnavigation of Africa by Eudoxus is referred to, as well as some geographical errors of Polybius. He makes observations of his own on the form and size of the earth in general, as well as of the inhabited portion of it, describing the method of representing it on a spherical or plane surface. A short outline is given of seas, countries, [Pg xiii]and nations; and he concludes with remarks on the system of climates,[16]and on the shadows projected by the sun.

The 3rd book commences with Iberia, and the subject of Europe is continued to the end of the 10th book. His references are the Periplus of Artemidorus, Polybius, and Poseidonius; all three of whom wrote as eye-witnesses. For descriptions and measurement of distances, Artemidorus is chiefly depended upon. The information possessed by Eratosthenes of these countries was meagre and uncertain. For the nations of southern Iberia, he adopts the account of Asclepiades of Myrlea, who had lived and been educated there. Some statements also are borrowed from Roman authors.

The 4th book contains Gallia, according to the four divisions then existing, viz. Gallia Narbonensis, Acquitanensis, Lugdunensis, and the Belgæ; also Britain, with Ierne, and Thule; and lastly, the Alps.

Here Eratosthenes and Ephorus are of little service. His chief guide is Julius Cæsar, whom he frequently quotes *verbatim*. Polybius is his guide for the Alps. Pytheas is the source of some scanty information respecting Ierne and Thule. Throughout his description he adds accounts obtained at Rome from travellers.

4

The 5th book commences with a general sketch of Italy, and refers principally to northern Italy. Dividing its history into ancient and modern, his chief reference for the former is Polybius, and for the latter we are indebted to the observations of the author himself, or to accounts received from others. Still the description of Upper Italy is poor and unsatisfactory, from the author not sufficiently availing himself of Roman resources. Then follows some account of Etruria with its neighbouring islands, Umbria, Samnium, Latium, and Rome, chiefly the result of the author's own researches and observations. The book concludes with some remarks on the inhabitants of the mountainous districts of Samnium and Campania.

The 6th book is a continuation of the same subject. Magna Græcia, Sicily, and the adjacent islands, are noticed, and the author concludes with a short discussion on the extent of the Roman Empire. Descriptions of some places are from his own observations; but the sources whence he takes his [Pg xiv]other account of Italy and the islands are the works of Polybius, Eratosthenes, Artemidorus, Ephorus, Fabius Pictor, Cæcilius (of Cale Acte in Sicily), and some others, besides an anonymous chorographer, supposed to be a Roman, from the circumstance of his distances being given, not in stadia, but in Roman miles.

The 7th book relates, first, to the people north of the Danube,—the Germans, Cimbri, Getæ, Dacians (particularly the European Scythians), and the Crimea; secondly, to the people south of the Danube, viz. those inhabiting Illyricum, Pannonia, Dalmatia, the eastern coast of Thrace to the Euxine, Epirus, Macedonia, Thrace, and the Hellespont. The latter part of this book is not preserved entire in any manuscript, but Kramer has, in his own opinion, succeeded in restoring from the epitomes left to us the greater part of what was wanting. Of Germany, Strabo had tolerable information, but he nowhere states whence it is derived; he may have been partly indebted to Asinius Pollio, whose work he had already examined for the Rhine. For the remaining northern countries, he had Poseidonius and the historians of the Mithridatic war. For the southern countries, he had a lost work of Aristotle on forms of government, Polybius, Poseidonius, and his chief disciples, Theopompus and Ephorus. Incidentally also he quotes Homer and his interpreters, and Philochorus.

The three following books are dedicated to the description of Greece, with the adjacent islands. The 8th comprises the Peloponnesus and its well-known seven provinces, Elis, Messenia, Laconia, Argolis, Corinthia with Sicyonia, Achaia, and Arcadia: the 9th, Attica, with Megaris, Bœotia, Phocis, both Locri and Thessaly: the 10th, Eubœa, Ætolia, and Acarnania, with the islands. After a long digression on the subject of the Curetes, the description of Europe closes with some account of Crete and the islands of the Ægean Sea. The design and construction of these three books differ considerably from the preceding. Homer is adopted as the foundation of his geographical descriptions; some things Strabo must have learnt as an eye-witness, but more from *vivâ voce* communications at Athens or at Corinth. All is interwoven together without any clear line of separation, and the result is some confusion. Athens, Corinth, Argos, and their neighbourhood, were the only parts of Greece our author saw. Heeren, indeed[Pg xv] maintains that he had seen the whole of it, and the Archipelago, but satisfactory proof of this is altogether wanting.

The 11th book commences with the description of the countries separated from Europe by the Tanaïs or Don. Asia is divided by our author (who here follows Eratosthenes) into two parts by the Taurus, which runs in a direction east and west. The northern part of Asia (or this side Taurus) is divided into four parts. The first part comprises the countries lying between the Don, the Sea of Azoff, the Euxine, and the Caspian; the second, the countries east of the Caspian; and the third, the countries south of Caucasus. These three parts of the first or northern division of Asia are contained in the 10th book; the remaining fourth part occupies the 12th, 13th, and 14th books.

The chief authorities for the first part are, besides information obtained from travellers and merchants at Amasia, Herodotus for the Don; Artemidorus and Eratosthenes for distances; Poseidonius and Theophanes of Mitylene, historians, of the Mithridatic war; Metrodorus of Skepsis; Hypsicrates of Amisus; and Cleitarchus for the digression on the Amazons.

For the second part, are principally Patrocles and Aristobulus, historians of the Asiatic campaigns of Alexander. For the third part, or Media and Armenia, are, Dellius, who

5

wrote a history of the war against the Parthians, in which he had served under Antony; Apollonides of Nicæa, who wrote a Periplus of Europe; and other writers before mentioned.

The 12th book commences with a detailed account of Anatolia, and contains the northern part. It was to have been expected that Strabo would have described most of these countries as an eye-witness, lying, as they do, so near his native country, Cappadocia. But this expectation vanishes, when we discover the meagreness of his account. With the exception of Pontus and Cappadocia, he had seen little of the rest, and depends upon historians and oral information. For earlier times, his authorities are Herodotus, Hellanicus, Theopompus, Ephorus, Artemidorus, Apollodorus, and Demetrius of Skepsis; for later times, historians of the wars of Mithridates and Pompey. For the ancient history of the Mysians and Phrygians, he is indebted to the celebrated Lydian historian Xanthus, and Menecrates.

[Pg xvi]

The 13th book continues the description of Anatolia. The greater part of the book is occupied with a dissertation on the Troad. Strabo had travelled over the country himself, but his great authority is Homer and Demetrius of Skepsis, the author of a work in twenty-six books, containing an historical and geographical commentary on that part of the second book of the Iliad, in which the forces of the Trojans are enumerated. A learned digression on the Leleges, Cilices, and Pelasgi, who preceded the Æolians and Ionians in the occupation of the country, is principally taken from Menecrates and Demetrius of Skepsis. The description then turns to the interior, and the account of the Æolian cities is probably due to Poseidonius. Throughout this book are evidences of great care and desire for accuracy.

The 14th book continues with the remainder of Anatolia, and an account of the islands Samos, Chios, Rhodes, and Cyprus. The authorities followed are, on the whole, the same as in the previous book—Herodotus, Thucydides, Ephorus, Artemidorus, Eratosthenes, and Poseidonius; besides Pherecydes of Syros, who wrote on the Ionian migration, and Anaximenes of Lampsacus, the author of a history in Greek of the Milesian colonies. For Caria, he had the historians of Alexander and an author named Philip, who wrote on the Leleges and Carians. For Cyprus he had Damastes and Eratosthenes.

The 15th and 16th books contain a description of the second portion of Asia, namely, the southern or the other side of Taurus. In the 15th book, Strabo describes India and Persia, the latter in two chief divisions, viz. Ariana or East Persia, and Persis or West Persia. These countries Strabo never saw; his description, therefore, is founded on the authority of travellers and historians. The topography of India is meagre, and limited to a few towns and rivers; but his account of the people of the country is more copious, he being supplied with materials from the historians of Alexander and of the campaigns of Seleucus in India. He looks on Megasthenes, Onesicritus, Deïmachus, and Cleitarchus as fabulous writers: but his confidence rests chiefly on Patrocles, Aristobulus (one of the companions and historians of Alexander), and Nearchus, the chief commander of Alexander's fleet. Artemidorus and Nicolaus of Damascus are occasionally consulted. For Ariana or East Persia, he had for his principal authority Eratosthenes;[Pg xvii] and for Persia Proper, he had, besides the above authors, Herodotus, Xenophon, and Polycletus of Larissa, an historian of Alexander.

In the 16th book, he describes the westerly half of south Asia, viz. Assyria with Babylonia, Mesopotamia, Syria, Phœnicia, and Palestine, the Persian and Arabian Gulfs, the coast of Ethiopia, and lastly, Arabia. For the three first countries (the old Assyrian kingdom), his chief authorities are, besides some of Alexander's historians, Eratosthenes, Poseidonius, and Herodotus; for the remainder he had, in addition to the same writers, Artemidorus, and probably also Nicolaus of Damascus. The account of Moses and the Jews, Heeren surmises, comes from Poseidonius, but it probably proceeds from oral communication had in Egypt; of these countries our author could describe nothing as an eye-witness, except the north-west of Syria. The accounts of Arabia, the Indian and the Red Seas, are from Agatharchides; and much that he describes of Arabia was obtained from his friends, Ælius Gallus and the Stoic, Athenodorus.

The 17th book concludes the work with the description of Egypt, Ethiopia, and the north coast of Africa. Strabo had travelled through the whole of Egypt, as far as Syene and Philæ, and writes with the decided tone of an eye-witness. Much verbal information, also, he

6

collected at Alexandria. His most important written authorities are, for the Nile, Eratosthenes (who borrowed from Aristotle), Eudoxus, and Aristo. For the most remarkable events of Egyptian history, he had Polybius, and for later times probably Poseidonius, besides *vivâ voce* accounts.

For the oracle at Ammon, he had the historians of Alexander; for Ethiopia, the accounts of Petronius, who had carried on war there, Agatharchides, and Herodotus. Of Libya or Africa Proper he had nothing new or authentic to say. Besides Eratosthenes, Artemidorus, and Poseidonius, his chief authorities, he had Iphicrates, who wrote on the plants and animals of Libya. The whole concludes with a short notice of the Roman Empire.

The dates at which particular books were written, as attempted to be given by Groskurd and Coraÿ, must be received with caution.

In book iv. c. vi. § 9, Strabo says that the Carni and Taurisci[Pg xviii] had quietly paid tribute for thirty-three years; and both these tribes were reduced to subjection by Tiberius and Drusus, B. C. 14. This book was therefore written in A. D. 19.

In book vi. c. iv. § 2, Cæsar Germanicus is spoken of as still living. He died in Syria, A. D. 20 (19). This book was therefore written before that year.

In book xii. c. viii. § 11, Strabo says that Cyzicus was still a free state. It lost its liberty A. D. 25. This book was therefore written before A. D. 25. Whether Strabo was alive or dead at this date, we have no means of determining.

The codices or manuscripts which exist of Strabo's work appear to be copies of a single manuscript existing in the middle ages, but now lost. From the striking agreement of errors and omissions in all now extant (with such differences only as can be accounted for, arising from the want of ability or carelessness of the copyist), it appears most probable that to this single manuscript we are indebted for the preservation of the work. Strabo himself describes the carelessness of bad scribes both at Rome and Alexandria,[12] in the following expressive language: "Some vendors of books, also, employed bad scribes and neglected to compare the copies with the originals. This happens in the case of other books, which are copied for sale both here and at Alexandria." After what Kramer has done for the text, we can hope for little improvement, unless, what is beyond all expectation, some other manuscript should be discovered which is either derived from another source, or is a more correct copy.

The following is some account of those in existence:—

Codices in the Imperial Library, Paris:

1. No. 1397 of the catalogue. This is the principal codex existing in the Imperial Library, and was written in the 12th century. It was formerly in the Strozzi Palace at Rome, and was brought to Paris by Maria de Medici. Not only are parts of the leaves, but even whole leaves of the 9th book, damaged or destroyed by damp, mice, bad binding, and careless attempts at correction. This codex contains the first nine books; the second part, containing the last eight, is lost. Collated by Kramer, and partly for Falconer, by Villebrune.

2. No. 1393 of the catalogue. On this codex Brequigny chiefly depended for his edition. Montfaucon says that it is [Pg xix]of the 12th or 13th century; Kramer, however, judging from the character of the handwriting and contractions, maintains that it belongs to the end of the 13th or beginning of the 14th century. It contains the whole seventeen books of the Geography, and was written in the East (not, however, by the same hand throughout), and brought from Constantinople to Paris by the Abbé Servin in 1732, to whom it had been presented by a Greek named Maurocordato. Collated by Villebrune for Falconer, and partly by Kramer.

3. No. 1408 contains the seventeen books, and appears to have been written towards the end of the 15th century. In general, the geography of Strabo is divided by transcribers into two parts, the first containing nine books, the second, the last eight; but in this codex there is a blank leaf inserted between the 10th and 11th books, from which it would appear that there was also another division of the work, separating the subjects, Europe and Asia. Partly collated by Villebrune for Falconer.

4. No. 1394. This contains the seventeen books, and is very beautifully written, and illuminated with arabesque designs. It was made by the order of Lorenzo the Magnificent;

and its date, therefore, is after the middle of the 15th century. Collated, as before, by Villebrune.

5. No. 1396 contains the whole seventeen books, and was probably written about the end of the 15th or the beginning of the 16th century. The division of the work is into ten books and seven books. In the beginning, it is stated to be "the gift of Antony the Eparch to Francis the great and illustrious king of France." Partly collated by Kramer.

6. No. 1395 contains the whole seventeen books, and served for the Aldine edition of Strabo. The handwriting of this codex is excellent, but the order of the words is arbitrarily changed, and there are frequent omissions, sometimes even of whole lines: it is corrupt beyond description, and among the worst we possess. Collated in some parts by Kramer.

No. 1398, written about the end of the 15th century. It contains the epitome of the first ten books, by Gemistus Pletho, and the last seven books entire. It is a copy of No. 397, in the Library of St. Mark, Venice. Collated by Villebrune.

Codices in the Vatican:

No. 1329 of the catalogue. This codex dates from the beginning,[Pg xx] probably, of the 14th century, and is remarkable for being the work of thirteen different transcribers. It is much to be lamented, that the greater part of it is lost; it begins from the end of the 12th book only, and a part of the last leaf of the 17th book is also destroyed; what remains to us surpasses all others in correctness of the text. The whole has been collated for the first time by Kramer.

No. 174 is of the 15th century, and contains the seventeen books: the first nine books are written by one transcriber, the last eight by another hand. The first nine books have been collated by Kramer.

No. 173 contains the first ten books, and is of the middle of the 15th century. It is badly and incorrectly written. The last seven books, which would complete the codex, are, as Kramer conjectures from the paper and handwriting, in the Library of the Grand Duchy of Parma. From a note in Greek at the end of the 10th book, it appears to have been brought to Rome A. D. 1466. Books 1, 2, 4, and 5, collated by Kramer.

No. 81 is tolerably well and correctly written. It contains the last eight books, and is of the end of the 15th century. It appears to be a copy of, or served as a copy to, the codex in the Laurentian Library, No. 19. Partially collated by Kramer.

Medicean Codices, in the Laurentian Library, Florence:

Codex 5 is elegantly and correctly written; it is of the beginning, probably, of the 15th century, and contains the first ten books. The 8th and 9th books are not entire; passages are curtailed, and much is omitted, to which the attention is not drawn, the lines being run on without spaces left to mark omissions. Errors of the first transcriber are corrected by a later hand, and noticed in the margin or between the lines. Collated by Bandini for Falconer, and almost the whole by Kramer.

Codex 40 contains the first ten books; a copy, probably, of the former. It was written after the middle of the 15th century.

Codex 15 is of the middle of the 15th century, and contains the last seven books. It is not in any way remarkable.

Codex 19, written at the end of the 15th century. It contains the last eight books, and resembles No. 81 of the Vatican. Collated by Bandini for Falconer.

[Pg xxi]
Venetian Codices:

No. 377 of the catalogue contains the first twelve books, and is written in the 15th century. Formerly the property of Cardinal Bessarion.

No. 378 contains the seventeen books, of which the first twelve are apparently copies of the above, No. 377; the remaining five are transcribed from some other codex. This was, also, formerly the property of Cardinal Bessarion.

No. 640 contains the last eight books. It was written, as appears from a note A. D.1321, by different hands. A great part of the 14th book is wanting; eight blank pages are left for the completion of it; but this was not done by the transcriber to whom this portion was assigned. It is placed by Kramer in the first class of manuscripts, and was wholly collated by him.

No. 379 is of the 15th century. It contains the Epitome of Gemistus Pletho of the first ten books, and the whole of the last seven books. It is the codex which served for the copy, No. 1398, in the Imperial Library at Paris. Formerly the property of Cardinal Bessarion.

No. 606 contains the last eight books, and was written towards the end of the 15th century. It contains nothing which is not to be found in other manuscripts.

Codices in the Ambrosian Library, Milan:

Codex M. 53 contains all but book ii., and is of the 15th century. The books are not written on paper of the same size, nor in consecutive order, although by the same hand. Book ii. is to be found in Codex N. 289, together with portions of other authors, written by a different transcriber, no doubt with the intention of completing this codex. According to Kramer, the first ten books are copied from Codex 5 of the Medici MS. The 13th, 14th, 12th books (the order in which they stand) from the Medici MS. 19, and the 11th, 15th, 16th, 17th, from the Medici MS. 15. Partly collated by Kramer.

Codex G. 53 contains the seventeen books, and is of the end of the 15th century. Five leaves at the beginning, and two at the end, are destroyed by damp, traces of which are to be seen throughout. Partly collated by Kramer.

In the Library of Eton College is a codex containing the first ten books; it was written at Constantinople. Kramer,[Pg xxii] who, however, did not see it, conjectures that the Medici MS., Codex 15, containing the last eight books, was formerly united to it, and completed the whole work. Collated for Falconer.

In the Library of the Escurial is a codex completed, as we are informed by a note at the end, A. D. 1423. Collated by Bayer for Falconer.

The Madrid Library possesses a codex written in the latter part of the 15th century, containing the seventeen books.

In the Library at Moscow is a codex containing the seventeen books; it was written at the end of the 15th or beginning of the 16th century. The first nine books resemble the Paris Codex, No. 1397; the last eight, the Venetian Codex, No. 640. It came from one of the monasteries of Mount Athos, and was not destroyed, as Groskurd suspects, in the great fire of 1812, but is still to be found in the Library of the Holy Synod, under No. 204 (Matt. ccv.), as I am informed by the Archimandrite Sabba, who dates from the Kremlin, April 4th, 1857.

A codex also is yet to be found in one of the monasteries of Mount Athos. From the accounts of learned travellers (Zacharias, *Reise in den Orient*, and Fallmerayer, in the *Allgem. Zeitg. 5 Jun. 1843*), it contains nothing which can supply the deficiencies of those MSS. with which we are acquainted.

Besides the above codices, there exist four epitomes of the Geography of Strabo, of which,

1. The Epitome Palatina, in the Heidelberg Library, is the oldest of all MSS. of this work. It is probably of the beginning of the 10th century, although Dodwell places it between 976 and 996. The codex from which it was copied appears to have been perfect, and contained the whole of the 7th book, which is imperfect in all other codices. It is, however, to be regretted that the author did not confine himself to following the text of Strabo; he has not only indulged in curtailing, transposing, and changing the words and sentences of the original, but has sometimes also added expressions of his own.

2. The Vatican Epitome is of more value than the preceding; the extracts are more copious, the author seldom wanders from the text of Strabo, and in no instance inserts language of his own. The codex which served as the basis for the Epitome contained the 7th book entire, and from this[Pg xxiii] and the Palatine Epitome Kramer collected the fragments of the last part of the 7th book, which appear for the first time in his edition (see vol. i. of the Translation, p. 504). This codex was written in the middle of the 14th century, and has suffered much by time and carelessness; several leaves are lost, and lines of the text at the top and bottom of the pages have been cut off in the binding.

3. The Parisian Epitome, on which no great value is placed by Kramer.

4. The Epitome of Gemistus Pletho, referred to above, is of great value, and held in the highest estimation by all editors.

9

The first appearance of Strabo's work in print was a Latin translation by Guarini, of Verona, and Gregorio of Tiferno. Of this, thirteen editions were printed, the first in 1469 or 1471, the twelfth in 1559, and the last in 1652. It is not known from what manuscripts the translation was taken, nor whether they now exist; but, though the translation itself is barbarous, and in many passages erroneous, its fidelity to the original is so apparent, that all editors to the present time have consulted it as a manuscript.

The first edition of the Greek text was printed at Venice by Aldus in 1516, and was taken from so corrupt a manuscript that Coraÿ compares it to the Augean stable. The second edition was a repetition of the Aldine, accompanied by the Latin translation of Guarini, and was published by Hopper and Heresbach, at Bâsle, in 1549. The third edition, by Xylander, in 1570, was also a repetition of the text of Aldus; but a new Latin translation accompanied it. The fourth and fifth editions, which do not essentially differ, were published in 1587 and 1620, by Isaac Casaubon. He collated for his edition four manuscripts, which he obtained from his father-in-law, H. Stephens, and was the first to add a commentary; but it is not known what manuscripts were made use of. The edition of Almeloveen, 1707, being a reprint of Casaubon, with notes, and an edition commenced by Brequigny, Paris, 1763, but not continued beyond the first three books, can scarcely be placed among the number of new editions. Brequigny left a French translation in manuscript and notes in Latin, which were consulted by the French translators.

The seventh edition was that of Thomas Falconer of Chester,[Pg xxiv] and of Brasenose College, published in 2 vols. folio, at Oxford, 1807. For the first time since Casaubon's last edition, nearly 200 years before, manuscripts were collated for this edition, namely, those of Eton, Moscow, the Escurial, and the Laurentian library; the conjectural emendations of Tyrwhitt, and notes of the editor and others, are added. "It has everything that is valuable in Casaubon's edition, besides having corrected numberless typographical errors. In the account given of it, the public are as much wronged as we are abused; for no view whatever is laid before them of its nature or its merits."[18] Thos. Falconer, having prepared the greater part of the work for the press, died in 1792. A little more than the two first books were edited by John Parsons, Bishop of Peterborough, and formerly Master of Balliol College, Oxford; but the whole work was, ultimately, in 1802 given up to Thomas Falconer (nephew of the former), of Corpus Christi College, Oxford, who completed it and wrote the preface. A complete revisal of the text, however, was not attempted.

The eighth edition was that of Professor J. P. Siebenkees, of which great expectations were formed. The deficiencies of his performance are strongly commented on by Kramer. Siebenkees lived to complete only the first six books; the remainder of the work was undertaken by Tzchucke, and conducted with greater skill and ability than by his predecessor. It was published in 1811, 6 vols. 8vo.

The ninth edition is that by Coraÿ, Paris, 1815-1818, 4 vols. 8vo. Kramer passes an unfavourable opinion on it. The editor, according to him, did not possess an aptitude for discriminating the value of the different manuscripts he collated, and considered more what he thought ought to have been written than what were really the author's words. Hence, although he was successful in restoring the true readings of many passages, he corrupted not a few, and left untouched many errors. Yet he was a very able scholar, and has the merit of attempting the first critical edition of Strabo.

The tenth edition is that by Professor Gustavus Kramer, in 3 vols. 8vo, the first of which appeared in 1844, the last in 1852. The editor has brought to his task great ability and [Pg xxv]unwearied labour; of the many years spent in the preparation of it, were passed in Italy for the purpose of collating manuscripts. This edition surpasses all others in completeness, and little is left for correction by subsequent editors.

A. Meineke published at Leipsic, in 3 vols., 1852, a reprint of Kramer's text, with some emendations of his own contained in his work, "Vindiciarum Straboniarum Liber." Berlin, 1852.

C. Müller and F. Dübner have also published the first vol., Paris, 1852, of a reprint of Kramer's text, with Meineke's corrections. It is accompanied by a new Latin translation, of which the first six books are by Dübner, and the remainder by Müller.

In modern languages, we have a translation by Alfonso Buonacciuoli, of Ferrara, in Italian, 2 vols. 8vo, Venice, 1552. It is a very literal translation from a manuscript, and is frequently quoted by the French translators. Also a translation in German by Abr. Penzel, in 4 vols., Lemgow, 1775. It is not literal, and abounds with wilful additions and alterations of the author's meaning.

A translation in French was published at Paris in five vols. 4to, from the year 1805 to 1819. The first three books are translated by De la Porte du Theil and Coraÿ together. The 4th, 7th, 8th, 12th, 13th, 14th, and 15th books are by Coraÿ; the 5th, 6th, 9th, 10th, and 11th, by De la Porte du Theil; on the death of the latter, Letronne undertook the translation of the 16th and 17th books. The whole is accompanied by very copious notes by the translators, and others on geographical and mathematical subjects, by Gossellin. As might be expected from the united labour of such distinguished men, this translation, which was undertaken at the command of Napoleon I., has been held in high estimation. De la Porte du Theil, for the purpose of conveying to the reader a more accurate idea of the state of the text of the ninth book than could be given by description or notes, has prefaced his translation by a copy, page for page and line for line, of the original manuscript. The number of mutilated passages amounts to two thousand. For the purpose of restoring the text, recourse has been had to other manuscripts, to conjectures, to extracts from the Epitomes, and to quotations of Strabo's work contained in the Geographical Lexicon of Stephanus of Byzantium, composed before the seventh century,[Pg xxvi] and in the Commentaries of Eustathius on Homer, which were written towards the end of the twelfth century. It is an example of Kramer's just remark, that no work of any ancient author, which has descended to our time, has suffered more from various causes.

A translation by F. Ambrosoli, forming part of the "Collana degli Antichi Storici Greci," was published in 1832, 4 vols. 8vo, Milan, and is founded on the French translation. A translation of the third book (Spain) by Lopez, was published at Madrid, 1788, and is well spoken of. The best translation of the whole work—and too much cannot be said in praise of it—is in German, by Groskurd, 4 vols. 8vo, Berlin, 1831-1834. The last volume contains a very copious index.

In conclusion, I have to acknowledge considerable obligations to the notes and prefaces of Groskurd, Kramer, the French translators, and others.

The part of the translation for which I am responsible commences at chap. iv. book vii., vol. i. p. 473, to the end of the work, and is partly based on an incomplete translation in MS. by my late father, the Rev. Dr. T. Falconer. The previous portion is the work of Hans C. Hamilton, Esq., F.S.A., to whom I am indebted for his continued interest in the translation throughout, for his care in correcting the press, and for valuable suggestions.

A complete index, which concludes the third volume, has been compiled with the greatest regard to accuracy, by a gentleman of tried skill and ability. It contains every geographical name mentioned by Strabo; and the modern names, printed in italics, are also added, as far as can be ascertained: they are not given with perfect confidence in all cases; discussion on doubtful points would have exceeded the limits of this work; and reference may be advantageously made, where more minute detail is required, to the able articles in Dr. W. Smith's Dictionary of Greek and Roman Geography.

W. FALCONER.

Rectory,　　　　　　　　　　*Bushey,*　　　　　　　　　　*Herts.*
September 1, 1857.

[Pg 1]

STRABO'S GEOGRAPHY.

BOOK XIV.
SUMMARY.

The Fourteenth Book contains an account of the Cyclades islands and the region opposite to them, Pamphylia, Isauria, Lycia, Pisidia, Cilicia as far as Seleucia of Syria, and that part of Asia properly called Ionia.

CHAPTER I.

1. THERE remain to be described Ionia, Caria, and the sea-coast beyond the Taurus, which is occupied by Lycians, Pamphylians, and Cilicians.[19] We shall thus finish the description of the whole circuit of the peninsula, the isthmus of which, we have said, consists of the tract between the Euxine and the Sea of Issus.

2. The navigation around Ionia along the coast is about 3430 stadia. It is a considerable distance, on account of the gulfs, and of the peninsular form for the most part of the country, but the length in a straight line is not great. The distance, for example, from Ephesus to Smyrna is a journey in a straight line of 320 stadia; to Metropolis[20] is 120 stadia, and the remainder to Smyrna; but this distance by sea is little less than 2200 stadia. The extent of the Ionian coast is reckoned from Poseidium,[21] belonging to the Milesians,[Pg 2] CAS. 632 and the boundaries of Caria, as far as Phocæa,[22] and the river Hermus.[23]

3. According to Pherecydes, Miletus, Myus,[24] Mycale, and Ephesus, on this coast, were formerly occupied by Carians; the part of the coast next in order, as far as Phocæa, and Chios, and Samos, of which Ancæus was king, were occupied by Leleges, but both nations were expelled by the Ionians, and took refuge in the remaining parts of Caria.

Pherecydes says that the leader of the Ionian, which was posterior to the Æolian migration, was Androclus, a legitimate son of Codrus king of the Athenians, and that he was the founder of Ephesus, hence it was that it became the seat of the royal palace of the Ionian princes. Even at present the descendants of that race are called kings, and receive certain honours, as the chief seat at the public games, a purple robe as a symbol of royal descent, a staff instead of a sceptre, and the superintendence of the sacrifices in honour of the Eleusinian Ceres.

Neleus, of a Pylian family, founded Miletus. The Messenians and Pylians pretend that there is some affinity between them; in reference to which later poets say that even Nestor was a Messenian, and that many Pylians accompanied Melanthus, the father of Codrus, to Athens, and that all this people sent out the colony in common with the Ionians. There is also to be seen on the promontory Poseidium an altar erected by Neleus.

Myus was founded by Cydrelus, a spurious son of Codrus; Lebedos[25] by Andropompus, who took possession of a place called Artis; Colophon by Andræmon, a Pylian, as Mimnermus mentions in his poem of Nanno;[26] Priene by Æpytus, son of Neleus; and afterwards by Philotas, who brought a colony from Thebes; Teos by Athamas, its first founder, whence Anacreon calls the city Athamantis, but at the time of the Ionian migration of the colony it received settlers from Nauclus, a spurious son of Codrus, and after this from Apœcus and Damasus, who were Athenians, and from Geres, a Bœotian; Erythræ was founded by Cnopus, who also was a spurious[Pg 3] son of Codrus; Phocæa by Athenians, who accompanied Philogenes; Clazomenæ by Paralus; Chios by Egertius, who brought with him a mixed body of colonists; Samos by Tembrion, and afterwards by Procles.

4. These are the twelve Ionian cities. At a subsequent period Smyrna also was added to the Ionian association at the instance of the Ephesians, for anciently they inhabited the same city, at which time Ephesus was called Smyrna. Callinus somewhere gives it this name, and calls the Ephesians Smyrnæans in the address to Jupiter:

"And pity the Smyrnæans;"

and in another passage,

"remember now, if ever, the beautiful thighs of the oxen [which the Smyrnæans burnt in sacrifice]."

Smyrna was an Amazon, who got possession of Ephesus; from her the inhabitants and the city had their name, in the same manner as some Ephesians were called Sisyrbitæ from Sisyrba; and a certain spot in Ephesus was called Smyrna, as Hipponax testifies:

"He lived in Smyrna, at the back of the city between Tracheia and Lepre Acta."

The mountain Prion was called Lepre Acta; it overhangs the present city, and has on it a portion of the wall. Even now the farms at the back of the Prion retain the name in the term Opistholepria. The country along the foot of the mountain about Coressus was called Tracheia. The city was anciently built about the Athenæum, which is now beyond the city, at the (fountain) Hypelæus. Smyrna therefore was situated near the present gymnasium, at the back of the present city, but between Tracheia and Lepre Acta. The Smyrnæans, upon quitting the Ephesians, marched to the place where Smyrna now stood, and which was in the possession of Leleges. They expelled these people and founded the ancient Smyrna, which is distant from the present city about 20 stadia. They were themselves afterwards expelled by Æolians, and took refuge at Colophon; they then returned with a body of men from the latter place, and recovered their own city, Smyrna. Mimnermus relates this in his poem of Nanno, and says of Smyrna, that it was always a subject of contention;

CAS. 634

"after leaving Pylus, the lofty city of Neleus, we came in our voyage to the long wished-for Asia, and settled at Colophon, and hastening thence from the river Astëeis, by the will of the gods we took Æolian Smyrna."

So much then on this subject.

We must, however, again describe each place in particular, beginning with the principal cities, from which the first settlements originated, I mean Miletus and Ephesus, for these are superior to all others, and the most celebrated.

5. Next after the Poseidium of the Milesians, at the distance of 18[27] stadia from the sea-coast, is the oracle of Apollo Didymeus among the Branchidæ. This, as well as the other temples, except that at Ephesus, was burnt by the order of Xerxes.[28] The Branchidæ delivered up the treasures of the god to the Persian king, and accompanied him in his flight, in order to avoid the punishment of sacrilege and treachery.

The Milesians afterwards built a temple, which exceeded in size all others, but it remained without a roof on account of its magnitude. The circuit of the sacred enclosure contained within it a village with a magnificent grove, which also extended beyond it; other sacred enclosures contain the oracle, and what belongs to the worship of the god.

Here is laid the scene of the fable of Branchus, and Apollo's love for him. The temple is adorned with the most costly offerings, the productions of ancient art.

Thence to the city the journey is not long either by land or sea.[29]

6. Ephorus relates that Miletus was first founded and fortified by the Cretans on the spot above the sea-coast where at present the ancient Miletus is situated, and that Sarpedon conducted thither settlers from the Miletus in Crete,[30] and gave it the same name; that Leleges were the former occupiers of the country, and that afterwards Neleus built the present city.

[Pg 5]The present city has four harbours, one of which will admit a fleet of ships.[31] The citizens have achieved many great deeds, but the most important is the number of colonies which they established. The whole Euxine, for instance, and the Propontis, and many other places, are peopled with their settlers.

Anaximenes of Lampsacus says, that the Milesians colonized both the island Icarus and Lerus, and Limnæ on the Hellespont, in the Chersonesus; in Asia, Abydus, Arisba, and Pæsus; on the island of the Cyziceni, Artace and Cyzicus; in the interior of the Troad, Scepsis. We have mentioned, in our particular description of places, other cities which this writer has omitted.

Both the Milesians and Delians invoke Apollo Ulius, as dispensing health and curing diseases; for οὔλειν[32] is to be in health, whence οὐλή,[33] a wound healed, and the phrase in Homer,[34] Οὖλέ τε καὶ μέγα χαῖρε, "health and good welcome;" for Apollo is a healer, and Artemis has her name from making persons ἀρτεμέας, or sound. The sun, also, and moon are associated with these deities, since they are the causes of the good qualities of the air; pestilential diseases, also, and sudden death are attributed to these deities.

7. Illustrious persons, natives of Miletus, were Thales, one of the seven wise men, the first person who introduced among the Greeks physiology and mathematics; his disciple Anaximander, and Anaximenes the disciple of Anaximander. Besides these, Hecatæus the historian;[35] and of our time, Æschines the orator, who was banished for having spoken with too great freedom before Pompey the Great, and died in exile.

Miletus shut her gates against Alexander, and experienced the misfortune of being taken by storm, which was also the fate of Halicarnassus; long before this time it was captured by the Persians. Callisthenes relates, that Phrynichus the tragic writer was fined a thousand drachmæ by the Athenians for composing a play entitled "The taking of Miletus by Darius." [Pg 6]

[CAS. 635]The island Lade lies close in front of Miletus, and small islands about Tragææ,[36] which afford a shelter for pirates.

8. Next follows the Gulf of Latmus, on which is situated "Heracleia under Latmus,"[37] as it is called, a small town with a shelter for vessels. It formerly had the same name as the mountain above, which Hecatæus thinks was the same as that called by the poet[38] the mountain of the Phtheiri, for he says that the mountain of the Phtheiri was situated below Latmus; but some say that it was Grium, as being parallel to Latmus, and extending from the Milesian territory towards the east, through Caria, as far as Euromus and Chalcetores. However, the mountain rises up in sight of[39] the city.

At a little distance further, after crossing a small river near Latmus, there is seen in a cave the sepulchre of Endymion. Then from Heracleia to Pyrrha, a small city, is about 100 stadia by sea, but a little more from Miletus to Heracleia, if we include the winding of the bays.

9. From Miletus to Pyrrha, in a straight line by sea, is 30 stadia; so much longer is the journey by sailing near the land.

10. When we are speaking of celebrated places, the reader must endure with patience the dryness of such geographical descriptions.

From Pyrrha to the mouth of the Mæander are 50 stadia. The ground about it is marshy and a swamp. In sailing up the river in vessels rowed by oars to the distance of 30 stadia, we come to Myus,[40] one of the twelve Ionian cities, which, on account of its diminished population, is now incorporated with Miletus. Xerxes is said to have given this city to Themistocles to supply him with fish, Magnesia with bread, and Lampsacus with wine.[41]

11. At four stadia from Myus is Thymbria, a Carian village, near which is Aornum; this is a sacred cave called Charonium,[Pg 7] which emits destructive vapours. Above it is Magnesia[42] on the Mæander, a colony of the Magnesians of Thessaly and Crete. We shall speak of it very soon.

12. After the mouths of the Mæander follows the shore of Priene. Above it is Priene,[43] and the mountain Mycale,[44] which abounds with animals of the chace, and is covered with forests. It is situated above the Samian territory, and forms towards it, beyond the promontory Trogilium,[45] a strait of above 7 stadia in width. Priene is called by some writers Cadme, because Philotus, its second founder, was a Bœotian. Bias, one of the seven wise men, was a native of Priene, of whom Hipponax uses this expression;

"More just in pleadings than Bias of Priene."

13. In front of Trogilium lies an island of the same name. Thence, which is the nearest way, is a passage across to Sunium of 1600 stadia. At the commencement of the voyage, on the right hand are Samos, Icaria, and the Corsiæ islands;[46] on the left, the Melantian rocks.[47] The remainder of the voyage lies through the middle of the Cyclades islands. The promontory Trogilium itself may be considered as a foot of the mountain Mycale. Close to Mycale is another mountain, the Pactyas, belonging to the Ephesian territory, where the Mesogis terminates.

14. From Trogilium to Samos are 40 stadia. Both this and the harbour, which has a station for vessels, have a southern aspect. A great part of it is situated on a flat, and is overflowed by the sea, but a part also rises towards the mountain which overhangs it. On the right hand, in sailing towards the city, is the Poseidium, a promontory, which forms towards Mycale the strait of 7 stadia. It has upon it a temple of Neptune. In front is a small island, Narthecis; on the left, near the Heræum, is the suburb, and the river Imbrasus, and the Heræum, an ancient temple, and a large nave, which at present is a repository for paintings. Besides the great number of paintings in the Heræum, there are other repositories and some small chapels, filled with works of ancient art. The Hypæthrum also is full of the best statues. Of these, three of colossal size, the work of Myron, stand [Pg 8] [CAS. 637]upon the same base. Antony took them all away, but Augustus Cæsar replaced

two, the Minerva and the Hercules, upon the same base. He transported the Jupiter to the Capitol, having built a chapel for its reception.

15. The voyage round the island Samos is 600 stadia.[48] Formerly, when the Carians inhabited it, it was called Parthenia, then Anthemus, then Melamphylus,[49]then Samos, either from the name of some native hero, or from some one who conducted a colony thither from Ithaca and Cephallenia. In it is a promontory looking towards Drepanum in Icaria, which has the name of Ampelos, (the Vine,) but the whole mountain, which spreads over the island, has the same name. The island is not remarkable for good wine,[50] although the islands around, as Chios, Lesbos, Cos, and almost all the adjacent continent, produce wines of the best kind. The Ephesian and the Metropolites are good wines, but the Mesogis, the Tmolus, the Catacecaumene, Cnidos, Smyrna, and other more obscure places, are distinguished for the excellence of their wines, whether for gratification or dietetic purposes.

Samos is not very fortunate as regards the production of wine, but in general it is fertile, as appears from its possession being a subject of warlike contention, and from the language of its panegyrists, who do not hesitate to apply to it the proverb,

"It produces even birds' milk,"

as Menander somewhere says. This was the cause also of the tyrannies established there, and of the enmity of the Athenians.

16. The tyrannies were at their height in the time of Polycrates and his brother Syloson. The former was distinguished for his good fortune, and the possession of such a degree of power as made him master of the sea. It is related as an instance of his good fortune, that having purposely thrown into the sea his ring, which was of great value both on account of [Pg 9]the stone and the engraving, a short time afterwards a fisherman caught the fish which had swallowed it, and on cutting the fish open, the ring was discovered. When the king of Egypt was informed of this, he declared, it is said, with a prophetic spirit, that Polycrates, who had been elevated to such a height of prosperity, would soon end his life unfortunately; and this was actually the case, for he was taken by the Persian satrap by stratagem, and crucified. Anacreon, the lyric poet, was his contemporary, and all his poetry abounds with the praises of Polycrates.

It is said that in his time Pythagoras, observing the growing tyranny, left the city, and travelled to Egypt and Babylon, with a view to acquire knowledge. On his return from his travels, perceiving that the tyranny still prevailed, he set sail for Italy, and there passed the remainder of his life.

So much respecting Polycrates.

17. Syloson was left by his brother in a private station. But he made a present to Darius, son of Hystaspes, of a robe which the latter saw him wearing, and very much desired to possess. Darius was not king at this time, but when he became king, Polycrates received as a compensation the tyranny of Samos. He governed with so much severity, that the city was depopulated, which gave occasion to the proverb,

"By the pleasure of Syloson there is room enough."

18. The Athenians formerly sent Pericles their general, and with him Sophocles the poet, who harassed with the evils of a siege the refractory Samians. Afterwards[51]they sent thither a colony of two thousand citizens, among whom was Neocles the father of Epicurus, and, according to report, a school-master. It is said, that Epicurus was educated here and at Teos, and was admitted among the ephebi at Athens, having as his comrade in that class Menander the comic poet. Creophylus was a native of Samos,[52] who, it is said, once entertained Homer as his guest, and received, in return, his poem entitled "The taking of Œchalia." Callimachus, on the contrary, intimates in an epigram that it was the composition of [Pg   10]
[CAS. 639]Creophylus, but ascribed to Homer on account of the story of his hospitable entertainment by Creophylus:

"I am the work of the Samian, who once entertained in his house, as a guest, the divine Homer. I grieve for the sufferings of Eurytus, and mourn for the yellow-haired Ioleia. I am called Homer's writing. O Jupiter, how glorious this for Creophylus."

Some say that he was Homer's master; according to others, it was not Creophylus, but Aristeas of Proconnesus.

19. The island of Icaria, from which the Icarian Sea has its name, is near Samos. The island has its name from Icarus, the son of Dædalus, who, it is said, having accompanied his father in his flight, when both of them, furnished with wings, set out from Crete, fell on that island, unable to sustain his flight. He had mounted too near the sun, and the wings dropped off on the melting of the wax [with which they were fastened].

The whole island is 300 stadia in circumference; it has no harbours, but only anchorages, the best of which is called Histi. A promontory stretches towards the west. There is also on the island a temple of Diana, called Tauropolium, and a small town Œnoë; and another, Dracanum,[53] of the same name as the promontory on which it stands, with an anchorage for vessels. The promontory is distant from the promontory of the Samians, called Cantharius, 80 stadia, which is the shortest passage from one to the other. The Samians occupy it at present in its depopulated state, chiefly for the sake of pasture which it affords for cattle.

20. Next to the Samian strait at Mycale, on the right hand on the voyage to Ephesus, is the sea-coast of the Ephesians, a part of which even the Samians possess. First on the sea-coast is the Panionium,[54] distant from the sea three stadia, where the Panionia, a common festival of the Ionians, is celebrated, and a sacrifice is performed in honour of the Heliconian Neptune. The priests are Prienians. We have spoken of them in the description of Peloponnesus.

Then follows Neapolis, which formerly belonged to the Ephesians, but now belongs to the Samians, having exchanged Marathesium[55] for it, the more distant for the nearer place. Next is Pygela, a small town, containing a temple of Diana Munychia. It was founded by Agamemnon, and colonized [Pg 11]by some of his soldiers, who had a disease in the buttocks, and were called Pygalgeis; as they laboured under this complaint, they settled there, and the town had the appropriate name of Pygela.[56]

Next is a harbour called Panormus, with a temple of the Ephesian Diana; then the city.

On the same coast, at a little distance from the sea, is Ortygia, a fine wood with trees of all kinds, but the cypress in the greatest abundance. Through this wood flows the river Cenchrius, in which Latona is said to have bathed after the birth of her child. For here is laid the scene of the birth of the child, the cares of the nurse Ortygia, the cave in which the birth took place, the neighbouring olive tree under which the goddess first reposed when the pains of child-birth had ceased.

Above the wood is the mountain Solmissus, where, it is said, the Curetes stationed themselves, and with the noise of their arms perplexed and terrified Juno, who was enviously watching in secret the delivery of Latona, who was thus assisted in concealing the birth of the child.

There are many temples in the place, some of which are ancient, others of later times; in the former are ancient statues; in the latter are works of Scopas, Latona holding a sceptre, and Ortygia standing by her with a child in each arm.

A convention and festival are celebrated there every year. It is the custom for young men to vie with each other, particularly in the splendour of their convivial entertainments. The body of Curetes celebrate their Symposia at the same time, and perform certain mystic sacrifices.

21. The city of Ephesus was inhabited both by Carians and Leleges. After Androclus had expelled the greatest part of the inhabitants, he settled his companions about the Athenæum, and the Hypelæum, and in the mountainous tract at the foot of the Coressus. It was thus inhabited till the time of Crœsus. Afterwards, the inhabitants descended from the mountainous district, and settled about the present temple, and continued there to the time of Alexander. Then Lysimachus built a wall round the present temple, and, perceiving the inhabitants [Pg   12]
[CAS. 640]unwilling to remove thither, took advantage of a heavy storm of rain which he saw approaching, and obstructed the drains so as to inundate the city, and the inhabitants were glad to leave it for another place.

He called the city Arsinoë, after the name of his wife, but the old name prevailed. A body of elders was enrolled, with whom were associated persons called Epicleti, who administered all the affairs of the city.

22. Chersiphron[57] was the first architect of the temple of Diana; another afterwards enlarged it, but when Herostratus set fire to it,[58] the citizens constructed one more magnificent. They collected for this purpose the ornaments of the women, contributions from private property, and the money arising from the sale of pillars of the former temple. Evidence of these things is to be found in the decrees of that time. Artemidorus says, that Timæus of Tauromenium, in consequence of his ignorance of these decrees, and being otherwise a calumniator and detractor, (whence he had the name of Epitimæus, or Reviler,) avers that the Ephesians restored the temple by means of the treasure deposited there by the Persians. But at that time no treasure was deposited, and if any had been deposited there, it must have been consumed together with the temple: after the conflagration, when the roof was destroyed, who would wish to have a deposit lying there, with the sacred enclosure exposed to the air?

Besides, Artemidorus says, that Alexander promised to defray the expense of its restoration, both what had been and what would be incurred, on condition that the work should be attributed to him in the inscription, but the Ephesians refused to accede to this; much less, then, would they be disposed to acquire fame by sacrilege and spoliation. He praises also the reply of an Ephesian to the king, "that it was not fit that a god should provide temples in honour of gods."

23. After the completion of the temple, which, he says, was the work of Cheirocrates (the same person who built Alexandria, and also promised Alexander that he would form [Pg 13]Mount Athos into a statue of him, which should represent him as pouring a libation into a dish out of an ewer; that he would build two cities, one on the right hand of the mountain, and another on the left, and a river should flow out of the dish from one to the other,)[59]— after the completion of the temple, he says that the multitude of other sacred offerings were purchased by the Ephesians, at the value set on them by artificers, and that the altar was almost entirely full of the works of Praxiteles. They showed us also some of the performances of Thraso, namely, the Hecatesium, a Penelope,[60] and the old woman Eurycleia.

The priests were eunuchs, who were called Megabyzi. It was the practice to send to various places for persons worthy of this office, and they were held in high honour. They were obliged to appoint virgins as their colleagues in their priesthood. At present some of their rites and customs are observed, and some are neglected.

The temple was formerly, and is at present, a place of refuge, but the limits of the sanctity of this asylum have been frequently altered; Alexander extended them to the distance of a stadium. Mithridates discharged an arrow from the angle of the roof, and supposed that it fell a little beyond the distance of a stadium. Antonius doubled this distance, and included within the range of the sanctuary a certain portion of the city. This was attended with much evil, as it placed the city in the power of criminals and malefactors. On this account Augustus Cæsar abolished the privilege.

24. The city has an arsenal and a harbour. The entrance of the harbour was made narrow, by order of the king Attalus Philadelphus, who, together with the persons that constructed it, was disappointed at the result. The harbour was formerly shallow, on account of the embankment of earth accumulated by the Caÿster; but the king, supposing that there would be [Pg 14]
[CAS. 641]deep water for the entrance of large vessels of burden, if a mole were thrown up before the mouth of the river, which was very wide, gave orders for the construction of a mole; but the contrary effect took place, for the mud, being confined within the harbour, made the whole of it shallow to the mouth. Before the construction of the mole, the flow and ebb of the sea cleared the mud away entirely, by forcing it outwards.

Such then is the nature of the harbour.

The city, by the advantages which it affords, daily improves, and is the largest mart in Asia within the Taurus.

25. Among illustrious persons in ancient times natives of Ephesus were Heracleitus, surnamed Scoteinus, or the Obscure, and Hermodorus, of whom Heracleitus himself says:

"The Ephesians, youths and all, deserve hanging, for expelling Hermodorus, an honest citizen,[61] a citizen distinguished for his virtues, and saying, let there be no such amongst us; if there be, let it be in another place and among other people."

Hermodorus seems to have compiled laws for the Romans. Hipponax the poet was an Ephesian, and the painters Parrhasius and Apelles.

In more recent times was Alexander the orator, surnamed Lychnus, or the Lamp;[62] he was an administrator of state affairs, a writer of history, and left behind him poems which contain a description of the heavenly phenomena and a geographical account of the continents, each of which forms the subject of a distinct poem.

26. Next to the mouth of the Caÿster is a lake called Selinusia, formed by the overflowing of the sea. It is succeeded by another, which communicates with this. They afford a large revenue, of which the kings, although it was sacred, deprived the goddess, but the Romans restored it; then the [Pg 15]tax-gatherers seized upon the tribute by force, and converted it to their own use. Artemidorus, who was sent on an embassy to Rome, as he says, recovered possession of the lakes for the goddess, and also of the territory of Heracleotis, which was on the point of separating from Ephesus, by proceeding in a suit at Rome. In return for these services, the city erected in the temple to his honour a statue of gold.

In the most retired part of the lake is a temple of a king, built, it is said, by Agamemnon.

27. Next follows the mountain Gallesius, and Colophon, an Ionian city, in front of which is the grove of Apollo Clarius, where was once an ancient oracle.[63] It is said that the prophet Calchas came hither on foot, on his return from Troy with Amphilochus, the son of Amphiaraus, and that meeting at Clarus with a prophet superior to himself, Mopsus, the son of Mantus, the daughter of Teiresias, he died of vexation.

Hesiod relates the fable somewhat in this manner: Calchas propounds to Mopsus something of this kind:

"I am surprised to see how large a quantity of figs there is on this small tree; can you tell me the number?"

Mopsus answered:

"There are ten thousand; they will measure a medimnus, and there is one over, which you cannot comprehend."

Thus he spoke; the number and measure were exact. Then Calchas closed his eyes in the sleep of death.

But Pherecydes says, that Calchas proposed a question respecting a pregnant sow, and asked how many young she had; the other answered, "three, one of which is a sow." Upon his giving the true answer, Calchas died of vexation. According to others, Calchas propounded the question of the sow, and Mopsus that of the fig-tree; that Mopsus returned the true answer, and that Calchas was mistaken, who died of vexation, according to some oracular prophecy.

Sophocles, in his "Helen Claimed," says that he was destined by fate to die when he should meet with a prophet superior to himself. But this writer transfers the scene of the rivalry, and of the death of Calchas, to Cilicia.

These are ancient traditions.

[Pg   16]
[CAS. 643]

28. The Colophonians once possessed a considerable armament, consisting both of ships and of cavalry. In the latter they were so much superior to other nations, that in any obstinate engagement, on whichever side the Colophonian horse were auxiliaries, they decided it; whence came the proverb, "he put the Colophon to it," when a person brought any affair to a decisive issue.[64]

Among some of the remarkable persons born at Colophon were Mimnermus, a flute-player and an elegiac poet; Xenophanes, the natural philosopher, who composed Silli in verse. Pindar mentions one Polymnastus also, a Colophonian, as distinguished for his skill in music:

"Thou knowest the celebrated strains of Polymnastus, the Colophonian:"

and some writers affirm that Homer was of that city. The voyage from Ephesus in a straight line is 70 stadia, and including the winding of the bays, 120.

29. Next to Colophon is the mountain Coracium, and a small island sacred to Artemis, to which it is believed that the hinds swim across to bring forth their young.

Then follows Lebedos,[65] distant from Colophon 120 stadia. This is the place of meeting and residence[66] of the Dionysiac artists (who travel about) Ionia as far as the Hellespont. In Ionia a general assembly is held, and games are celebrated every year in honour of Bacchus. These artists formerly inhabited Teos,[67] a city of the Ionians, next in order after Colophon, but on the breaking out of a sedition they took refuge at Ephesus; and when Attalus settled them at Myonnesus,[68] between Teos and Lebedos, the Teians sent a deputation to request the Romans not to permit Myonnesus to be fortified, as it would endanger their safety. They migrated to Lebedos, and the Lebedians were glad to receive them, on account of their own scanty population.

Teos is distant from Lebedos 120 stadia. Between these two places is the island Aspis,[69] which some writers call Arconnesus.[Pg 17] Myonnesus is situated upon high ground resembling a peninsula.

30. Teos is situated upon a peninsula, and has a port. Anacreon, the lyric poet, was a native of this place; in his time, the Teians, unable to endure the insults and injuries of the Persians, abandoned Teos, and removed to Abdera, whence originated the verse—

"Abdera, the beautiful colony of the Teians."

Some of them returned in after-times to their own country. We have said that Apellicon was of Teos, and Hecataeus also, the historian.

There is another port to the north, at the distance of 30 stadia from the city, Gerrhaeidae.[70]

31. Next follows Chalcideis, and the isthmus of the peninsula[71] of the Teians and Erythraeans; the latter inhabit the interior of the isthmus. The Teians and Clazomenians are situated on the isthmus itself. The Teians occupy the southern side of the isthmus, namely, Chalcideis;[72] the Clazomenians, the northern side, whence they are contiguous to the Erythraean district. At the commencement of the isthmus is Hypocremnus, having on this side the Erythraean, and on the other, the Clazomenian territory. Above Chalcideis is a grove, dedicated to Alexander, the son of Philip, and a festival called Alexandreia is proclaimed and celebrated there by the common body of the Ionians.

The passage across the isthmus from the Alexandrine grove and Chalcideis, as far as the Hypocremnus, is 50 stadia (150?). The circuit round by sea is more than 1000 stadia. Somewhere about the middle of the voyage is Erythrae,[73] an Ionian city, with a port, having in front four small islands, called Hippoi (the Horses).

32. But before we come to Erythrae, the first place we meet with is Erae,[74] a small city belonging to the Teians.

Next is Corycus, a lofty mountain; and below it, Casystes, a port;[75] then another, called the port of Erythrae, and afterwards many others.

[Pg   18]

[CAS. 644]It is said that the whole sea-coast along the Corycus was the haunt of pirates, who were called Corycaeans, and who had contrived a new mode of attacking vessels. They dispersed themselves among the ports, and went among the merchants who had just arrived, and listened to their conversation respecting the freight of their ships, and the places whither they were bound. The pirates then collected together, attacked the merchants at sea, and plundered the vessels. Hence all inquisitive persons and those who listen to private and secret conversation we call Corycaeans, and say proverbially,

"The Corycaean must have overheard it,"

when any one thinks that he has done or said anything not to be divulged, but is betrayed by spies or persons anxious to be informed of what does not concern them.

33. Next to Corycus is Halonnesus, a small island, then the Argennum,[76] a promontory of the Erythræan territory, situated close to Poseidium, belonging to the Chians, and forming a strait of about 60 stadia in width. Between Erythræ and Hypocremnus is Mimas,[77] a lofty mountain, abounding with beasts of chase, and well wooded. Then follows Cybelia, a village, and a promontory called Melæna,[78](or Black,) which has a quarry whence millstones are obtained.

34. Erythræ was the native place of the Sibyl, an ancient inspired prophetess. In the time of Alexander there was another Sibyl, who was also a prophetess, whose name was Athenais, a native of the same city; and in our age there was Heracleides the Herophilian physician, a native of Erythræ, a fellow-student of Apollonius surnamed Mus.

35. The coasting circumnavigation of Chios is 900 stadia. It has a city[79] with a good port, and a station for eighty vessels. In the voyage round the island, a person sailing from the city, with the island on his right hand, first meets with Poseidium,[80]then Phanæ,[81] a deep harbour, and a temple of Apollo, and a grove of palm trees; then Notium, a part of the coast affording a shelter for vessels; next Laïus,[82] which is also a place of [Pg 19]shelter for vessels; hence to the city is an isthmus of 60 stadia. The circumnavigation is 360 stadia, as I have before described it. Next, the promontory Melæna,[83] opposite to which is Psyra,[84] an island distant from the promontory 50 stadia, lofty, with a city of the same name. The island is 40 stadia in circumference. Next is the rugged tract, Ariusia, without harbours, about 30 stadia in extent. It produces the best of the Grecian wines. Then follows Pelinæum,[85] the highest mountain in the island. In the island is a marble quarry.

Among illustrious natives of Chios were Ion[86] the tragic writer, Theopompus the historian, and Theocritus the sophist. The two latter persons were opposed to each other in the political parties in the state. The Chians claim Homer as a native of their country, alleging as a proof the Homeridæ, as they are called, descendants from his family, whom Pindar mentions:

"Whence also the Homeridæ, the chanters of the rhapsodies, most frequently begin their song."[87]

The Chians once possessed a naval force, and aspired to the sovereignty of the sea, and to liberty.[88]

From Chios to Lesbos is a voyage of about 400 stadia, with a south wind.

[Pg 20]

[CAS. 645]36. After the Hypocremnus is Chytrium, a place where Clazomenæ[89] formerly stood; then the present city, having in front eight small islands, the land of which is cultivated by husbandmen.

Anaxagoras, the natural philosopher, was a distinguished Clazomenian; he was a disciple of Anaximenes the Milesian, and master of Archelaus the natural philosopher, and of Euripides the poet.

Next is a temple of Apollo, and hot springs, the bay of Smyrna, and the city Smyrna.

37. Next is another bay, on which is situated the ancient Smyrna, at the distance of 20 stadia from the present city. After Smyrna had been razed by the Lydians, the inhabitants continued for about four hundred years to live in villages. It was then restored by Antigonus, and afterwards by Lysimachus, and at present it is the most beautiful city in Ionia.

One portion of Smyrna is built up on a hill, but the greater part is in the plain near the harbour, the Metroum, and the Gymnasium. The division of the streets is excellent, and as nearly as possible in straight lines. There are paved roads, large quadrangular porticos, both on a level with the ground and with an upper story.

There is also a library, and the Homereium, a quadrangular portico, which has a temple of Homer and a statue. For the Smyrnæans, above all others, urge the claims of their city to be the birth-place of Homer, and they have a sort of brass money, called Homereium.[90]

The river Meles flows near the walls. Besides other conveniences with which the city is furnished, there is a close harbour.

There is one, and not a trifling, defect in the work of the architects, that when they paved the roads, they did not make drains beneath them; the filth consequently lies on the surface, and, during rains particularly, the receptacles of the filth spread it over the streets.

It was here that Dolabella besieged and slew Trebonius, one of the murderers of divus Cæsar; he also destroyed many parts of the city.

38. Next to Smyrna is Leucæ,[21] a small city, which Aristonicus[Pg 21] caused to revolt, after the death of Attalus, the son of Philometor,[22] under pretence of being descended from the royal family, but with the intention of usurping the kingdom. He was, however, defeated in a naval engagement by the Ephesians, near the Cumæan district, and expelled. But he went into the interior of the country, and quickly collected together a multitude of needy people and slaves, who were induced to follow him by the hope of obtaining their freedom, whom he called Heliopolitæ. He first surprised Thyateira,[23] he then got possession of Apollonis, and had an intention of making himself master of other fortresses, but he did not maintain his ground long. The cities sent immediately a large body of troops against him, and were supported by Nicomedes the Bithynian and the kings of Cappadocia. Afterwards five deputies of the Romans came, then an army, and the consul Publius Crassus. These were followed by M. Perperna, who took Aristonicus prisoner, sent him to Rome, and thus put an end to the war. Aristonicus died in prison; Perperna died of some disease, and Crassus fell near Leucæ, in a skirmish with some people who had attacked him from an ambuscade. Manius Aquillius the consul came afterwards, with ten lieutenants; he regulated the affairs of the province, and established that form of government which continues at present.

After Leucæ follows Phocæa,[24] situated on a bay. I have mentioned this place in the description of Massalia.[25] Then follow the confines of the Ionians and the Æolians. I have already spoken of these.[26]

In the interior of the Ionian maritime territory there remain to be described the places about the road leading from Ephesus, as far as Antioch[27] and the Mæander.

This tract is occupied by a mixed population of Lydians, Carians, and Greeks.

39. The first place after Ephesus is Magnesia, an Æolian city, and called Magnesia on the Mæander, for it is situated near it; but it is still nearer the Lethæus, which discharges itself into the Mæander. It has its source in Pactyes, a mountain in the Ephesian district. There is another Lethæus in [Pg 22] [CAS. 647]Gortyne, a third near Tricca, where Asclepius is said to have been born, and the fourth among the Hesperitæ Libyans.[98]

Magnesia lies in a plain, near a mountain called Thorax,[99] on which it is said Daphitas the grammarian was crucified, for reviling the kings in a distich—

"O slaves, with backs purpled with stripes, filings of the gold of Lysimachus, you are the kings of Lydia and Phrygia."

An oracle is said to have warned Daphitas to beware of the Thorax.[100]

40. The Magnesians appear to be the descendants of Delphians who inhabited the Didymæan mountains in Thessaly, and of whom Hesiod says,

"or, as the chaste virgin, who inhabits the sacred Didymæan hills in the plain of Dotium, opposite Amyrus, abounding with vines, and bathes her feet in the lake Bœbias—"

At Magnesia also was the temple of Dindymene, the mother of the gods. Her priestess, according to some writers, was the daughter, according to others, the wife, of Themistocles. At present there is no temple, because the city has been transferred to another place. In the present city is the temple of Artemis Leucophryene, which in the size of the nave and in the number of sacred offerings is inferior to the temple at Ephesus; but, in the fine proportion and the skill exhibited in the structure of the enclosure, it greatly surpasses the Ephesian temple; in size it is superior to all the temples in Asia, except that at Ephesus and that at Didymi.

Anciently the Magnetes were utterly extirpated by Treres, a Cimmerian tribe, who for a long period made successful inroads. Subsequently Ephesians got possession of the place.[101] Callinus speaks of the Magnetes as still in a flourishing state, and successful in the war against the Ephesians. But Archilochus[Pg 23] seems to have been acquainted with the calamities which had befallen them:

"bewail the misfortunes of the Thasians, not of the Magnetes;"

whence we may conjecture that Archilochus was posterior to Callinus. Yet Callinus mentions some other earlier inroad of the Cimmerians, when he says—

"and now the army of the daring Cimmerians is advancing,"

where he is speaking of the capture of Sardis.

41. Among the illustrious natives of Magnesia were Hegesias the orator, who first introduced the Asiatic fervour, as it was called, and corrupted the established Attic style of eloquence; Simon (Simus?) the lyric poet, who also corrupted the system and plan of former lyric poets, by introducing the Simodia; it was still more corrupted by the Lysiodi and Magodi;[102] Cleomachus the pugilist, who was enamoured of a certain cinædus, and a female servant, who was maintained by the cinædus, imitated the sort of dialect and the manners of the cinædi. Sotades was the first person that employed the language of the cinædi, and he was followed by Alexander the Ætolian; but these were only prose writers. Lysis added verse, but this had been done before his time by Simus.

The theatres had raised the reputation of Anaxenor, the player on the cithara, but Antony elevated him as high as possible, by appointing him receiver of the tribute from four cities, and by giving him a guard of soldiers for the protection of his person. His native country also augmented his dignity, by investing him with the sacred purple of Jupiter Sosipolis, as is represented in the painted figure in the forum. There is also in the theatre a figure in brass, with this inscription:

[Pg 24]

[CAS. 648]

"It is truly delightful to listen to a minstrel such as he is, whose voice is like that of the gods."[103]

The artist who engraved the words was inattentive to the space which they would occupy, and omitted the last letter of the second verse, AYΔHI, (voice,) the breadth of the base not being large enough to allow its insertion; this afforded an occasion of accusing the citizens of ignorance, on account of the ambiguity of the inscription; for it is not clear whether the nominative AYΔH, or the dative AYΔHI, is to be understood, for many persons write the dative cases without the I, and reject the usage, as not founded on any natural reason.

42. After Magnesia is the road to Tralles;[104] travellers have on the left hand Mesogis,[105] and on the right hand, and from the road itself, the plain of the Mæander, which is occupied in common by Lydians, Carians, Ionians, Milesians, Mysians, and the Æolians of Magnesia.

The character of the sites of places is the same even as far as Nysa[106] and Antioch.

The city of Tralles is built upon ground in the shape somewhat of a trapezium. It has a citadel strongly fortified, and the places around are well defended. It is as well peopled as any of the cities in Asia, and its inhabitants are wealthy; some of them constantly occupy chief stations in the province, and are called Asiarchs. Among the latter was Pythodorus, originally a native of Nysa; but, induced by the celebrity of the place, he migrated hither. He was one of the few friends of Pompey who were fortunate. His wealth was kingly, and consisted of more than two thousand talents, which he redeemed when it was confiscated by divus Cæsar, on account of his attachment to Pompey, and left it undiminished to his children. Pythodoris, who is at present queen in Pontus, and whom we have mentioned before, is his daughter. Pythodorus flourished in our times, and also Menodorus, an eloquent man, and a person of dignified and grave demeanour; he was priest of Jupiter Larisæus. He was circumvented by the adherents of Domitius Ænobarbus, who, on the credit of [Pg 25]informers, put him to death, for attempting, as was supposed, the revolt of his fleet.

Tralles produced also celebrated orators, Dionysocles, and after him Damasus, surnamed Scombrus.

It is said to have been founded by Argives and a body of Tralli Thracians,[107] from whom it had its name. It was governed for a short time by tyrants, sons of Cratippus, about the period of the Mithridatic war.

43. Nysa is situated near the Mesogis, resting for the most part against the mountain. It is as it were a double town, for a kind of torrent watercourse divides it into two parts, and forms a valley, one part of which has a bridge over it, connecting the two towns; the other is adorned with an amphitheatre; underneath it is a passage through which the waters of the torrents flow out of sight.

Near the theatre are situated[108] two heights; below one lies the gymnasium for the young men; below the other is the forum, and a place of exercise for older persons. To the south below the city lies the plain, as at Tralles.

44. On the road between Tralles and Nysa is a village of the Nysæans, not far from the city Acharaca, in which is the Plutonium, to which is attached a large grove, a temple of Pluto and Proserpine, and the Charonium, a cave which overhangs the grove, and possesses some singular physical properties. The sick, it is said, who have confidence in the cures performed by these deities, resort thither, and live in the village near the cave, among experienced priests, who sleep at night in the open air, on behoof of the sick, and direct the modes of cure by their dreams. The priests invoke the gods to cure the sick, and frequently take them into the cave, where, as in a den, they are placed to remain in quiet without food for several days. Sometimes the sick themselves observe their own dreams, but apply to these persons, in their character of priests and guardians of the mysteries, to interpret them, and to counsel what is to be done. To others the place is interdicted and fatal.

An annual festival, to which there is a general resort, is celebrated at Acharaca, and at that time particularly are to be [Pg 26] [CAS. 650]seen and heard those who frequent it, conversing about cures performed there.[109] During this feast the young men of the gymnasium and the ephebi, naked and anointed with oil,[110] carry off a bull by stealth at midnight, and hurry it away into the cave. It is then let loose, and after proceeding a short distance falls down and expires.

45. Thirty stadia from Nysa, as you cross the Mesogis towards the southern parts of Mount Tmolus,[111] is a place called Leimon, or the Meadow, to which the Nysæans and all the people around repair when they celebrate a festival. Not far from this plain is an aperture in the ground, sacred to the same deities, which aperture is said to extend as far as Acharaca. They say that the poet mentions this meadow, in the words,

"On the Asian mead,"[112]

and they show a temple dedicated to two heroes, Caÿstrius and Asius, and the Caÿster flowing near it.

46. Historians relate that three brothers, Athymbrus, Athymbradus, and Hydrelus, coming hither from Lacedæmon, founded (three?) cities, to which they gave their own names; that the population of these towns afterwards declined, but that out of these jointly Nysa was peopled. The Nysæans at present regard Athymbrus as their founder.

47. Beyond the Mæander and in the neighbourhood are considerable settlements, Coscinia[113] and Orthosia, and on this side the river, Briula, Mastaura,[114] Acharaca, and above the city on the mountain, Aroma; the letter *o* is shortened in the pronunciation. From this latter place is obtained the Aromeus, the best Mesogitian wine.

48. Among illustrious natives of Nysa were Apollonius the Stoic philosopher, the most eminent of the disciples of Panætius, and of Menecrates, the disciple of Aristarchus; Aristodemus, the son of Menecrates, whom, when I was a very young man, I heard lecturing on philosophy, in extreme old [Pg 27]age, at Nysa; Sostratus, the brother of Aristodemus, and another Aristodemus, his cousin, the master of Pompey the Great, were distinguished grammarians. My master taught rhetoric also at Rhodes, and in his own country he had two schools; in the morning he taught rhetoric, in the evening grammar. When he superintended the education of the children of Pompey at Rome, he was satisfied with teaching a school of grammar.

CHAPTER II.

1. THE places beyond the Mæander, which remain to be described, belong to the Carians. The Carians here are not intermixed with Lydians, but occupy the whole country by

themselves, if we except a small portion of the sea-coast, of which the Milesians and Mysians have taken possession.

Caria[115] begins on the sea-coast opposite to Rhodes, and ends at Poseidium,[116]belonging to the Milesians. In the interior are the extremities of Taurus, which extend as far as the Mæander. For the mountains situated above the Chelidonian islands,[117] as they are called, which lie in front of the confines of Pamphylia and Lycia, are, it is said, the beginning of the Taurus; for the Taurus has there some elevation, and indeed a mountainous ridge of Taurus separates the whole of Lycia towards the exterior and the southern part from Cibyra and its district, as far as the country opposite to Rhodes. Even there a mountainous tract is continued; it is, however, much lower in height, and is not considered as any longer belonging to Taurus, nor is there the distinction of parts lying within and parts lying without the Taurus, on account of the eminences and depressions being scattered about through the whole country both in breadth and length, and not presenting anything like a separation-wall.

The whole voyage along the coast, including the winding [Pg 28] [CAS. 651]of the bays, is 4900 stadia, and that along the country opposite to Rhodus 1500 stadia.

2. The beginning of this tract is Dædala,[118] a stronghold; and ends at the mountain Phœnix,[119] as it is called, both of which belong to the Rhodian territory. In front, at the distance of 120 stadia from Rhodes, lies Eleussa.[120] In sailing from Dædala towards the west in a straight line along Cilicia, Pamphylia, and Lycia, in the midway is a bay called Glaucus, with good harbours; then is the promontory Artemisium, and a temple; next, the grove sacred to Latona; above this, and at the distance of 60 stadia, is Calynda, a city; then Caunus,[121] and a deep river near it, the Calbis,[122] which may be entered by vessels; between these is Pisilis.

3. The city Caunus has a naval arsenal and a close harbour. Above the city upon a height is Imbrus, a stronghold. Although the country is fertile, yet the city is allowed by all to be unhealthy in summer, on account of the heat, and in autumn, from the abundance of fruits.

Stories of the following kind are related respecting the city. Stratonicus, the player on the cithara, seeing the Caunians somewhat dark and yellow,[123] said that this was what the poet meant in the line,

"As are the leaves, so is the race of men."[124]

When he was accused of ridiculing the unhealthiness of the city, he answered, "Can I be so bold as to call that city unhealthy, where even the dead walk about?"

The Caunians once revolted from the Rhodians, but, by a decision of the Romans, they were received again by the Rhodians into favour. There is in existence an oration of Molo against the Caunians.

It is said that they speak the same language as the Carians, that they came from Crete, and retained their own laws and customs.[125]

[Pg 29]4. Next is Physcus,[126] a small town; it has a port and a grove sacred to Latona: then Loryma, a rugged line of sea-coast, and a mountain, the highest of any in that quarter, on the summit of which is Phœnix, a stronghold, of the same name as the mountain. In front is the island Eleussa, at the distance of 4 stadia. Its circumference is about 8 stadia.

5. The city of the Rhodians is on the eastern promontory. With regard to harbours, roads, walls, and other buildings, it so much surpasses other cities, that we know of none equal, much less superior to it.

Their political constitution and laws were excellent, and the care admirable with which they administered affairs of state generally, and particularly those relative to their marine. Hence being for a long period masters of the sea, they put an end to piracy, and became allies of the Romans, and of those kings who were well affected to the Romans and the Greeks; hence also the city was suffered to preserve her independence, and was embellished with many votive offerings. These are distributed in various places, but the greatest part of them are deposited in the Dionysium and in the gymnasium. The most remarkable is the Colossus of the Sun, which, the author of the iambics says, was "seventy cubits in height, the work of Chares of Lindus."

It now lies on the ground, having been thrown down by an earthquake, and is broken off at the knees. An oracle prohibited its being raised again. This is the most remarkable of the votive offerings, and it is allowed to be one of the seven wonders of the world.[127] There were also the pictures by Protogenes,[128] the Ialysus, and the Satyr, who was represented [Pg 30] [CAS. 652]standing by a pillar. On the top of the pillar was a partridge. The bird strongly attracted, as was natural, the gaping admiration of the people, when the picture was first hung up in public, and they were so much delighted, that the Satyr, although executed with great skill, was not noticed. The partridge-breeders were still more struck with the picture of the bird. They brought tame partridges, which, when placed opposite to the picture, made their call, and drew together crowds of people. When Protogenes observed that the principal had become the subordinate part of his work, he obtained permission of the curators of the temple to efface the bird, which he did.

The Rhodians, although their form of government is not democratic, are attentive to the welfare of the people, and endeavour to maintain the multitude of poor. The people receive allowances of corn, and the rich support the needy, according to an ancient usage. There are also public offices in the state, the object of which is to procure and distribute provisions,[129] so that the poor may obtain subsistence, and the city not suffer for want of persons to serve her, especially in manning her fleets.

Some of the dockyards are kept private, and the multitude are prohibited from seeing them. If any person should be found inspecting, or to have entered them, he would be punished with death. As at Massalia and Cyzicus,[130] so here particularly, everything relating to architects, the manufacture of engines, stores of arms, and of other materials, is administered with peculiar care, much more so than in other places.

6. Like the people of Halicarnassus,[131] Cnidus, and Cos, the Rhodians are of Doric origin. Some of the Dorians, who founded Megara after the death of Codrus, remained there; others associated themselves with the colony which went to Crete under the conduct of Althæmenes the Argive; the rest were distributed at Rhodus, and among the cities just mentioned.

But these migrations are more recent than the events related[Pg 31] by Homer. For Cnidus and Halicarnassus were not then in existence. Rhodes and Cos existed, but were inhabited by Heracleidæ. Tlepolemus, when he attained manhood,

"slew the maternal uncle of his father, the aged Licymnius. He immediately built ships, and, collecting a large body of people, fled away with them:"[132]

and adds afterwards—

"after many sufferings on the voyage, he came to Rhodes; they settled there according to their tribes, in three bodies:"

and mentions by name the cities then existing[133]—

"Lindus, Ialysus, and the white Cameirus,"

the city of the Rhodians not being yet founded.

Homer does not here mention Dorians by name, but means Æolians and Bœotians, since Hercules and Licymnius lived in Bœotia. If however, as others relate, Tlepolemus set out from Argos and Tiryns, even so the colony would not be Dorian, for it was settled before the return of the Heracleidæ.

And of the Coans also Homer says—

"their leaders were Pheidippus and Antiphus, two sons of Thessalus the King, an Heracleid;"[134]

and these names designate rather an Æolian than a Dorian origin.

7. Rhodes was formerly called Ophiussa and Stadia, then Telchinis, from the Telchines, who inhabited the island.[135]

These Telchines are called by some writers charmers and enchanters, who besprinkle animals and plants, with a view to destroy them, with the water of the Styx, mingled with sulphur. Others on the contrary say, that they were persons who excelled in certain mechanical arts, and that they were calumniated by jealous rivals, and thus acquired a bad reputation; that they came from Crete, and first landed at Cyprus, and then removed to Rhodes. They were the first workers in iron and brass, and were the makers of Saturn's scythe.

I have spoken of them before, but the variety of fables [Pg 32][CAS. 654] which are related of them induces me to resume their history, and to supply what may have been omitted.

8. After the Telchines, the Heliadæ[136] were said, according to fabulous accounts, to have occupied the island. One of these Heliadæ, Cercaphus, and his wife Cydippe had children, who founded the cities called after their names—

"Lindus, Ialysus, and the white Cameirus."[137]

Others say, that Tlepolemus founded them, and gave to them the names of some of the daughters of Danaüs.

9. The present city was built during the Peloponnesian war, by the same architect,[138] it is said, who built the Piræus. The Piræus, however, does not continue to exist, having formerly sustained injuries from the Lacedæmonians, who threw down the walls, and then from Sylla, the Roman general.

10. It is related of the Rhodians that their maritime affairs were in a flourishing state, not only from the time of the foundation of the present city, but that many years before the institution of the Olympic festival, they sailed to a great distance from their own country for the protection of sailors. They sailed as far as Spain, and there founded Rhodus, which the people of Marseilles afterwards occupied; they founded Parthenope[139] among the Opici, and Elpiæ in Daunia, with the assistance of Coans. Some authors relate, that after their return from Troy they colonized the Gymnasian islands. According to Timæus, the greater of these islands is the largest known,[140] next the seven following, Sardinia, Sicily, Cyprus, Crete, Eubœa,[141] Corsica, and Lesbos; but this is a mistake, for these others are much larger. It is said, that gymnetes (or light-armed soldiers[142]) are called by the Phœnicians balearides, and that from hence the Gymnasian islands were called Balearides.

Some of the Rhodians settled in the neighbourhood of [Pg 33]Sybaris, in the Chonian territory.[143] Homer seems to bear evidence of the former prosperity of the Rhodians, from the very foundation of the three cities;

"they settled according to their tribes, in three companies, and were the favourites of Jupiter, who showered upon them great wealth."[144]

Other writers have applied these verses to a fable, according to which, at the birth of Minerva, it rained gold on the island from the head of Jupiter, as Pindar has said.[145]

The island is 920 stadia in circumference.

11. In sailing from the city, and leaving the island on the right hand, the first place we meet with is Lindus,[146] a city situated on a mountain extending far towards the south, and particularly towards Alexandreia (in Egypt).[147] There is here a celebrated temple of the Lindian Diana, built by the Danaides. Formerly, the Lindians, like the inhabitants of Cameirus,[148] and Ialyssus, formed an independent state, but afterwards they all settled at Rhodes.

Cleobulus, one of the seven wise men, was a native of Lindus.

12. Next to Lindus is Ixia,[149] a stronghold, and Mnasyrium; then the Atabyris,[150]the highest mountain in the island, sacred to Jupiter Atabyrius; then Cameirus; then Ialysus a village, and above it is an acropolis called Ochyroma (the Fortification); then, at the distance of about 80 stadia, the city of the Rhodians. Between these is the Thoantium, a sort of beach, immediately in front of which are situated the Sporades islands lying about Chalcis, which we have mentioned before.[151]

13. There have been many remarkable persons, natives of Rhodes, both generals and athletæ, among whom were the ancestors of Panætius the philosopher. Among statesmen, orators, and philosophers, were Panætius, Stratocles, Andronicus the Peripatetic, Leonides the Stoic, and long before the time of these persons, Praxiphanes, Hieronymus, and Eudēmus. Poseidonius was concerned in the administration of the affairs of state, and taught philosophy at Rhodes, (but he was a native of Apameia in Syria,) as did Apollonius Malacus,

[CAS. 655]Molon, who were natives of Alabanda, and disciples of Menecles the rhetorician. Apollonius had resided at Rhodes long before, but Molon came late; whence the former said to him "late comer," Ὀψὲ μολών, instead of ἐλθών.[152] Peisander, a Rhodian poet, author of the Heracleia; Simmias the grammarian, and Aristocles, of our time. Dionysius the Thracian, and Apollonius, author of the Argonautics, although natives of Alexandreia, were called Rhodians.

This is sufficient on the subject of the island of Rhodes.

14. There is a bend of the Carian coast opposite to Rhodes, immediately after Eleus[153] and Loryma, towards the north, and then the ship's course is in a straight line to the Propontis,[154] and forms as it were a meridian line of about 500 stadia in length, or somewhat less. Along this line are situated the remainder of Caria, Ionians, Æolians, Troy, and the parts about Cyzicus and Byzantium. Next to Loryma is the Cynossema, or dogs' monument,[155] and the island Syme.[156]

15. Then follows Cnidus,[157] which has two harbours, one of which is a close harbour, fit for receiving triremes, and a naval station for 20 vessels. In front of Cnidus is an island, in circumference about 7 stadia; it rises high, in the form of a theatre, and is united by a mole to the continent, and almost makes Cnidus a double city, for a great part of the inhabitants occupy the island, which shelters both harbours. Opposite to it, far out at sea, is Nisyrus.[158]

Illustrious natives of Cnidus were, first, Eudoxus the mathematician, a disciple of Plato's; Agatharchides, the Peripatetic philosopher and historian; Theopompus, one of the most powerful of the friends of divus Cæsar, and his son Artemidorus. Ctesias also, the physician of Artaxerxes, was a native of this place. He wrote a history of Assyria and Persia.

Next after Cnidus are Ceramus[159] and Bargasa, small towns overlooking the sea.

16. Then follows Halicarnasus, formerly called Zephyra, the royal seat of the dynasts of Caria. Here is the sepulchre of Mausolus, one of the seven wonders of the world;[160] [Pg 35]Artemisia erected it, in honour of her husband. Here also is the fountain Salmacis, which has a bad repute, for what reason I know not, for making those who drink of it effeminate. Mankind, enervated by luxury, impute the blame of its effects to different kinds of air and water, but these are not the causes of luxury, but riches and intemperance.

There is an acropolis at Halicarnasus. In front of it lies Arconnesus.[161] It had, among others, as its founders, Anthes and a body of Trœzenians.[162]

Among the natives of Halicarnasus were Herodotus the historian, who was afterwards called Thurius, because he was concerned in sending out the colony to Thurii; Heracleitus the poet, the friend of Callimachus; and in our time, Dionysius the historian.

17. Halicarnasus suffered, when it was taken by storm by Alexander. Hecatomnus, who was then king of the Carians, had three sons, Mausolus, Hidrieus, and Pixodarus, and two daughters. Mausolus, the eldest son, married Artemisia, the eldest daughter; Hidrieus, the second son, married Ada, the other sister. Mausolus came to the throne, and, dying without children, left the kingdom to his wife, by whom the above-mentioned sepulchre was erected. She pined away for grief at the loss of her husband. Hidrieus succeeded her; he died a natural death, and was succeeded by his wife Ada. She was ejected by Pixodarus, the surviving son of Hecatomnus. Having espoused the party of the Persians, Pixodarus sent for a satrap to share the kingdom with him. After the death of Pixodarus, the satrap became master of Halicarnasus. But upon the arrival of Alexander, he sustained a siege. His wife was Ada, daughter of Pixodarus, and Aphneïs, a woman of Cappadocia. But Ada, the daughter of Hecatomnus, whom Pixodarus ejected, entreated Alexander, and endeavoured to prevail upon him to reïnstate her in the kingdom of which she had been deprived; she promised (in return) her assistance in reducing to obedience the parts of the country which had revolted; for the persons who were in possession of them [Pg 36] [CAS. 657]were her relations and subjects. She also delivered up Alinda where she herself resided. Alexander granted her request and proclaimed her queen, after the city was taken, but not the acropolis, which was doubly fortified. He assigned to Ada the siege of the acropolis, which was taken in a short time afterwards, the besiegers having attacked it with fury and exasperation at the resistance of the besieged.

18. Next is Termerium,[163] a promontory of the Myndians, opposite to which lies Scandaria, a promontory of Cos, distant 40 stadia from the continent. There is also above the promontory a fortress, Termerum.

19. The city of the Coans was formerly called Astypalæa, and was built in another place, but is at present on the sea-coast. Afterwards, on account of a sedition, they migrated to the present city, near Scandarium, and changed the name to that of the island, Cos. The city is not large, but beautifully built, and a most pleasing sight to mariners who are sailing by the coast. The island is about 550 stadia in circumference. The whole of it is fertile, and produces, like Chios and Lesbos, excellent wine. It has, towards the south, the promontory Laceter,[164] from which to Nicyrus is 60 stadia, and near Laceter is Halisarna, a stronghold; on the west is Drecanum, and a village called Stomalimne. Drecanum is distant about 200 stadia from the city. The promontory Laceter adds to the length of the navigation 35 stadia. In the suburb is the celebrated temple Asclepieium, full of votive offerings, among which is the Antigonus of Apelles. It formerly contained the Venus Anadyomene, (Venus emerging from the sea,) but that is now at Rome, dedicated to divus Cæsar by Augustus, who consecrated to his father the picture of her who was the author of his family. It is said that the Coans obtained, as a compensation for the loss of this painting, an abatement, amounting to a hundred talents, of their usual tribute.

It is said, that Hippocrates learned and practised the dietetic part of medicine from the narrative of cures suspended in the temple. He is one of the illustrious natives of Cos. Simus, also, the physician, Philetas the poet and critic, Nicias of our time, who was tyrant of Cos; Ariston, the disciple and heir of Ariston the Peripatetic philosopher; and Theomnestus, a minstrel of name, who was of the opposite political party to Nicias.

[Pg 37]20. On the coast of the continent opposite to the Myndian territory is Astypalæa a promontory, and Zephyrium. The city Myndus follows immediately after, which has a harbour; then the city Bargylia. In the intervening distance is Caryanda[165] a harbour, and an island of the same name, occupied by Caryandians. Scylax the ancient historian was a native of this island. Near Bargylia is the temple of Artemis Cindyas, round which the rain falls, it is believed, without touching it. There was once a strong place called Cindya.

Among the distinguished natives of Bargylia was Protarchus the Epicurean; Demetrius surnamed Lacon was his disciple.

21. Next follows Iasus, situated upon an island,[166] on the side towards the continent. It has a port, and the inhabitants derive the greatest part of their subsistence from the sea, which abounds with fish, but the soil is very barren. Stories of the following kind are related of Iasus.

As a player on the cithara was displaying his art in public, every one listened to him attentively till the market bell rung for the sale of fish, when he was deserted by all except one man, who was quite deaf. The minstrel coming up to him said, "Friend, I am much obliged to you for the honour you have done me, and I admire your love of music, for all the others have left me at the sound of the bell."—"What say you, has the bell rung?"—"Yes, he replied?"—"Good bye to you," said the man, and away he also went.

Diodorus the Dialectician was a native of this place. He was surnamed Cronus (or Old Time); the title was not properly his from the first; it was his master Apollonius who (in the first instance) had received the surname of Cronus, but it was transferred to Diodorus on account of the want of celebrity in the true Cronus.

22. Next to Iasus is Cape Poseidium[167] of the Milesians. In the interior are three considerable cities, Mylasa,[168] Stratoniceia,[169] and Alabanda.[170] The others are guard forts to these or to the maritime towns, as Amyzon, Heracleia, Euromus, Chalcetor. But we make little account of these.

23. Mylasa is situated in a very fertile plain; a mountain, containing a very beautiful marble quarry, overhangs the city; and it is no small advantage to have stone for building [Pg 38]
[CAS. 659]in abundance and near at hand, particularly for the construction of temples and other public edifices; consequently, no city is embellished more beautifully than this with portico and temples. It is a subject of surprise, however, that persons should be guilty of the absurdity of building the city at the foot of a perpendicular and lofty precipice. One of the

governors of the province is reported to have said, when he expressed his astonishment at this circumstance, "If the founder of the city had no fear, he had no shame."

The Mylasians have two temples, one of Jupiter called Osogo, and another of Jupiter Labrandenus. The former is in the city. Labranda is a village on the mountain, near the passage across it from Alabanda to Mylasa, at a distance from the city. At Labranda is an ancient temple of Jupiter, and a statue of Jupiter Stratius, who is worshipped by the neighbouring people and by the inhabitants of Mylasa. There is a paved road for a distance of about 60 stadia from the temple to the city; it is called the Sacred Way, along which the sacred things are carried in procession. The most distinguished citizens are always the priests, and hold office during life. These temples belong peculiarly to the city. There is a third temple of the Carian Jupiter, common to all the Carians, in the use of which the Lydians, also, and Mysians participate, as being brethren.

Mylasa is said to have been anciently a village, but the native place and royal residence of Hecatomnus and the Carians. The city approaches nearest to the sea at Physcus, which is their naval arsenal.

24. Mylasa has produced in our time illustrious men, who were at once orators and demagogues, Euthydemus and Hybreas. Euthydemus inherited from his ancestors great wealth and reputation. He possessed commanding eloquence, and was regarded as a person of eminence, not only in his own country, but was thought worthy of the highest honours even in Asia. The father of Hybreas, as he used to relate the circumstance in his school, and as it was confirmed by his fellow-citizens, left him a mule which carried wood, and a mule driver. He was maintained for a short time by their labour, and was enabled to attend the lectures of Diotrephes of Antioch. On his return he held the office of superintendent of the market. But here being harassed, and gaining but[Pg 39] little profit, he applied himself to the affairs of the state, and to attend to the business of the forum. He quickly advanced himself and became an object of admiration, even during the lifetime of Euthydemus, and still more after his death, as the leading person in the city. Euthydemus possessed great power, and used it for the benefit of the city, so that if some of his acts were rather tyrannical, this character was lost in their public utility.

The saying of Hybreas, at the conclusion of an harangue to the people, is applauded: "Euthydemus, you are an evil necessary to the city; for we can live neither with thee nor without thee."[171]

Hybreas, although he had acquired great power, and had the reputation of being both a good citizen and an excellent orator, was defeated in his political opposition to Labienus. For the citizens, unarmed, and disposed to peace, surrendered to Labienus, who attacked them with a body of troops and with Parthian auxiliaries, the Parthians being at that time masters of Asia. But Zeno of Laodiceia and Hybreas, both of them orators, did not surrender, but caused their own cities to revolt. Hybreas provoked Labienus, an irritable and vain young man, by saying, when the youth announced himself emperor of the Parthians, "Then I shall call myself emperor of the Carians." Upon this Labienus marched against the city, having with him cohorts drafted from the Roman soldiery stationed in Asia. He did not however take Hybreas prisoner, who had retreated to Rhodes, but plundered and destroyed his house, which contained costly furniture, and treated the whole city in the same manner. After Labienus had left Asia, Hybreas returned, and restored his own affairs and those of the city to their former state.

This then on the subject of Mylasa.

25. Stratoniceia is a colony of Macedonians. It was embellished by the kings with costly edifices. In the district of the Stratoniceians are two temples. The most celebrated, that of Hecate, is at Lagina, where every year great multitudes assemble at a great festival. Near the city is the temple of Jupiter Chrysoreus,[172] which is common to all the Carians, and whither they repair to offer sacrifice, and to deliberate on their common interests. They call this meeting the Chrysaoreon, [Pg 40][Cas. 660] which is composed of villages. Those who represent the greatest number of villages have the precedency in voting, like the Ceramiētæ. The Stratoniceians, although they are not of Carian race, have a place in this assembly, because they possess villages included in the Chrysaoric body.

29

In the time of our ancestors there flourished at Stratoniceia a distinguished person, Menippus the orator, surnamed Catocas, whom Cicero[173] commends in one of his writings above all the Asiatic orators whom he had heard, comparing him to Xenocles, and to those who flourished at that time.

There is another Stratoniceia, called Stratoniceia at the Taurus, a small town adjacent to the mountain.

26. Alabanda lies at the foot of two eminences, in such a manner as to present the appearance of an ass with panniers. On this account Apollonius Malacus ridicules the city, and also because it abounds with scorpions; he says, it was an ass, with panniers full of scorpions.

This city and Mylasa, and the whole mountainous tract between them, swarm with these reptiles.

The inhabitants of Alabanda are addicted to luxury and debauchery. It contains a great number of singing girls.

Natives of Alabanda, distinguished persons, were two orators, brothers, Menecles, whom we mentioned a little above, and Hierocles, Apollonius, and Molo; the two latter afterwards went to Rhodes.

27. Among the various accounts which are circulated respecting the Carians, the most generally received is that the Carians, then called Leleges, were governed by Minos, and occupied the islands. Then removing to the continent, they obtained possession of a large tract of sea-coast and of the interior, by driving out the former occupiers, who were, for the most part, Leleges and Pelasgi. The Greeks again, Ionians and Dorians, deprived the Carians of a portion of the country.

As proofs of their eager pursuit of war, the handles of shields, badges, and crests, all of which are called Carian, are alleged. Anacreon says,

"Come, grasp the well-made Caric handles;"

and Alcæus—

"Shaking a Carian crest."

[Pg 41]28. But when Homer uses these expressions, "Masthles commanded the Carians, who speak a barbarous language,"[174] it does not appear why, when he was acquainted with so many barbarous nations, he mentions the Carians alone as using a barbarous language, but does not call any people Barbarians. Nor is Thucydides right, who says that none were called Barbarians, because as yet the Greeks were not distinguished by any one name as opposed to some other. But Homer himself refutes this position that the Greeks were not distinguished by this name:

"A man whose fame has spread through Greece and Argos;"[175]

and in another place—

"But if you wish to go through Hellas and the middle of Argos."[176]

But if there was no such term as Barbarian, how could he properly speak of people as Barbarophonoi (i. e. speaking a barbarous language)?

Neither is Thucydides nor Apollonius the grammarian right, because the Greeks, and particularly the Ionians, applied to the Carians a common term in a peculiar and vituperative sense, in consequence of their hatred of them for their animosity and continual hostile incursions. Under these circumstances he might call them Barbarians. But we ask, why does he call them Barbarophonoi, but not once Barbarians? Because, replies Apollonius, the plural number does not fall in with the metre; this is the reason why Homer does not call them Barbarians. Admitting then that the genitive case (βαρβάρων) does not fall in with the measure of the verse, the nominative case (βάρβαροι) does not differ from that of Dardani (Δάρδανοι);

"Trojans, Lycians, and Dardani;"

and of the same kind is the word Troïi[177] in this verse,

"Like the Troïi horses" (Τρώιοι ἵπποι).

Nor is the reason to be found in the alleged excessive harshness of the Carian language, for it is not extremely harsh; and besides, according to Philippus, the author of a history of Caria, their language contains a very large mixture of Greek words.

[Pg 42][Cas. 661] I suppose that the word "barbarian" was at first invented to designate a mode of pronunciation which was embarrassed, harsh, and rough; as we use the words battarizein, traulizein, psellizein,[178] to express the same thing. For we are naturally very much disposed to denote certain sounds by names expressive of those sounds, and characteristic of their nature; and hence invented terms abound, expressive of the sounds which they designate, as kelaryzein, clange, psophos, boe, krotos,[179] most of which words are at present used in an appropriate sense.

As those who pronounce their words with a thick enunciation are called Barbarians, so foreigners, I mean those who were not Greeks, were observed to pronounce their words in this manner. The term Barbarians was therefore applied peculiarly to these people, at first by way of reproach, as having a thick and harsh enunciation; afterwards the term was used improperly, and applied as a common gentile term in contradistinction to the Greeks. For after a long intimacy and intercourse had subsisted with the Barbarians, it no longer appeared that this peculiarity arose from any thickness of enunciation, or a natural defect in the organs of the voice, but from the peculiarities of their languages.

But there was in our language a bad and what might be called a barbarous utterance, as when any person speaking Greek should not pronounce it correctly, but should pronounce the words like the Barbarians, who, when beginning to learn the Greek language, are not able to pronounce it perfectly, as neither are we able to pronounce perfectly their languages.

This was peculiarly the case with the Carians. For other nations had not much intercourse with the Greeks, nor were disposed to adopt the Grecian manner of life, nor to learn our language, with the exception of persons who by accident and singly had associated with a few Greeks; but the Carians were dispersed over the whole of Greece, as mercenary soldiers. Then the barbarous pronunciation was frequently met with among them, from their military expeditions into Greece; and afterwards it spread much more, from the time that they occupied the islands together with the Greeks: not even when [Pg 43]driven thence into Asia, could they live apart from Greeks, when the Ionians and Dorians arrived there.

Hence arose the expression, "to barbarize," for we are accustomed to apply this term to those whose pronunciation of the Greek language is vicious, and not to those who pronounce it like the Carians.

We are then to understand the expressions, "barbarous speaking" and "barbarous speakers," of persons whose pronunciation of the Greek language is faulty. The word "to barbarize" was formed after the word "to Carize," and transferred into the books which teach the Greek language; thus also the word "to solœcize" was formed, derived either from Soli or some other source.

29. Artemidorus says that the journey from Physcus, on the coast opposite to Rhodes, towards Ephesus, as far as Lagina is 850 stadia; thence to Alabanda 250 stadia; to Tralles 160. About halfway on the road to Tralles the Mæander is crossed, and here are the boundaries of Caria. The whole number of stadia from Physcus to the Mæander, along the road to Ephesus, is 1180 stadia. Again, along the same road, from the Mæander of Ionia to Tralles 80 stadia, to Magnesia 140 stadia, to Ephesus 120, to Smyrna 320, to Phocæa and the boundaries of Ionia, less than 200 stadia; so that the length of Ionia in a straight line would be, according to Artemidorus, a little more than 800 stadia.

But as there is a public frequented road by which all travellers pass on their way from Ephesus to the east, Artemidorus thus describes it. [From Ephesus] to Carura, the boundary of Caria towards Phrygia, through Magnesia and Tralles, Nysa, Antioch, is a journey of 740 stadia. From Carura, the first town in Phrygia, through Laodiceia, Apameia, Metropolis, and Chelidoniæ,[180] to Holmi, the beginning of the Paroreius, a country lying at the foot of the mountains, about 920 stadia; to Tyriæum,[181] the termination towards Lycaonia of the Paroreius,[182] through Philomelium[183] is little more than 500 stadia. Next is Lycaonia as far as Coropassus,[184] through Laodiceia in the Catacecaumene, 840 stadia; from Coropassus [Pg 44][Cas. 662] in Lycaonia to Garsaüra,[185] a small city of Cappadocia, situated on its borders,

120 stadia; thence to Mazaca,[186] the metropolis of the Cappadocians, through Soandus and Sadacora, 680 stadia; thence to the Euphrates, as far as Tomisa, a stronghold in Sophene, through Herphæ,[187] a small town, 1440 stadia.

The places in a straight line with these, as far as India, are described in the same manner by Artemidorus and Eratosthenes. Polybius says, that with respect to those places we ought chiefly to depend upon Artemidorus. He begins from Samosata in Commagene, which is situated at the passage, and the Zeugma of the Euphrates, to Samosata across the Taurus, from the mountains of Cappadocia about Tomisa, he says is a distance of 450 stadia.

CHAPTER III.

1. AFTER the part of the coast opposite[188] to Rhodes, the boundary of which is Dædala, in sailing thence towards the east, we come to Lycia, which extends to Pamphylia; next is Pamphylia, extending as far as Cilicia Tracheia, which reaches as far as the Cilicians, situated about the Bay of Issus. These are parts of the peninsula, the isthmus of which we said was the road from Issus as far as Amisus,[189] or, according to some authors, to Sinope.

[Pg 45]The country beyond the Taurus consists of the narrow line of sea-coast extending from Lycia to the places about Soli, the present Pompeiopolis. Then the sea-coast near the Bay of Issus, beginning from Soli and Tarsus, spreads out into plains.

The description of this coast will complete the account of the whole peninsula. We shall then pass to the rest of Asia without the Taurus, and lastly we shall describe Africa.

2. After Dædala of the Rhodians there is a mountain of Lycia, of the same name, Dædala, and here the whole Lycian coast begins, and extends 1720 stadia. This maritime tract is rugged, and difficult to be approached, but has very good harbours, and is inhabited by a people who are not inclined to acts of violence. The country is similar in nature to that of Pamphylia and Cilicia Tracheia. But the former used the places of shelter for vessels for piratical purposes themselves, or afforded to pirates a market for their plunder and stations for their vessels.

At Side,[190] a city of Pamphylia, the Cilicians had places for building ships. They sold their prisoners, whom they admitted were freemen, by notice through the public crier.

But the Lycians continued to live as good citizens, and with so much restraint upon themselves, that although the Pamphylians had succeeded in obtaining the sovereignty of the sea as far as Italy, yet they were never influenced by the desire of base gain, and persevered in administering the affairs of the state according to the laws of the Lycian body.

3. There are three and twenty cities in this body, which have votes. They assemble from each city at a general congress, and select what city they please for their place of meeting. Each of the largest cities commands three votes, those of intermediate importance two, and the rest one vote. They contribute in the same proportion to taxes and other public charges. The six largest cities, according to Artemidorus, are Xanthus,[191] Patara,[192] Pinara,[193] Olympus, Myra, Tlos,[194] which is situated at the pass of the mountain leading to Cibyra.

At the congress a lyciarch is first elected, then the other officers of the body. Public tribunals are also appointed for [Pg 46][Cas. 665] the administration of justice. Formerly they deliberated about war and peace, and alliances, but this is not now permitted, as these things are under the control of the Romans. It is only done by their consent, or when it may be for their own advantage.

Thus judges and magistrates are elected according to the proportion of the number of votes belonging to each city.[195] It was the fortune of these people, who lived under such an excellent government, to retain their liberty under the Romans, and the laws and institutions of their ancestors; to see also the entire extirpation of the pirates, first by Servilius Isauricus, at the time that he demolished Isaura, and afterwards by Pompey the Great, who burnt more than 1300 vessels, and destroyed their haunts and retreats. Of the survivors in these contests he transferred some to Soli, which he called Pompeiopolis; others to Dyme, which had a deficient population, and is now occupied by a Roman colony.

The poets, however, particularly the tragic poets, confound nations together; for instance, Trojans, Mysians, and Lydians, whom they call Phrygians, and give the name of Lycians to Carians.

4. After Dædala is a Lycian mountain, and near it is Telmessus,[196] a small town of the Lycians, and Telmessis, a promontory with a harbour. Eumenes took this place from the Romans in the war with Antiochus, but after the dissolution of the kingdom of Pergamus, the Lycians recovered it again.

5. Then follows Anticragus, a precipitous mountain, on which is Carmylessus,[197] a fortress situated in a gorge; next is Mount Cragus, with eight peaks,[198] and a city of the same name. The neighbourhood of these mountains is the scene of the fable of the Chimæra; and at no great distance is Chimæra, a sort of ravine, extending upwards from the shore. Below the Cragus in the interior is Pinara, which is one of the largest cities of Lycia. Here Pandarus is worshipped, of the same name perhaps as the Trojan Pandarus;

[Pg 47]

"thus the pale nightingale, daughter of Pandarus;"[199]

for this Pandarus, it is said, came from Lycia.

6. Next is the river Xanthus, formerly called Sirbis.[200] In sailing up it in vessels which ply as tenders, to the distance of 10 stadia, we come to the Letoum, and proceeding 60 stadia beyond the temple, we find the city of the Xanthians, the largest in Lycia. After the Xanthus follows Patara, which is also a large city with a harbour, and containing a temple of Apollo. Its founder was Patarus. When Ptolemy Philadelphus repaired it, he called it the Lycian Arsinoë, but the old name prevailed.

7. Next is Myra, at the distance of 20 stadia from the sea, situated upon a lofty hill; then the mouth of the river Limyrus, and on ascending from it by land 20 stadia, we come to the small town Limyra. In the intervening distance along the coast above mentioned are many small islands and harbours. The most considerable of the islands is Cisthene, on which is a city of the same name.[201] In the interior are the strongholds Phellus, Antiphellus, and Chimæra, which I mentioned above.

8. Then follow the Sacred Promontory[202] and the Chelidoniæ, three rocky islands, equal in size, and distant from each other about 5, and from the land 6 stadia. One of them has an anchorage for vessels. According to the opinion of many writers, the Taurus begins here, because the summit is lofty, and extends from the Pisidian mountains situated above Pamphylia, and because the islands lying in front exhibit a [Pg 48] [CAS. 666]remarkable figure in the sea, like a skirt of a mountain. But in fact the mountainous chain is continued from the country opposite Rhodes to the parts near Pisidia, and this range of mountains is called Taurus.

The Chelidoniæ islands seem to be situated in a manner opposite to Canopus,[203]and the passage across is said to be 4000 stadia.

From the Sacred Promontory to Olbia[204] there remain 367 stadia. In this distance are Crambusa,[205] and Olympus[206] a large city, and a mountain of the same name, which is called also Phœnicus;[207] then follows Corycus, a tract of sea-coast.

9. Then follows Phaselis,[208] a considerable city, with three harbours and a lake. Above it is the mountain Solyma[209] and Termessus,[210] a Pisidic city, situated on the defiles, through which there is a pass over the mountain to Milyas. Alexander demolished it, with the intention of opening the defiles.

About Phaselis, near the sea, are narrow passes through which Alexander conducted his army. There is a mountain called Climax. It overhangs the sea of Pamphylia, leaving a narrow road along the coast, which in calm weather is not covered with water, and travellers can pass along it, but when the sea is rough, it is in a great measure hidden by the waves. The pass over the mountains is circuitous and steep, but in fair weather persons travel on the road along the shore. Alexander came there when there was a storm, and trusting generally to fortune, set out before the sea had receded, and the soldiers marched during the whole day up to the middle of the body in water.

Phaselis also is a Lycian city, situated on the confines of Pamphylia. It is not a part of the Lycian body, but is an independent city.

10. The poet distinguishes the Solymi from the Lycians, when he despatches Bellerophon by the king of the Lycians to this second adventure;

"he encountered the brave Solymi;"[211]

[Pg 49]other writers say that the Lycians were formerly called Solymi, and afterwards Termilæ, from the colonists that accompanied Sarpedon from Crete; and afterwards Lycians, from Lycus the son of Pandion, who, after having been banished from his own country, was admitted by Sarpedon to a share in the government; but their story does not agree with Homer. We prefer the opinion of those who say that the poet called the people Solymi who have now the name of Milyæ, and whom we have mentioned before.

CHAPTER IV.

1. AFTER Phaselis is Olbia; here Pamphylia begins. It is a large fortress. It is followed by the Cataractes,[212] as it is called, a river which descends violently from a lofty rock, with a great body of water, like a winter torrent, so that the noise of it is heard at a great distance.

Next is Attaleia,[213] a city, so called from its founder Attalus Philadelphus, who also settled another colony at Corycus, a small city near Attaleia, by introducing other inhabitants, and extending the circuit of the walls.

It is said, that between Phaselis and Attaleia, Thebe and Lyrnessus[214] are shown; for, according to Callisthenes, a part of the Trojan Cilicians were driven from the plain of Thebe into Pamphylia.

2. Next is the river Cestrus;[215] on sailing up its stream 60 stadia we find the city Perge,[216] and near it upon an elevated place, the temple of the Pergæan Artemis, where a general festival is celebrated every year.

Then at the distance of about 40 stadia from the sea is [Syllium],[217] on an elevated site, and visible at Perge. Next is Capria, a lake of considerable extent; then the river Eurymedon;[218] sailing up it to the distance of 60 stadia, we come to Aspendus,[219] a well-peopled city, founded by Argives. Above it is Petnelissus;[220]then another river, and many small islands [Pg 50] [CAS. 668]lying in front; then Side, a colony of the Cymæans, where there is a temple of Minerva. Near it is the coast of the Little Cibyratæ; then the river Melas,[221] and an anchorage for vessels; then Ptolemais[222]a city; next the borders of Pamphylia, and Coracesium,[223] where Cilicia Tracheia begins. The whole of the voyage along the coast of Pamphylia is 640 stadia.

3. Herodotus says,[224] that the Pamphylians are descendants of the people who accompanied Amphilochus and Calchas from Troy, a mixture of various nations. The majority of them settled here, others were dispersed over different countries. Callinus says that Calchas died at Clarus, but that some of the people who, together with Mopsus, crossed the Taurus, remained in Pamphylia, and that others were scattered in Cilicia and Syria, and as far even as Phœnicia.

CHAPTER V.

1. OF Cilicia without the Taurus one part is called Cilicia Tracheia, the rugged; the other, Cilicia Pedias, the flat or plain country.

The coast of the Tracheia is narrow, and either has no level ground or it rarely occurs; besides this, the Taurus overhangs it, which is badly inhabited as far even as the northern side, about Isaura and the Homonadeis as far as Pisidia. This tract has the name of Tracheiotis, and the inhabitants that of Tracheiotæ. The flat or plain country extends from Soli and Tarsus as far as Issus, and the parts above, where the Cappadocians are situated on the northern side of the Taurus. This tract consists chiefly of fertile plains.

I have already spoken of the parts within the Taurus; I shall now describe those without the Taurus, beginning with the Tracheiotæ.

2. The first place is Coracesium,[225] a fortress of the Cilicians, [Pg 51]situated upon an abrupt rock. Diodotus surnamed Tryphon used it as a rendezvous at the time that he caused Syria to revolt from her kings, and carried on war against them with various success. Antiochus, the son of Demetrius, obliged him to shut himself up in one of the fortresses, and there he killed himself.

Tryphon was the cause of originating among the Cilicians a piratical confederacy. They were induced also to do this by the imbecility of the kings who succeeded each other on the thrones of Syria and Cilicia. In consequence of his introduction of political changes, others imitated his example, and the dissensions among brothers exposed the country to the attacks of invaders.

The exportation of slaves was the chief cause of inducing them to commit criminal acts, for this traffic was attended with very great profit, and the slaves were easily taken. Delos was at no great distance, a large and rich mart, capable of receiving and transporting, when sold, the same day, ten thousand slaves; so that hence arose a proverbial saying,

"Merchant, come into port, discharge your freight—everything is sold."

The Romans, having acquired wealth after the destruction of Carthage and Corinth, employed great numbers of domestic slaves, and were the cause of this traffic. The pirates, observing the facility with which slaves could be procured, issued forth in numbers from all quarters, committing robbery and dealing in slaves.

The kings of Cyprus and of Egypt, who were enemies of the Syrians, favoured their marauding enterprises; the Rhodians were no less hostile to the Syrians, and therefore afforded the latter no protection. The pirates, therefore, under the pretence of trading in slaves, continued without intermission their invasions and robbery.

The Romans paid little attention to the places situated without the Taurus; they sent, however, Scipio Æmilianus, and afterwards some others, to examine the people and the cities. They discovered that the evils arose from negligence on the part of the sovereigns, but they were reluctant to deprive the family of Seleucus Nicator of the succession, in which he had been confirmed by themselves.

For the same reason the Parthians, who occupied the parts[Pg 52] [CAS. 669] beyond the Euphrates, became masters of the country; and lastly the Armenians, who also gained possession of the country without the Taurus as far as Phœnicia. They used their utmost to extirpate the power of the kings and all their descendants, but surrendered the command of the sea to the Cilicians.

The Romans were subsequently compelled to reduce the Cilicians, after their aggrandizement, by war and expeditions, whose progress, however, and advancement they had not obstructed; yet it would be improper to accuse the Romans of neglect, because, being engaged with concerns nearer at hand, they were unable to direct their attention to more distant objects.

I thought proper to make these remarks in a short digression from my subject.

3. Next to the Coracesium is the city Syedra;[226] then Hamaxia,[227] a small town upon a hill, with a harbour, to which is brought down timber for ship-building; the greatest part of it consists of cedar. This country seems to produce this tree in abundance. It was on this account that Antony assigned it to Cleopatra, as being capable of furnishing materials for the construction of her fleet.

Then follows Laertes a fortress, situated upon the crest of a hill, of a pap-like form; a port belongs to it; next, the city Selinus,[228] then Cragus, a precipitous rock on the sea-coast; then Charadrus[229] a fortress, which has a port (above it is the mountain Andriclus[230]) and a rocky shore, called Platanistus, next Anemurium[231] a promontory, where the continent approaches nearest to Cyprus, towards the promontory Crommyum,[232] the passage across being 350 stadia.

From the boundaries of Pamphylia to Anemurium, the voyage along the Cilician coast is 820 stadia; the remainder of it as far as Soli[233] is about 500 stadia (1500?). On this coast, after Anemurium, the first city is Nagidus, then Arsinoë,[234] with a small port; then a place called Melania,[235] and Celenderis[236] a city, with a harbour.

[Pg 53]Some writers,[237] among whom is Artemidorus, consider this place as the commencement of Cilicia, and not Coracesium. He says, that from the Pelusiac mouth to Orthosia are 3900 stadia, and to the river Orontes[238] 1130 stadia; then to the gates of Cilicia 525 stadia, and to the borders of Cilicia 1260 stadia.[239]

4. Next is Holmi,[240] formerly inhabited by the present Seleucians; but when Seleucia on the Calycadnus was built, they removed there. On doubling the coast, which forms a

promontory called Sarpedon,[241] we immediately come to the mouth of the Calycadnus.[242] Zephyrium[243] a promontory is near the Calycadnus. The river may be ascended as far as Seleucia, a city well peopled, and the manners of whose inhabitants are very different from those of the people of Cilicia and Pamphylia.

In our time there flourished at that place remarkable persons of the Peripatetic sect of philosophers, Athenæus and Xenarchus. The former was engaged in the administration of the affairs of state in his own country, and for some time espoused the party of the people; he afterwards contracted a friendship with Murena, with whom he fled, and with whom he was captured, on the discovery of the conspiracy against Augustus Cæsar; but he established his innocence, and was set at liberty by Cæsar. When he returned from Rome, he addressed the first persons who saluted him, and made their inquiries, in the words of Euripides—

"I come from the coverts of the dead, and the gates of darkness."[244]

He survived his return but a short time, being killed by the fall, during the night, of the house in which he lived.

Xenarchus, whose lectures I myself attended, did not long remain at home, but taught philosophy at Alexandreia, Athens, and Rome. He enjoyed the friendship of Areius, and afterwards of Augustus Cæsar; he lived to old age, honoured and respected. Shortly before his death he lost his sight, and died a natural death.

[Pg 54]
[CAS. 670]

5. After the Calycadnus, is the rock called Pœcile,[245] which has steps, like those of a ladder, cut in the rock, on the road to Seleucia. Then follows the promontory Anemurium,[246] of the same name with the former, Crambusa an island, and then Corycus[247] a promontory, above which, at the distance of 20 stadia, is the Corycian cave, where grows the best saffron. It is a large valley of a circular form, surrounded by a ridge of rock, of considerable height all round. Upon descending into it, the bottom is irregular, and a great part of it rocky, but abounding with shrubs of the evergreen and cultivated kind. There are interspersed spots which produce the saffron. There is also a cave in which rises a river of pure and transparent water. Immediately at its source the river buries itself in the ground, and continues its subterraneous course till it discharges itself into the sea. The name of (Pikron Hydor) "bitter water" is given to it.

6. After Corycus, is the island Elæussa,[248] lying very near the continent. Here Archelaus resided, and built a palace, after having become master of the whole of Cilicia Tracheiotis, except Seleucia, as Augustus had been before, and as at a still earlier period it was held by Cleopatra. For as the country was well adapted by nature for robbery both by sea and land, (by land, on account of the extent of the mountains, and the nations situated beyond them, who occupy plains, and large tracts of cultivated country easy to be overrun; by sea, on account of the supply of timber for ship-building, the harbours, fortresses, and places of retreat,) for all these reasons the Romans thought it preferable that the country should be under the government of kings, than be subject to Roman governors sent to administer justice, but who would not always be on the spot, nor attended by an army. In this manner Archelaus obtained possession of Cilicia Tracheia, in addition to Cappadocia. Its boundaries between Soli and Elæussa are the river Lamus,[249] and a village of the same name.[250]

7. At the extremity of the Taurus is Olympus a [Pg 55]mountain,[251] the piratical hold of Zenicetus, and a fortress of the same name. It commands a view of the whole of Lycia, Pamphylia, and Pisidia. When the mountain was taken by (Servilius) Isauricus, Zenicetus burnt himself, with all his household. To this robber belonged Corycus, Phaselis, and many strongholds in Pamphylia, all of which were taken by (Servilius) Isauricus.

8. Next to Lamus is Soli,[252] a considerable city, where the other Cilicia, that about Issus, commences. It was founded by Achæans, and by Rhodians from Lindus. Pompey the Great transferred to this city, which had a scanty population, the survivors of the pirates, whom he thought most entitled to protection and clemency, and changed its name to Pompeiopolis.

Chrysippus the Stoic philosopher, the son of an inhabitant of Tarsus, who left it to live at Soli; Philemon the comic poet; and Aratus, who composed a poem called "the Phænomena," were among the illustrious natives of this place.

9. Next follows Zephyrium,²⁵³ of the same name as that near Calycadnus; then Anchiale, a little above the sea, built by Sardanapalus, according to Aristobulus. (According to the same author) the tomb of Sardanapalus is here, and a stone figure representing him with the fingers of his right hand brought together as in the act of snapping them, and the following inscription in Assyrian letters: "SARDANAPALUS, THE SON OF ANACYNDARAXES, BUILT ANCHIALE AND TARSUS IN ONE DAY. EAT, DRINK, BE MERRY; EVERYTHING ELSE IS NOT WORTH²⁵⁴ THAT"—the snapping of the fingers.

Chœrilus mentions this inscription, and the following lines are everywhere known:

"Meat and drink, wanton jests, and the delights of love, these I have enjoyed; but my great wealth I have left behind."²⁵⁴

10. Above Anchiale is situated Cyinda a fortress, where the Macedonian kings formerly kept their treasure. Eumenes, when he revolted from Antigonus, took it away. Further above this place and Soli, is a mountainous tract, where is situated Olbe a city, which has a temple of Jupiter, founded by Ajax, son of Teucer. The priest of this temple was master [Pg 56]

[CAS. 672]of the Tracheiotis. Subsequently many tyrants seized upon the country, and it became the retreat of robbers. After their extermination, the country was called, even to our times, the dominion of Teucer; and the priesthood, the priesthood of Teucer; indeed, most of the priests had the name of Teucer, or of Ajax. Aba, the daughter of Xenophanes, one of the tyrants, entered into this family by marriage, and obtained possession of the government. Her father had previously administered it as guardian, but Antony and Cleopatra afterwards conferred it upon Aba, as a favour, being ultimately prevailed upon to do so by her entreaties and attentions. She was afterwards dispossessed, but the government remained in the hands of the descendants of her family.

Next to Anchiale are the mouths of the Cydnus²⁵⁵ at the Rhegma, (the Rent,) as it is called. It is a place like a lake, and has ancient dockyards; here the Cydnus discharges itself, after flowing through the middle of Tarsus. It rises in the Taurus, which overhangs the city. The lake is a naval arsenal of Tarsus.

11. The whole of the sea-coast, beginning from the part opposite to Rhodes, extends to this place in the direction from the western to the eastern point of the equinoctial. It then turns towards the winter solstice, as far as Issus, and thence immediately makes a bend to the south to Phœnicia. The remainder towards the west terminates at the pillars (of Hercules).²⁵⁶

The actual isthmus of the peninsula, which we have described, is that which extends from Tarsus and the mouth of the Cydnus as far as Amisus, for this is the shortest distance from Amisus to the boundaries of Cilicia; from these to Tarsus are 120 stadia, and not more from Tarsus to the mouth of the Cydnus. To Issus, and the sea near it, there is no shorter road from Amisus than that leading through Tarsus, nor from Tarsus to Issus is there any nearer than that leading to Cydnus; so that it is clear, that, in reality, this is the isthmus. Yet it is pretended that the isthmus extending as far as the [Pg 57]Bay of Issus is the true isthmus, on account of its presenting remarkable points.

Hence, not aiming at exactness, we say that the line drawn from the country opposite to Rhodes, which we protracted as far as Cydnus, is the same as that extending as far as Issus, and that the Taurus extends in a straight direction with this line as far as India.

12. Tarsus is situated in a plain. It was founded by Argives, who accompanied Triptolemus in his search after Io. The Cydnus flows through the middle of it, close by the gymnasium of the young men. As the source is not far distant, and the stream passing through a deep valley, then flows immediately into the city, the water is cold and rapid in its course; hence it is of advantage to men and beasts affected with swellings of the sinews, fluxions, and gout.²⁵⁷

13. The inhabitants of this city apply to the study of philosophy and to the whole encyclical compass of learning with so much ardour, that they surpass Athens, Alexandreia,

and every other place which can be named where there are schools and lectures of philosophers.

It differs however so far from other places, that the studious are all natives, and strangers are not inclined to resort thither. Even the natives themselves do not remain, but travel abroad to complete their studies, and having completed them reside in foreign countries. Few of them return.

The contrary is the case in the other cities which I have mentioned, except Alexandreia; for multitudes repair to them, and reside there with pleasure; but you would observe that few of the natives travel abroad from a love of learning, or show much zeal in the pursuit of it on the spot. But both these things are to be seen at Alexandreia, a large number of strangers is received, (into their schools,) and not a few of their own countrymen are sent out to foreign countries (to study). They have schools of all kinds, for instruction in the liberal arts. In other respects Tarsus is well peopled, extremely powerful, and has the character of being the capital.[258]

[Pg 58]

[CAS. 674]14. The Stoic philosophers Antipater, Archedemus, and Nestor were natives of Tarsus: and besides these, the two Athenodori, one of whom, Cordylion, lived with Marcus Cato, and died at his house; the other, the son of Sandon, called Cananites, from some village, was the preceptor of Cæsar,[259] who conferred on him great honours. In his old age he returned to his native country, where he dissolved the form of government existing there, which was unjustly administered by various persons, and among them by Boëthus, a bad poet and a bad citizen, who had acquired great power by courting the favour of the people. Antony contributed to increase his importance by having in the first instance commended a poem which he had composed on the victory at Philippi; his influence was still augmented by the facility which he possessed (and it is very general among the inhabitants of Tarsus) of discoursing at great length, and without preparation, upon any given subject. Antony also had promised the people of Tarsus to establish a gymnasium; he appointed Boëthus chief director of it, and intrusted to him the expenditure of the funds. He was detected in secreting, among other things, even the oil, and when charged with this offence by his accusers in the presence of Antony, he deprecated his anger by this, among other remarks in his speech, that as Homer had sung the praises of "Achilles, Agamemnon, and Ulysses, so have I sung yours. I therefore ought not to be brought before you on such a charge." The accuser answered, "Homer did not steal oil from Agamemnon[260] nor Achilles; but you have stolen it from the gymnasium, and therefore you shall be punished." Yet he contrived to avert the displeasure of Antony by courteous offices, and continued to plunder the city until the death of his protector.

Athenodorus found the city in this state, and for some time attempted to control Boëthus and his accomplices by argument; but finding that they continued to commit all kinds of injustice, he exerted the power given to him by Cæsar, condemned them to banishment, and expelled them. They had previously caused to be written upon the walls, "Action for the young, counsel for the middle-aged, discharging wind for the [Pg 59]old;" but Athenodorus, accepting it as a jest, gave orders to inscribe by the side of it, "Thunder for the old." Some one, however, in contempt for his good manners, having a lax state of body, bespattered the gate and wall of his house as he passed by it at night. Athenodorus, in an assembly of the people, accusing persons of being factiously disposed, said, "We may perceive the sickly condition of the city, and its bad habit of body, from many circumstances, but particularly from its discharges."

Those men were Stoics, but Nestor, of our time, the tutor of Marcellus, son of Octavia, the sister of Cæsar, was of the Academic sect. He was also at the head of the government, having succeeded Athenodorus, and continued to be honoured both by the Roman governors and by the citizens.

15. Among the other philosophers,

"Those whom I know, and could in order name,"[261]

were Plutiades and Diogenes, who went about from city to city, instituting schools of philosophy as the opportunity occurred. Diogenes, as if inspired by Apollo, composed and rehearsed poems, chiefly of the tragic kind, upon any subject that was proposed. The grammarians of Tarsus, whose writings we have, were Artemidorus and Diodorus. But the best writer of tragedy, among those enumerated in "The Pleiad," was Dionysides. Rome is best able to inform us what number of learned men this city has produced, for it is filled with persons from Tarsus and Alexandreia.

Such then is Tarsus.

16. After the Cydnus follows the Pyramus,[262] which flows from Cataonia. We have spoken of it before. Artemidorus says, that from thence to Soli is a voyage in a straight line of 500 stadia. Near the Pyramus is Mallus,[263] situated upon a height; it was founded by Amphilochus, and Mopsus, the son of Apollo, and Mantus, about whom many fables are related. I have mentioned them in speaking of Calchas, and of the contest between Calchas and Mopsus respecting their skill in divination. Some persons, as Sophocles, transfer the scene of this contest to Sicily, which, after the custom of tragic poets, they call Pamphylia, as they call Lycia, Caria, and [Pg 60] [CAS. 676]Troy and Lydia, Phrygia. Sophocles, among other writers, says that Calchas died there. According to the fable, the contest did not relate to skill in divination only, but also to sovereignty. For it is said, that Mopsus and Amphilochus, on their return from Troy, founded Mallus; that Amphilochus afterwards went to Argos, and being dissatisfied with the state of affairs there, returned to Mallus, where, being excluded from a share in the government, he engaged with Mopsus in single combat. Both were killed, but their sepulchres are not in sight of each other. They are shown at present at Magarsa, near the Pyramus.

Crates the grammarian was a native of this place, and Panætius is said to have been his disciple.

17. Above this coast is situated the Aleïan plain, over which Philotas conducted Alexander's cavalry, he himself leading the phalanx from Soli along the sea-coast and the territory of Mallus to Issus, against the forces of Darius. It is said that Alexander performed sacrifices in honour of Amphilochus, on account of their common affinity to Argos. Hesiod says that Amphilochus was killed by Apollo at Soli; according to others, at the Aleïan plain; and others again say, in Syria, upon his quitting the Aleïan plain on account of the quarrel.

18. Mallus is followed by Ægæ, a small town[264] with a shelter for vessels; then the Amanides Gates, (Gates of Amanus,[265]) with a shelter for vessels. At these gates terminates the mountain Amanus,[266] which extends from the Taurus, and lies above Cilicia towards the east. It was successively in the possession of several tyrants, who had strongholds; but, in our time, Tarcondimotus, who was a man of merit, became master of all; for his good conduct and bravery, he received from the Romans the title of King, and transmitted the succession to his posterity.

19. Next to Ægæ is Issus, a small town with a shelter for vessels, and a river, the Pinarus.[267] At Issus the battle was fought between Alexander and Darius. The bay is called the Issic Bay. The city Rhosus[268] is situated upon it, [Pg 61]as also the city Myriandrus, Alexandreia,[269] Nicopolis, Mopsuestia,[270] and the Gates,[271] as they are called, which are the boundary between Cilicia and Syria.

In Cilicia are the temple of the Sarpedonian Artemis and an oracle. Persons possessed with divine inspiration deliver the oracles.

20. After Cilicia, the first Syrian city is Seleucia-in-Pieria;[272] near it the river Orontes[273] empties itself. From Seleucia to Soli is a voyage in a straight line of nearly 1000 stadia.

21. Since the Cilicians of the Troad, whom Homer mentions, are situated at a great distance from the Cilicians without the Taurus, some writers declare that the leaders of the latter colony were Cilicians of the Troad, and point to Thebe and Lyrnessus in Pamphylia, places bearing the same name as those in the Troad; other authors are of a contrary opinion, and (considering the Cilicians of the Troad as descendants of those from beyond the Taurus) point to an Aleïan plain (in support of their hypothesis).

22. Having described the parts of the before-mentioned Chersonesus without the Taurus, I must add these particulars.

Apollodorus, in his work on the catalogue of the ships mentioned in Homer, relates, that all the allies of the Trojans, who came from Asia, inhabited, according to the poet, the peninsula of which at its narrowest part is the isthmus between the innermost recess of the bay at Sinope and Issus. The exterior sides (of this peninsula), which is of a triangular shape, are unequal. Of these, one extends from Cilicia to Chelidoniæ, (islands,) another thence to the mouth of the Euxine, and the third from the mouth of the Euxine to Sinope.

The assertion that the allies were only those who occupied the peninsula may be proved to be erroneous by the same arguments by which we before showed that those who lived within the Halys were not the only allies. For the places about Pharnacia, where we said the Halizoni lived, are situated without the Halys, and also without the isthmus, for they [Pg 62]

[CAS. 677]are without the line drawn from Sinope to Issus;²⁷⁴ and not only without this line, but also without the true line of the isthmus drawn from Amisus to Issus; for Apollodorus incorrectly describes the isthmus and the line of its direction, substituting one line for another (the line drawn from Sinope to Issus for the line drawn from Amisus to Issus).

But the greatest absurdity is this, that after having said that the peninsula was of a triangular shape, he speaks of three *exterior* sides. For in speaking of *exterior* sides, he seems to except the line of the isthmus itself, considering it still a side, although not an *exterior* side, from its not being upon the sea. But if this line were so shortened that the extremities of the (*exterior*) sides falling upon Issus and Sinope nearly coincided, the peninsula might in that case be said to be of a triangular shape; but as his own line (from Sinope to Issus) is 3000 stadia in length, it would be ignorance, and not a knowledge of chorography, to call such a four-sided figure a triangle. Yet he published a work on Chorography, in the metre of comedy, (Iambic metre,) entitled "The Circuit of the Earth."

He is still liable to the same charge of ignorance, even if we should suppose the isthmus to be contracted to its least dimensions, and follow writers who erroneously estimate the distance at one-half of the sum, namely 1500 stadia, to which it is reduced by Artemidorus; but even this would not by any means reduce the thus contracted space to the figure of a triangle.

Besides, Artemidorus has not correctly described the exterior sides; one side, he says, extends from Issus to the Chelidoniæ islands, although the whole Lycian coast, and the country opposite to Rhodes as far as Physcus, lies in a straight line with, and is a continuation of it; the continent then makes a bend at Physcus, and forms the commencement of the second or western side, extending to the Propontis and Byzantium.

23. Ephorus had said that this peninsula was inhabited by sixteen tribes, three of which were Grecian, and the rest barbarous, with the exception of the mixed nations; he placed [Pg 63]on the sea-coast Cilicians, Pamphylians, Lycians, Bithynians, Paphlagonians, Mariandyni, Troes, and Carians; and in the interior, Pisidians, Mysians, Chalybes, Phrygians, and Milyæ.²⁷⁵ Apollodorus, when discussing this position, says there is a seventeenth tribe, the Galatians, who are more recent than the time of Ephorus; that of the sixteen tribes mentioned, the Greeks were not settled (in the peninsula) at the period of the Trojan war, and that time has produced great intermixture and confusion among the barbarous nations. Homer, he continues, recites in his Catalogue the Troes, and those now called Paphlagonians, Mysians, Phrygians, Carians, Lycians, Meionians, instead of Lydians and other unknown people, as Halizoni and Caucones; nations besides not mentioned in the Catalogue but elsewhere, as Ceteii, Solymi, the Cilicians from the plain of Thebe, and Leleges. But the Pamphylians, Bithynians, Mariandyni, Pisidians, and Chalybes, Milyæ, and Cappadocians are nowhere mentioned by the poet; some because they did not then inhabit these places, and some because they were surrounded by other tribes, as Idrieis and Termilæ by Carians, Doliones and Bebryces by Phrygians.

24. But Apollodorus does not seem to have carefully examined the statements of Ephorus, for he confounds and misrepresents the words of Homer. He ought first to have inquired of Ephorus why he placed the Chalybes within the peninsula, who were situated at a great distance from Sinope, and Amisus towards the east. Those who describe the isthmus of this peninsula to be on the line drawn from Issus to the Euxine, lay down this line as a sort of meridian line, which some suppose to pass through Sinope, others through Amisus;

but no one through the Chalybes, for such a line would be altogether an oblique line. For the meridian passing through the Chalybes, drawn through the Lesser Armenia, and the Euphrates, would comprise (on the east) the whole of Cappadocia, Commagene, Mount Amanus, and the Bay of Issus. [Pg 64] [CAS. 678]But if we should grant (to Ephorus) that this oblique line is the direction of the isthmus, most of these places, Cappadocia in particular, would be included, and (the kingdom of) Pontus, properly so called, which is a part of Cappadocia on the Euxine; so that if we were to admit the Chalybes to be a part of the peninsula, with more reason we ought to admit the Cataonians, the two nations of Cappadocians, and the Lycaonians, whom even he himself has omitted. But why has he placed in the interior the Chalybes, whom the poet, as we have shown, calls Halizoni? It would have been better to divide them, and to place one portion of them on the sea-coast, and another in the inland parts. The same division ought to be made of the Cappadocians and Cilicians. But Ephorus does not even mention the former, and speaks only of the Cilicians on the sea-coast. The subjects, then, of Antipater of Derbe, the Homonadeis, and many other tribes contiguous to the Pisidians,

"men, who know not the sea, nor have ever eaten food seasoned with salt,"[276]

where are they to be placed? Nor does he say whether the Lydians and the Meonians are two nations or the same nation, or whether they live separately by themselves or are comprehended in another tribe. For it was impossible for Ephorus to be ignorant of so celebrated a nation, and does he not, by passing it over in silence, appear to omit a most important fact?

25. But who are "the mixed nations"? For we cannot say that he either named or omitted others, besides those already mentioned, whom we should call mixed nations. Nor, indeed, should we say that they were a part of those nations whom he has either mentioned or omitted. For if they were a mixed people, still the majority constituted them either Greeks or Barbarians. We know nothing of a third mixed people.

26. But how (according to Ephorus) are there three tribes of Greeks who inhabit the peninsula? Is it because anciently the Athenians and Ionians were the same people? In that case the Dorians and the Æolians should be considered as the same nation, and then there would be (only) two tribes (and not three, inhabiting the peninsula). But if, following modern [Pg 65]practice, we are to distinguish nations according to dialects, there will be four nations, as there are four dialects. But this peninsula is inhabited, especially if we adopt the division by Ephorus, not only by Ionians, but also by Athenians, as we have shown in the account of each particular place.

It was worth while to controvert the positions of Ephorus, Apollodorus however disregards all this, and adds a seventeenth to the sixteen nations, namely, the Galatians; although it is well to mention this, yet it is not required in a discussion of what Ephorus relates or omits; Apollodorus has assigned as the reason of the omission, that all these nations settled in the peninsula subsequently to the time of Ephorus.

27. Passing then to Homer, Apollodorus is correct in saying that there was a great intermixture and confusion among the barbarous nations, from the Trojan war to the present time, on account of the changes which had taken place; for some nations had an accession of others, some were extinct or dispersed, or had coalesced together.

But he is mistaken in assigning two reasons why the poet does not mention some nations, namely, either because the place was not then occupied by the particular people, or because they were comprehended in another tribe. Neither of these reasons could induce him to be silent respecting Cappadocia or Cataonia, or Lycaonia itself, for we have nothing of the kind in history relating to these countries. It is ridiculous to be anxious to find excuses why Homer has omitted to speak of Cappadocia [Cataonia] and Lycaonia, and not to inform us why Ephorus omitted them, particularly as the proposed object of Apollodorus was to examine and discuss the opinions of Ephorus; and to tell us why Homer mentions Mæonians instead of Lydians, and also not to remark that Ephorus has not omitted to mention either Lydians or Mæonians.[277]

28. Apollodorus remarks, that Homer mentions certain unknown nations, and he is right in specifying Caucones, Solymi, Ceteii, Leleges, and the Cilicians from the plain of

41

Thebe; but the Halizones are a fiction of his own, or rather of those who, not knowing who the Halizones were, frequently altered the mode of writing the name, and invented the existence of [Pg 66] [CAS. 680]mines of silver and of many other mines, all of which are abandoned.

With this vain intention they collected the stories related by the Scepsian, (Demetrius,) and taken from Callisthenes and other writers, who did not clear them from false notions respecting the Halizones; for example, the wealth of Tantalus and of the Pelopidæ was derived, it is said, from the mines about Phrygia and Sipylus; that of Cadmus from the mines about Thrace and Mount Pangæum; that of Priam from the gold mines at Astyra, near Abydos (of which at present there are small remains, yet there is a large quantity of matter ejected, and the excavations are proofs of former workings); that of Midas from the mines about Mount Bermium; that of Gyges, Alyattes, and Crœsus, from the mines in Lydia and the small deserted city between Atarneus and Pergamum, where are the sites of exhausted mines.[278]

29. We may impute another fault to Apollodorus, that although he frequently censures modern writers for introducing new readings at variance with the meaning of Homer, yet in this instance he not only neglects his own advice, but actually unites together places which are not so represented (by Homer).

(For example), Xanthus the Lydian says, that after the Trojan times the Phrygians came from Europe (into Asia) and the left (western) side of the Euxine, and that their leader Scamandrius conducted them from the Berecynti and Ascania. Apollodorus adds, that Homer mentions the same Ascania as Xanthus,

"Phorcys and the divine Ascanius led the Phrygians from the distant Ascania."[279]

If this be so, the migration (from Europe to Asia) must be later than the Trojan war; but in the Trojan war the auxiliaries mentioned by the poet came from the opposite continent, from the Berecynti and Ascania. Who then were the Phrygians,

"who were then encamped on the banks of the Sangarius,"

when Priam says,

"And I joined them with these troops as an auxiliary"?[280]

[Pg 67]And how came Priam to send for the Phrygians from among the Berecynti, between whom and himself no compact existed, and pass over the people who were contiguous to him, and whose ally he formerly had been?

Apollodorus, after having spoken of the Phrygians in this manner, introduces an account concerning the Mysians which contradicts this. He says that there is a village of Mysia called Ascania, near a lake of the same name,[281] out of which issues the river Ascanius, mentioned by Euphorion:[282]

"near the waters of the Mysian Ascanius;"

and by Alexander of Ætolia:

"they who dwell on the stream of Ascanius, on the brink of the Ascanian lake, where lived Dolion, the son of Silenus and Melia."

The district, he says, about Cyzicus, on the road to Miletopolis, is called Dolionis and Mysia.

If this is the case, and if it is confirmed by existing places and by the poets, what prevented Homer, when he mentioned this Ascania, from mentioning the Ascania also of which Xanthus speaks?

I have already spoken of these places in the description of Mysia and Phrygia, and shall here conclude the discussion.

CHAPTER VI.

1. IT remains for me to describe the island Cyprus, which adjoins this peninsula on the south. I have already said, that the sea comprised between Egypt, Phœnice, Syria, and the remainder of the coast as far as that opposite to Rhodes, consists, [Pg 68] [CAS. 681]so to say, of the Egyptian and Pamphylian seas and the sea along the Bay of Issus.

In this sea lies the island Cyprus, having its northern side approaching to Cilicia Tracheia, and here also it approaches nearest to the continent; on the east it is washed by the Bay of Issus, on the west by the Pamphylian sea, and on the south by that of Egypt. The latter sea is confluent on the west with the Libyan and Carpathian seas. On its southern and eastern parts is Egypt, and the succeeding tract of coast as far as Seleucia and Issus. On the north is Cyprus, and the Pamphylian sea.

The Pamphylian sea is bounded on the north by the extremities of Cilicia Tracheia, of Pamphylia, and of Lycia as far as the territory opposite to Rhodes; on the west, by the island of Rhodes; on the east, by the part of Cyprus near Paphos, and the Acamas; on the south, it unites with the Egyptian sea.

2. The circumference of Cyprus is 3420 stadia, including the winding of the bays. Its length from Cleides[283] to the Acamas,[284] to a traveller on land proceeding from east to west, is 1400 stadia.

The Cleides are two small islands lying in front of Cyprus on the eastern side, at the distance of 700 stadia from the Pyramus.[285]

The Acamas is a promontory with two paps, and upon it is a large forest. It is situated at the western part of the island, but extends towards the north, approaching very near Selinus in Cilicia Tracheia, for the passage across is only 1000 stadia; to Side in Pamphylia the passage is 1600 stadia, and to the Chelidoniæ (islands) 1900 stadia.

The figure of the whole island is oblong, and in some places on the sides, which define its breadth, there are isthmuses.

We shall describe the several parts of the island briefly, beginning from the point nearest to the continent.

3. We have said before, that opposite to Anemyrium, a promontory of Cilicia Tracheia, is the extremity of Cyprus, namely, the promontory of Crommyon,[286] at the distance of 350 stadia.

From the cape, keeping the island on the right hand, and [Pg 69]continent on the left, the voyage to the Cleides in a straight line towards north and east is a distance of 700 stadia.

In the interval is the city Lapathus,[287] with a harbour and dockyards; it was founded by Laconians and Praxander. Opposite to it was Nagidus. Then follows Aphrodisium;[288] here the island is narrow, for over the mountains to Salamis[289] are 70 stadia. Next is the sea-beach of the Achæans; here Teucer, the founder of Salamis in Cyprus, being it is said banished by his father Telamon, first disembarked. Then follows the city Carpasia,[290] with a harbour. It is situated opposite to the promontory Sarpedon.[291] From Carpasia there is a transit across the isthmus of 30 stadia to the Carpasian islands and the southern sea; next are a promontory and a mountain. The name of the promontory is Olympus, and upon it is a temple of Venus Acræa, not to be approached nor seen by women.

Near and in front lie the Cleides, and many other islands; next are the Carpasian islands, and after these Salamis, the birth-place of Aristus the historian; then Arsinoë, a city with a harbour; next Leucolla, another harbour; then the promontory Pedalium, above which is a hill, rugged, lofty, and table-shaped, sacred to Venus; to this hill from Cleides are 680 stadia. Then to Citium[292] the navigation along the coast is for the greater part difficult and among bays. Citium has a close harbour. It is the birth-place of Zeno, the chief of the Stoic sect, and of Apollonius the physician. Thence to Berytus are 1500 stadia. Next is the city Amathus,[293] and between Citium and Berytus, a small city called Palæa, and a pap-shaped mountain, Olympus; then follows Curias,[294] a promontory of a peninsular form, to which from Throni[295] are 700 stadia; then the city Curium,[296]with a harbour, founded by Argives.

Here we may observe the negligence of the author, whether Hedylus, or whoever he was, of the elegiac lines which begin,

"We hinds, sacred to Phœbus, hither came in our swift course; we traversed the broad sea, to avoid the arrows of our pursuers."

He says, that the hinds ran down from the Corycian heights, [Pg 70] [CAS. 683]and swam across from the Cilician coast to the beach near Curias, and adds,

"That it was a cause of vast surprise to men to think how we scoured the trackless waves, aided by the vernal Zephyrs."

For it is possible (by doubling the cape) to sail round from Corycus to the beach of Curias, but not with the assistance of the west wind, nor by keeping the island on the right, but on the left hand; and there is no (direct) passage across.

At Curium is the commencement of the voyage towards the west in the direction of Rhodes; then immediately follows a promontory, whence those who touch with their hands the altar of Apollo are precipitated. Next are Treta,[297] Boosura,[298] and Palæpaphus, situated about 10 stadia from the sea, with a harbour and an ancient temple of the Paphian Venus; then follows Zephyria,[299] a promontory with an anchorage, and another Arsinoë, which also has an anchorage, a temple, and a grove. At a little distance from the sea is Hierocepis.[300] Next is Paphos, founded by Agapenor, with a harbour and temples, which are fine buildings. It is distant from Palæpaphus 60 stadia by land. Along this road the annual sacred processions are conducted, when a great concourse both of men and women resort thither from other cities. Some writers say, that from Paphos to Alexandreia are 3600 stadia. Next after Paphos is the Acamas; then after the Acamas the voyage is easterly to Arsinoë a city, and to the grove of Jupiter; then Soli[301] a city, where there is a harbour, a river, and a temple of Venus and Isis. It was founded by Phalerus and Acamas, who were Athenians. The inhabitants are called Solii. Stasanor, one of the companions of Alexander, was a native of Soli, and was honoured with a chief command. Above Soli in the interior is Limenia a city, then follows the promontory of Crommyon.

4. But why should we be surprised at poets, and those particularly who study modes of expression only, when we compare them with Damastes? The latter gives the length of the island from north to south, from Hierocepia, as he says, to Cleides.

Nor does even Eratosthenes give it exactly. For, when [Pg 71]he censures Damastes, he says that Hierocepia is not on the north, but on the south. Yet neither is it on the south, but the west, since it lies on the western side, where are situated Paphos and Acamas.

Such then is the position of Cyprus.

5. It is not inferior in fertility to any one of the islands, for it produces good wine and oil, and sufficient corn to supply the wants of the inhabitants. At Tamassus there are abundant mines of copper, in which the calcanthus is found, and rust of copper, useful for its medicinal properties.

Eratosthenes says, that anciently the plains abounded with timber, and were covered with forests, which prevented cultivation; the mines were of some service towards clearing the surface, for trees were cut down to smelt the copper and silver. Besides this, timber was required for the construction of fleets, as the sea was now navigated with security and by a large naval force; but when even these means were insufficient to check the growth of timber in the forests, permission was given to such as were able and inclined, to cut down the trees and to hold the land thus cleared as their own property, free from all payments.

6. Formerly the Cyprian cities were governed by tyrants, but from the time that the Ptolemaïc kings were masters of Egypt, Cyprus also came into their power, the Romans frequently affording them assistance. But when the last Ptolemy that was king, brother of the father of Cleopatra, the queen of Egypt in our time, had conducted himself in a disorderly manner, and was ungrateful to his benefactors, he was deposed, and the Romans took possession of the island, which became a Prætorian province by itself.

The chief author of the deposition of the king was Pub. Claudius Pulcher, who having fallen into the hands of the Cilician pirates, at that time at the height of their power, and a ransom being demanded of him, despatched a message to the king, entreating him to send it for his release. The king sent a ransom, but of so small an amount, that the pirates disdained to accept it, and returned it, but they dismissed Pulcher without any payment. After his escape, he remembered what he owed to both parties; and when he became tribune of the people, he had sufficient influence to have Marcus[Pg 72] [CAS. 684] Cato sent to deprive the king of the possession of Cyprus. The latter put himself to death before the arrival of Cato, who, coming soon afterwards, took possession of Cyprus, sold the king's property, and conveyed the money to the public treasury of the Romans.

44

From this time the island became, as it is at present, a Prætorian province. During a short intervening period Antony had given it to Cleopatra and her sister Arsinoë, but upon his death all his arrangements were annulled.

[Pg 73]

BOOK XV.
SUMMARY.
The Fifteenth Book contains India and Persia.
CHAPTER I.

1. THE parts of Asia which remain to be described are those without the Taurus, except Cilicia, Pamphylia, and Lycia; extending from India to the Nile, and situated between the Taurus and the exterior Southern Sea.[302]

Next to Asia is Africa, which I shall describe hereafter. At present I shall begin from India, the first and the largest country situated towards the east.

2. The reader must receive the account of this country with indulgence, for it lies at a very great distance, and few persons of our nation have seen it; those also who have visited it have seen only some portions of it; the greater part of what they relate is from report, and even what they saw, they became acquainted with during their passage through the country with an army, and in great haste. For this reason they do not agree in their accounts of the same things, although they write about them as if they had examined them with the greatest care and attention. Some of these writers were fellow-soldiers and fellow-travellers, as those who belonged to the army which, under the command of Alexander, conquered Asia; yet they frequently contradict each other. If, then, they differ so much respecting things which they had seen, what must we think of what they relate from report?

3. Nor do the writers who, many ages since Alexander's time, have given an account of these countries, nor even those who at present make voyages thither, afford any precise information.

Apollodorus, for instance, author of the Parthian History, when he mentions the Greeks who occasioned the revolt of Bactriana from the Syrian kings, who were the successors of [Pg 74] [CAS. 686]Seleucus Nicator, says, that when they became powerful they invaded India. He adds no discoveries to what was previously known, and even asserts, in contradiction to others, that the Bactrians had subjected to their dominion a larger portion of India than the Macedonians; for Eucratidas (one of these kings) had a thousand cities subject to his authority. But other writers affirm that the Macedonians conquered nine nations situated between the Hydaspes[303] and the Hypanis,[304] and obtained possession of five hundred cities, not one of which was less than Cos Meropis,[305] and that Alexander, after having conquered all this country, delivered it up to Porus.

4. Very few of the merchants who now sail from Egypt by the Nile and the Arabian Gulf to India have proceeded as far as the Ganges; and, being ignorant persons, were not qualified to give an account of places they have visited. From one place in India, and from one king, namely, Pandion, or, according to others,[306]Porus, presents and embassies were sent to Augustus Cæsar. With the ambassadors came the Indian Gymno-Sophist, who committed himself to the flames at Athens,[307] like Calanus, who exhibited the same spectacle in the presence of Alexander.

5. If, then, we set aside these stories, and direct our attention to accounts of the country prior to the expedition of Alexander, we shall find them still more obscure. It is probable that Alexander, elated by his extraordinary good fortune, believed these accounts.

According to Nearchus, Alexander was ambitious of conducting his army through Gedrosia,[308] when he heard that Semiramis and Cyrus had undertaken expeditions against India (through this country), although both had abandoned the enterprise, the former escaping with twenty, and Cyrus with seven men only. For he considered that it would be a glorious achievement for him to lead a conquering army safe through the same nations and countries where Semiramis and Cyrus had suffered such disasters. Alexander, therefore, believed these stories.

45

6. But how can we place any just confidence in the accounts [Pg 75]of India derived from such expeditions as those of Cyrus and Semiramis? Megasthenes concurs in this opinion; he advises persons not to credit the ancient histories of India, for, except the expeditions of Hercules, of Bacchus, and the later invasion of Alexander, no army was ever sent out of their country by the Indians, nor did any foreign enemy ever invade or conquer it. Sesostris the Egyptian (he says), and Tearco the Ethiopian, advanced as far as Europe; and Nabocodrosor, who was more celebrated among the Chaldæans than Hercules among the Greeks, penetrated even as far as the Pillars,[309] which Tearco also reached; Sesostris conducted an army from Iberia to Thrace and Pontus; Idanthyrsus the Scythian overran Asia as far as Egypt; but not one of these persons proceeded as far as India, and Semiramis died before her intended enterprise was undertaken. The Persians had sent for the Hydraces[310] from India, a body of mercenary troops; but they did not lead an army into that country, and only approached it when Cyrus was marching against the Massagetæ.

7. Megasthenes, and a few others, think the stories respecting Hercules and Bacchus to be credible, but the majority of writers, among whom is Eratosthenes, regard them as incredible and fabulous, like the Grecian stories. Dionysus, in the Bacchæ of Euripides, makes this boasting speech:

[Pg   76]
[CAS. 687]

"But now from Lydia's field,

With gold abounding, from the Phrygian realm

And that of Persia scorch'd by torrid suns,

Pressing through Bactrian gates, the frozen land

Of Media, and through Araby the Blest,

With Asia's wide extended continent——"[311]

In Sophocles, also, a person is introduced speaking the praises of Nysa,[312] as being a mountain sacred to Bacchus:

"whence I beheld the famed Nysa, the resort of the Bacchanalian bands, which the horned Iacchus makes his most pleasant and beloved retreat, where no bird's clang is heard,"

and so on. [He is called also Merotraphes.][313]

Homer also mentions Lycurgus the Edonian in these words,

"who formerly pursued the nurses of the infuriate Bacchus along the sacred mountain Nysa."[314]

So much respecting Bacchus. But with regard to Hercules, some persons say, that he penetrated to the opposite extremities on the west only, while others maintain that he also advanced to those of the east.

8. From such stories as those related above, they gave the name of Nysæans to some imaginary nation, and called their city Nysa, founded by Bacchus; a mountain above the city they called Meron, alleging as a reason for imposing these names that the ivy and vine grow there, although the latter does not perfect its fruit; for the bunches of grapes, in consequence of excessive rains, drop off before they arrive at maturity.

They say, also, that the Sydracæ (Oxydracæ) are descendants of Bacchus, because the vine grows in their country, and because their kings display great pomp in setting out on their warlike expeditions, after the Bacchic manner; whenever they appear in public, it is with beating of drums, and are dressed in flowered robes, which is the common custom among the other Indians.

[Pg 77]When Alexander took, on the first assault, Aornos,[315] a fortress on a rock, the foot of which is washed by the Indus near its source, his flatterers exaggerated this act, and said that Hercules thrice assailed this rock and was thrice repulsed.

They pretended that the Sibæ[316] were descended from the people who accompanied Hercules in his expedition, and that they retained badges of their descent; that they wore skins like Hercules, and carried clubs, and branded with the mark of a club their oxen and mules. They confirm this fable with stories about Caucasus[317] and Prometheus, for they transferred hither from Pontus these tales, on the slight pretence that they had seen a sacred cave among the Paropamisadæ.[318]This they alleged was the prison of Prometheus, that Hercules came hither to release Prometheus, and that this mountain was the Caucasus, to which the Greeks represent Prometheus as having been bound.

9. That these are the inventions of the flatterers of Alexander is evident, first, because the writers do not agree with one another, some of whom speak of these things; others make no mention of them whatever. For it is not probable, that actions so illustrious, and calculated to foster pride and vanity, should be unknown, or if known, that they should not be thought worthy of record, especially by writers of the greatest credit.

Besides, the intervening people, through whose country the armies of Bacchus and Hercules must have marched in their [Pg 78] [CAS. 688]way to India, do not exhibit any proofs of their passage through the country. The kind of dress, too, of Hercules is much more recent than the memorials of Troy, an invention of those who composed the Heracleia (or exploits of Hercules,) whether it were Peisander or some one else who composed it. But the ancient wooden statues do not represent Hercules in that attire.

10. Under such circumstances, therefore, we must receive everything that approaches nearest to probability. I have already discussed this subject to the extent of my ability at the beginning of this work;[319] I shall now assume those opinions as clearly proved, and shall add whatever may seem to be required for the sake of perspicuity.

It appeared from the former discussion, that in the summary given by Eratosthenes, in the third book of his Geography, is contained the most credible account of the country considered as India at the time of its invasion by Alexander.

At that period the Indus was the boundary of India and of Ariana,[320] situated towards the west, and in the possession of the Persians, for afterwards the Indians occupied a larger portion of Ariana, which they had received from the Macedonians.

The account of Eratosthenes is as follows:—

11. The boundaries of India, on the north, from Ariana to the Eastern Sea,[321] are the extremities of Taurus, to the several parts of which the natives give, besides others, the names of Paropamisus, Emodus, and Imaus,[322] but the Macedonians call them Caucasus; on the west, the river Indus; the southern and eastern sides, which are much larger than the others, project towards the Atlantic Sea, and the figure of the country [Pg 79]becomes rhomboïdal,[323] each of the greater sides exceeding the opposite by 3000 stadia; and this is the extent of the extremity, common to the eastern and southern coast, and which projects beyond the rest of that coast equally on the east and south.

The western side, from the Caucasian mountains to the Southern Sea, is estimated at 13,000 stadia, along the river Indus to its mouth; wherefore the eastern side opposite, with the addition of the 3000 stadia of the promontory, will be 16,000 stadia in extent. This is both the smallest and greatest breadth of India.[324] The length is reckoned from west to east. The part of this extending (from the Indus) as far as Palibothra[325] we may describe more confidently; for it has been measured by Schœni,[326] and is a royal road of 10,000 stadia. The extent of the parts beyond depends upon conjecture derived from the ascent of vessels from the sea by the Ganges to Palibothra. This may be estimated at 6000 stadia.

The whole, on the shortest computation, will amount to 16,000 stadia, according to Eratosthenes, who says that he took it from the register of the Stathmi (or the several stages from place to place),[327] which was received as authentic, and Megasthenes agrees with him. But Patrocles says, that the sum of the whole is less by 1000 stadia. If again we add to this [Pg 80] [CAS. 689]distance the extent of the extremity which advances far towards the east, the greatest length of India will be 3000 stadia; this length is reckoned from the mouths of the river Indus along the coast, in a line with the mouths to the above-mentioned extremity and its eastern limits. Here the people called Coniaci[328] live.

12. From what has been said, we may perceive how the opinions of the other writers differ from one another. Ctesias says that India is not less than the rest of Asia; Onesicritus regards it as the third part of the habitable world; Nearchus says that it is a march of four months through the plain only. The computations of Megasthenes and Deïmachus are more

moderate, for they estimate the distance from the Southern Sea to Caucasus[329] at above 20,000 stadia. Deïmachus says that in some places it exceeds 30,000 stadia.

We have replied to these writers in the early part of this work.[330] At present it is sufficient to say that these opinions are in favour of the writers who, in describing India, solicit indulgence if they do not advance anything with confidence.

13. The whole of India is watered by rivers, some of which empty themselves into the two largest, the Indus and the Ganges; others discharge themselves into the sea by their own mouths. But all of them have their sources in the Caucasus. At their commencement their course is towards the south; some of them continue to flow in the same direction, particularly those which unite with the Indus; others turn to the east, as the Ganges. This, the largest of the Indian rivers, descends from the mountainous country, and when it reaches the plains, turns to the east, then flowing past Palibothra, a very large city, proceeds onwards to the sea in that quarter, and discharges its waters by a single mouth. The Indus falls into the Southern Sea, and empties itself by two mouths, encompassing the country called Patalene, which resembles the Delta of Egypt.

By the exhalation of vapours from such vast rivers, and by [Pg 81]the Etesian winds, India, as Eratosthenes affirms, is watered by summer rains, and the plains are overflowed. During the rainy season flax,[331] millet, sesamum, rice, and bosmorum[332] are sowed; and in the winter season, wheat, barley, pulse, and other esculent fruits of the earth with which we are not acquainted. Nearly the same animals are bred in India as in Ethiopia and Egypt, and the rivers of India produce all the animals of those countries, except the hippopotamus, although Onesicritus asserts that even this animal is found in them.

The inhabitants of the south resemble the Ethiopians in colour, but their countenances and hair are like those of other people. Their hair does not curl, on account of the humidity of the atmosphere. The inhabitants of the north resemble the Egyptians.

14. Taprobane[333] is said to be an island, lying out at sea, distant from the most southerly parts of India, which are opposite the Coniaci, seven days'[334] sail towards the south. Its length is about 8000 stadia in the direction of Ethiopia.[335] It produces elephants.

This is the account of Eratosthenes. The accounts of other writers, in addition to this, whenever they convey exact information, will contribute to form the description[336] (of India).

15. Onesicritus, for example, says of Taprobane, that its magnitude is 5000 stadia, without distinction of length or breadth, and that it is distant twenty days' sail from the continent, but that it was a voyage performed with difficulty and danger by vessels with sails ill constructed, and built with prows at each end, but without holds and keels;[337] that there are other islands between this and India, but that Taprobane lies farthest to the south; that there are found in the sea, about the island, animals of the cetaceous kind, in form like oxen, horses, and other land-animals.

16. Nearchus, speaking of the accretion of earth formed[Pg 82] [CAS. 691] by the rivers, adduces these instances. The plains of Hermes, Caÿster, Mæander, and Caïcus have these names, because they have been formed by the soil which has been carried over the plains by the rivers; or rather they were produced by the fine and soft soil brought down from the mountains; whence the plains are, as it were, the offspring of the rivers, and it is rightly said, that the plains belong to the rivers. What is said by Herodotus[338] of the Nile, and of the land about it, may be applied to this country, namely, that it is the gift of the Nile. Hence Nearchus thinks that the Nile had properly the synonym of Egypt.

17. Aristobulus, however, says, that rain and snow fall only on the mountains and the country immediately below them, and that the plains experience neither one nor the other, but are overflowed only by the rise of the waters of the rivers; that the mountains are covered with snow in the winter; that the rains set in at the commencement of spring, and continue to increase; that at the time of the blowing of the Etesian winds they pour down impetuously, without intermission, night and day till the rising of Arcturus,[339] and that the rivers, filled by the melting of the snow and by the rains, irrigate the flat grounds.

These things, he says, were observed by himself and by others on their journey into India from the Paropamisadæ. This was after the setting of the Pleiades,[340] and during their stay in the mountainous country in the territory of the Hypasii, and in that of Assacanus during the winter. At the beginning of spring they descended into the plains to a large city called Taxila,[341] thence they proceeded to the Hydaspes and the country of Porus. During the winter they saw no rain, but only snow. The first rain which fell was at Taxila. After their descent to the Hydaspes and the conquest of Porus, their progress was eastwards to the Hypanis, and thence again to the Hydaspes. At this time it rained continually, and particularly during the blowing of the Etesian winds, but at the rising of Arcturus the rains ceased. They remained at the Hydaspes while the ships were constructing, [Pg 83]and began their voyage not many days before the setting of the Pleiades, and were occupied during the whole autumn, winter, and the ensuing spring and summer, in sailing down the river, and arrived at Patalene[342] about the rising of the Dog-Star;[343] during the passage down the river, which lasted ten months, they did not experience rain at any place, not even when the Etesian winds were at their height, when the rivers were full and the plains overflowed; the sea could not be navigated on account of the blowing of contrary winds, but no land breezes succeeded.

18. Nearchus gives the same account, but does not agree with Aristobulus respecting the rains in summer, but says that the plains are watered by rain in the summer, and that they are without rain in winter. Both writers, however, speak of the rise of the rivers. Nearchus says, that the men encamped upon the Acesines[344]were obliged to change their situation for another more elevated, and that this was at the time of the rise of the river, and of the summer solstice.

Aristobulus gives even the measure of the height to which the river rises, namely, forty cubits, of which twenty would fill the channel beyond its previous depth up to the margin, and the other twenty are the measure of the water when it overflows the plains.

They agree also in saying that the cities placed upon mounds become islands, as in Egypt and Ethiopia, and that the inundation ceases after the rising of Arcturus, when the waters recede. They add, that the ground when half dried is sowed, after having been prepared by the commonest labourer, yet the plant comes to perfection, and the produce is good. The rice, according to Aristobulus, stands in water in an enclosure. It is sowed in beds. The plant is four cubits in height, with many ears, and yields a large produce. The harvest is about the time of the setting of the Pleiades, and the grain is beaten out like barley. It grows in Bactriana, Babylonia, Susis, and in the Lower Syria. Megillus says that it is sowed before the rains, but does not require irrigation or transplantation, being supplied with water from tanks.

The bosmorum, according to Onesicritus, is a kind of corn smaller than wheat, and grows in places situated between[Pg 84] [CAS. 692] rivers. After it is threshed out, it is roasted; the threshers being previously bound by an oath not to carry it away unroasted from the threshing floor; a precaution to prevent the exportation of the seed.

19. Aristobulus, when comparing the circumstances in which this country resembles, and those in which it differs from, Egypt and Ethiopia, and observing that the swelling of the Nile is occasioned by rains in the south, and of the Indian rivers by rains from the north, inquires why the intermediate places have no rain; for it does not rain in the Thebaïs as far as Syene, nor at the places near Meroë, nor in the parts of India from Patalene to the Hydaspes. But the country situated above these parts,[345] in which both rain and snow occur, is cultivated by the husbandman in the same manner as the country without India; for the rain and the snow supply the ground with moisture.

It is probable from what he relates that the country is subject to shocks of earthquakes, that the ground is loose and hollow by excess of moisture, and easily splits into fissures, whence even the course of rivers is altered.

He says that when he was despatched upon some business into the country, he saw a tract of land deserted, which contained more than a thousand cities with their dependent villages; the Indus, having left its proper channel, was diverted into another, on the left hand, much deeper, and precipitated itself into it like a cataract, so that it no longer watered the country by the (usual) inundation on the right hand, from which it had receded, and this was elevated above the level, not only of the new channel of the river, but above that of the (new) inundation.

20. The account of Onesicritus confirms the facts of the rising of the rivers and of the absence of land breezes. He says that the sea-shore is swampy, particularly near the mouths of rivers, on account of the mud, tides, and the force of the winds blowing from the sea.

Megasthenes also indicates the fertility of India by the circumstance of the soil producing fruits and grain twice a year. Eratosthenes relates the same facts, for he speaks of a winter and a summer sowing, and of the rain at the same [Pg 85]seasons. For there is no year, according to him, which is without rain at both those periods, whence ensues great abundance, the ground never failing to bear crops.

An abundance of fruit is produced by trees; and the roots of plants, particularly of large reeds, possess a sweetness, which they have by nature and by coction; for the water, both from rains and rivers, is warmed by the sun's rays. The meaning of Eratosthenes seems to be this, that what among other nations is called the ripening of fruits and juices, is called among these *coction*, and which contributes as much to produce an agreeable flavour as the coction by fire. To this is attributed the flexibility of the branches of trees, from which wheels of carriages are made, and to the same cause is imputed the growth upon some trees of wool.[346] Nearchus says that their fine clothes were made of this wool, and that the Macedonians used it for mattresses and the stuffing of saddles. The Serica[347] also are of a similar kind, and are made of dry byssus, which is obtained from some sort of bark of plants. He says that reeds[348] yield honey, although there are no bees, and that there is a tree from the fruit of which honey is procured, but that the fruit eaten fresh causes intoxication.

21. India produces many singular trees. There is one whose branches incline downwards, and whose leaves are not less in size than a shield. Onesicritus, describing minutely the country of Musicanus, which he says is the most southerly part[349] of India, relates, that there are some large trees the branches of which extend to the length even of twelve cubits. They then grow downwards, as though bent (by force), till they touch the earth, where they penetrate and take root like layers. They next shoot upwards and form a trunk. They again grow as we have described, bending downwards, and implanting one layer after another, and in the above order, so that one tree forms a long shady roof, like a tent, supported by many pillars. In speaking of the size of the trees, he says their trunks could scarcely be clasped by five men.[350]

Aristobulus also, where he mentions the Acesines, and its confluence with the Hyarotis, speaks of trees with their boughs bent downwards and of a size that fifty, but, according[Pg 86]
[CAS. 694] to Onesicritus, four hundred horsemen might take shelter at mid-day beneath the shade of a single tree.

Aristobulus mentions another tree, not large, bearing great pods, like the bean, ten fingers in length, full of honey,[351] and says that those who eat it do not easily escape with life. But the accounts of all these writers about the size of the trees have been exceeded by those who assert that there has been seen, beyond the Hyarotis,[352] a tree which casts a shade at noon of five stadia.

Aristobulus says of the wool-bearing trees, that the flower pod contains a kernel, which is taken out, and the remainder is combed like wool.

22. In the country of Musicanus there grows, he says, spontaneously grain resembling wheat, and a vine that produces wine, whereas other authors affirm that there is no wine in India. Hence, according to Anacharsis, they had no pipes, nor any musical instruments, except cymbals, drums, and crotala, which were used by jugglers.

Both Aristobulus and other writers relate that India produces many medicinal plants and roots, both of a salutary and noxious quality, and plants yielding a variety of colours. He adds, that, by a law, any person discovering a deadly substance is punished with death unless he also discover an antidote; in case he discovers an antidote, he is rewarded by the king.

Southern India, like Arabia and Ethiopia, produces cinnamon, nard, and other aromatics. It resembles these countries as regards the effect of the sun's rays, but it surpasses them in having a copious supply of water, whence the atmosphere is humid, and on this

50

account more conducive to fertility and fecundity; and this applies to the earth and to the water, hence those animals which inhabit both one and the other are of a larger size than are found in other countries. The Nile contributes to fecundity more than other rivers, and among other animals of large bulk, produces the amphibious kind. The Egyptian women also sometimes have four children at a birth, and Aristotle says that one woman had seven children at one birth.[353] He calls the Nile most fecundating and nutritive, on [Pg 87]account of the moderate coction effected by the sun's rays, which leave behind the nutritious part of substances, and evaporate that which is superfluous.

23. It is perhaps owing to this cause that the water of the Nile boils, as he says, with one half of the heat which other water requires. In proportion however, as the water of the Nile traverses in a straight line, a long and narrow tract of country, passing through a variety of climates and of atmosphere, while the Indian rivers are poured forth into wider and more extensive plains, their course being delayed a long time in the same climate, in the same degree the waters of India are more nutritious than those of the Nile; they produce larger animals of the cetaceous kind, and in greater number (than the Nile), and the water which descends from the clouds has already undergone the process of coction.

24. This would not be admitted by the followers of Aristobulus, who say that the plains are not watered by rain. Onesicritus, however, thinks that rain-water is the cause of the peculiar properties of animals, and alleges in proof, that the colour of foreign herds which drink of it is changed to that of the native animals.

This is a just remark; but it is not proper to attribute to the power of the water merely the cause of the black complexion and the woolly hair of the Ethiopians, and yet he censures Theodectes, who refers these peculiarities to the effects of the sun, in these words,

"Near these approaching with his radiant car,

The sun their skins with dusky tint doth dye,

And sooty hue; and with unvarying forms

Of fire, crisps their tufted hair."

There may be reason in this, for he says that the sun does not approach nearer to the Ethiopians than to other nations, but shines more perpendicularly, and that on this account the heat is greater; indeed, it cannot be correctly said that the sun approaches near to the Ethiopians, for he is at an equal distance from all nations. Nor is the heat the cause of the black complexion, particularly of children in the womb, who are out of the reach of the sun. Their opinion is to be preferred, who attribute these effects to the sun and to intense solar heat, causing a great deficiency of moisture on the surface[Pg 88] [CAS. 696] of the skin. Hence we say it is that the Indians have not woolly hair, nor is their colour so intensely[354] dark, because they live in a humid atmosphere.

With respect to children in the womb, they resemble their parents (in colour) according to a seminal disposition and constitution, on the same principle that hereditary diseases, and other likenesses, are explained.

The equal distance of the sun from all nations (according to Onesicritus) is an argument addressed to the senses, and not to reason. But it is not an argument addressed to the senses generally, but in the meaning that the earth bears the proportion of a point to the sun, for we may understand such a meaning of an argument addressed to the senses, by which we estimate heat to be more or less, as it is near or at a distance, in which cases it is not the same; and in this meaning, not in that of Onesicritus, the sun is said to be near the Ethiopians.

25. It is admitted by those who maintain the resemblance of India to Egypt and Ethiopia, that the plains which are not overflowed do not produce anything for want of water.

Nearchus says, that the old question respecting the rise of the Nile is answered by the case of the Indian rivers, namely, that it is the effect of summer rains; when Alexander saw crocodiles in the Hydaspes, and Egyptian beans in the Acesines, he thought that he had discovered the sources of the Nile, and was about to equip a fleet with the intention of sailing by this river to Egypt; but he found out shortly afterwards that his design could not be accomplished, "for in midway were vast rivers, fearful waters, and first the ocean,"[355]

into which all the Indian rivers discharge themselves; then Ariana, the Persian and Arabian Gulfs, all Arabia and Troglodytica.

The above is what has been said on the subject of winds and rains, the rising of rivers, and the inundation of plains.

26. We must describe these rivers in detail, with the particulars, which are useful for the purposes of geography, and which have been handed down to us by historians.

Besides this, rivers, being a kind of physical boundaries of the size and figures of countries, are of the greatest use in [Pg 89]every part of the present work. But the Nile and the rivers in India have a superiority above the rest, because the country could not be inhabited without them. By means of the rivers it is open to navigation and capable of cultivation, when otherwise it would not be accessible, nor could it be occupied by inhabitants.

We shall speak of the rivers deserving notice, which flow into the Indus, and of the countries which they traverse; with regard to the rest we know some particulars, but are ignorant of more. Alexander, who discovered the greatest portion of this country, first of all resolved it to be more expedient to pursue and destroy those who had treacherously killed Darius, and were meditating the revolt of Bactriana. He approached India therefore through Ariana, which he left on the right hand, and crossed the Paropamisus to the northern parts, and to Bactriana.[356] Having conquered all the country subject to the Persians, and many other places besides, he then entertained the desire of possessing India, of which he had received many, although indistinct, accounts.

He therefore returned, crossing over the same mountains by other and shorter roads, having India on the left hand; he then immediately turned towards it, and towards its western boundaries and the rivers Cophes and Choaspes.[357] The latter river empties itself into the Cophes,[358] near Plemyrium, after passing by another city Gorys, in its course through Bandobene and Gandaritis.[359]

He was informed that the mountainous and northern parts were the most habitable and fertile, but that the southern part was either without water, or liable to be overflowed by rivers at one time, or entirely burnt up at another, more fit to be the haunts of wild beasts than the dwellings of men. He resolved therefore to get possession of that part of India first which had been well spoken of, considering at the same time that the rivers which it was necessary to pass, and which flowed[Pg 90] [CAS. 697]transversely through the country which he intended to attack, would be crossed with more facility near their sources. He heard also that many of the rivers united and formed one stream, and that this more frequently occurred the farther they advanced into the country, so that from want of boats it would be more difficult to traverse. Being apprehensive of this obstruction, he crossed the Cophes, and conquered the whole of the mountainous country situated towards the east.

27. Next to the Cophes was the Indus, then the Hydaspes, the Acesines, the Hyarotis, and last, the Hypanis. He was prevented from proceeding farther, partly from regard to some oracles, and partly compelled by his army, which was exhausted by toil and fatigue, but whose principal distress arose from their constant exposure to rain. Hence we became acquainted with the eastern parts of India on this side the Hypanis, and whatever parts besides which have been described by those who, after Alexander, advanced beyond the Hypanis to the Ganges and Palibothra.

After the river Cophes, follows the Indus. The country lying between these two rivers is occupied by Astaceni, Masiani, Nysæi, and Hypasii.[360] Next is the territory of Assacanus, where is the city Masoga (Massaga?), the royal residence of the country. Near the Indus is another city, Peucolaïtis.[361] At this place a bridge which was constructed afforded a passage for the army.

28. Between the Indus and the Hydaspes is Taxila, a large city, and governed by good laws. The neighbouring country is crowded with inhabitants and very fertile, and here unites with the plains. The people and their king Taxiles received Alexander with kindness, and obtained in return more presents than they had offered to Alexander; so that the

Macedonians became jealous, and observed, that it seemed as if Alexander had found none on whom he could confer favours before he passed the Indus. Some writers say that this country is larger than Egypt.

Above this country among the mountains is the territory of Abisarus,[362] who, as the ambassadors that came from him [Pg 91]reported, kept two serpents, one of 80, and the other, according to Onesicritus, of 140 cubits in length. This writer may as well be called the master fabulist as the master pilot of Alexander. For all those who accompanied Alexander preferred the marvellous to the true, but this writer seems to have surpassed all in his description of prodigies. Some things, however, he relates which are probable and worthy of record, and will not be passed over in silence even by one who does not believe their correctness.

Other writers also mention the hunting of serpents in the Emodi mountains,[363] and the keeping and feeding of them in caves.

29. Between the Hydaspes and Acesines is the country of Porus,[364] an extensive and fertile district, containing nearly three hundred cities. Here also is the forest in the neighbourhood of the Emodi mountains in which Alexander cut down a large quantity of fir, pine, cedar, and a variety of other trees fit for ship-building, and brought the timber down the Hydaspes. With this he constructed a fleet on the Hydaspes, near the cities, which he built on each side of the river where he had crossed it and conquered Porus. One of these cities he called Bucephalia,[365] from the horse Bucephalus, which was [Pg 92] [CAS. 699]killed in the battle with Porus. The name Bucephalus[366] was given to it from the breadth of its forehead. He was an excellent war-horse, and Alexander constantly rode him in battle.

The other city he called Nicæa from the victory, NIKH (Nice), which he had obtained.

In the forest before mentioned it is said there is a vast number of monkeys,[367] and as large as they are numerous. On one occasion the Macedonians, seeing a body of them standing in array opposite to them, on some bare eminences, (for this animal is not less intelligent than the elephant,) and presenting the appearance of an army, prepared to attack them as real enemies, but being informed by Taxiles, who was then with the king, of the real fact, they desisted.

The chase of this animal is conducted in two different manners. It is an imitative creature, and takes refuge up among the trees. The hunters, when they perceive a monkey seated on a tree, place in sight a basin containing water, with which they wash their own eyes; then, instead of water, they put a basin of bird-lime, go away, and lie in wait at a distance. The animal leaps down, and besmears itself with the bird-lime, and when it winks, the eyelids are fastened together; the hunters then come upon it, and take it.

The other method of capturing them is as follows: the hunters dress themselves in bags like trowsers, and go away, leaving behind them others which are downy, with the inside smeared over with bird-lime. The monkeys put them on, and are easily taken.

30. Some writers place Cathaia[368] and the country of Sopeithes, one of the nomarchs, in the tract between the rivers (Hydaspes and Acesines); some, on the other side of the Acesines and of the Hyarotis, on the confines of the territory of the other Porus, the nephew of Porus who was taken prisoner by Alexander, and call the country subject to him Grandaris.

A very singular usage is related of the high estimation in which the inhabitants of Cathaia hold the quality of [Pg 93]beauty, which they extend to horses and dogs. According to Onesicritus, they elect the handsomest person as king. The child (selected), two months after birth, undergoes a public inspection, and is examined. They determine whether it has the amount of beauty required by law, and whether it is worthy to be permitted to live. The presiding magistrate then pronounces whether it is to be allowed to live, or whether it is to be put to death.

They dye their heads with various and the most florid colours, for the purpose of improving their appearance. This custom prevails elsewhere among many of the Indians, who pay great attention to their hair and dress; and the country produces colours of great beauty. In other respects the people are frugal, but are fond of ornament.

A peculiar custom is related of the Cathæi. The bride and the husband are respectively the choice of each other, and the wives burn themselves with their deceased husbands. The reason assigned for this practice is, that the women sometimes fell in love with young men, and deserted or poisoned their husbands. This law was therefore established in order to check the practice of administering poison; but neither the existence nor the origin of the law are probable facts.

It is said, that in the territory of Sopeithes there is a mountain composed of fossile salt, sufficient for the whole of India. Valuable mines also both of gold and silver are situated, it is said, not far off among other mountains, according to the testimony of Gorgus, the miner (of Alexander). The Indians, unacquainted with mining and smelting, are ignorant of their own wealth, and therefore traffic with greater simplicity.

31. The dogs in the territory of Sopeithes are said to possess remarkable courage: Alexander received from Sopeithes a present of one hundred and fifty of them. To prove them, two were set at a lion; when these were mastered, two others were set on; when the battle became equal, Sopeithes ordered a man to seize one of the dogs by the leg, and to drag him away; or to cut off his leg, if he still held on. Alexander at first refused his consent to the dog's leg being cut off, as he wished to save the dog. But on Sopeithes saying, "I will give you four in the place of it," Alexander consented; and he saw the dog permit his leg to be cut off by a slow incision, rather than loose his hold.

[Pg 94]
[CAS. 700]

32. The direction of the march, as far as the Hydaspes, was for the most part towards the south. After that, to the Hypanis, it was more towards the east. The whole of it, however, was much nearer to the country lying at the foot of the mountains than to the plains. Alexander therefore, when he returned from the Hypanis to the Hydaspes and the station of his vessels, prepared his fleet, and set sail on the Hydaspes.

All the rivers which have been mentioned (the last of which is the Hypanis) unite in one, the Indus. It is said that there are altogether fifteen[369] considerable rivers which flow into the Indus. After the Indus has been filled by all these rivers, so as to be enlarged in some places to the extent of a hundred stadia, according to writers who exaggerate, or, according to a more moderate estimate, to fifty stadia at the utmost, and at the least to seven, [and who speak of many nations and cities about this river,][370] it discharges itself by two mouths into the southern sea, and forms the island called Patalene.

Alexander's intention was to relinquish the march towards the parts situated to the east, first, because he was prevented from crossing the Hypanis; next, because he learnt by experience the falsehood of the reports previously received, to the effect that the plains were burnt up with fire, and more fit for the haunts of wild beasts than for the habitation of man. He therefore set out in this direction, relinquishing the other track; so that these parts became better known than the other.

33. The territory lying between the Hypanis and the Hydaspes is said to contain nine nations and five thousand cities, not less in size than Cos Meropis;[371] but the number seems to be exaggerated. We have already mentioned nearly all the nations deserving of notice, which inhabit the country situated between the Indus and the Hydaspes.

Below, and next in order, are the people called Sibæ, whom we formerly mentioned,[372] and the great nations, the Malli[373] and Sydracæ (Oxydracæ). It was among the Malli that Alexander[Pg 95] was in danger of losing his life, from a wound he received at the capture of a small city. The Sydracæ, we have said, are fabled to be allied to Bacchus.

Near Patalene is placed the country of Musicanus, that of Sabus,[374] whose capital is Sindomana, that of Porticanus, and of other princes who inhabited the country on the banks of the Indus. They were all conquered by Alexander; last of all he made himself master of Patalene, which is formed by the two branches of the Indus. Aristobulus says that these two branches are distant 1000 stadia from each other. Nearchus adds 800 stadia more to this number. Onesicritus reckons each side of the included island, which is of a triangular shape, at 2000 stadia; and the breadth of the river, where it is separated into two mouths, at about

200 stadia.^{375} He calls the island Delta, and says that it is as large as the Delta of Egypt; but this is a mistake. For the Egyptian Delta is said to have a base of 1300 stadia, and each of the sides to be less than the base. In Patalene is Patala, a considerable city, from which the island has its name.

34. Onesicritus says, that the greatest part of the coast in this quarter abounds with swamps, particularly at the mouths of the river, which is owing to the mud, the tides, and the want of land breezes; for these parts are chiefly under the influence of winds blowing from the sea.

He expatiates also in praise of the country of Musicanus, and relates of the inhabitants what is common to other Indian tribes, that they are long-lived, and that life is protracted even to the age of 130 years, (the Seres,^{376} however, are said by some [Pg 96] [CAS. 701]writers to be still longer lived,) that they are temperate in their habits and healthy; although the country produces everything in abundance.

The following are their peculiarities: to have a kind of Lacedæmonian common meal, where they eat in public. Their food consists of what is taken in the chase. They make no use of gold nor silver, although they have mines of these metals. Instead of slaves, they employed youths in the flower of their age, as the Cretans employ the Aphamiotæ, and the Lacedæmonians the Helots. They study no science with attention but that of medicine; for they consider the excessive pursuit of some arts, as that of war, and the like, to be committing evil. There is no process at law but against murder and outrage, for it is not in a person's own power to escape either one or the other; but as contracts are in the power of each individual, he must endure the wrong, if good faith is violated by another; for a man should be cautious whom he trusts, and not disturb the city with constant disputes in courts of justice.

Such are the accounts of the persons who accompanied Alexander in his expedition.

35. A letter of Craterus to his mother Aristopatra is circulated, which contains many other singular circumstances, and differs from every other writer, particularly in saying that Alexander advanced as far as the Ganges. Craterus says, that he himself saw the river, and the whales^{377} which it produces, and [his account] of its magnitude, breadth, and depth, far exceeds, rather than approximates, probability. For that the Ganges is the largest of known rivers in the three continents, it is generally agreed; next to this is the Indus; and, thirdly, the Danube; and, fourthly, the Nile. But different authors differ in their account of it, some assigning 30, others 3 stadia, as the least breadth. But Megasthenes says that its ordinary width is 100 stadia,^{378} and its least depth twenty orguiæ.^{379}

[Pg 97]

36. At the confluence of the Ganges and of another river (the Erannoboas^{380}) is situated (the city) Palibothra, in length 80, and in breadth 15 stadia. It is in the shape of a parallelogram, surrounded by a wooden wall pierced with openings through which arrows may be discharged. In front is a ditch, which serves the purpose of defence and of a sewer for the city. The people in whose country the city is situated are the most distinguished of all the tribes, and are called Prasii. The king, besides his family name, has the surname of Palibothrus, as the king to whom Megasthenes was sent on an embassy had the name of Sandrocottus.^{381}

Such also is the custom among the Parthians; for all have the name Arsacæ,^{382}although each has his peculiar name of Orodes, Phraates, or some other appellation.

37. All the country on the other side of the Hypanis is allowed to be very fertile, but we have no accurate knowledge of it. Either through ignorance or from its remote situation, everything relative to it is exaggerated or partakes of the wonderful. As, for example, the stories of myrmeces (or ants),^{383} which dig up gold; of animals and men with peculiar shapes, and possessing extraordinary faculties; of the longevity of the Seres, whose lives exceed the age of two hundred years. They speak also of an aristocratical form of government, consisting of five hundred counsellors, each of whom furnishes the state with an elephant.

According to Megasthenes, the largest tigers are found among the Prasii, almost twice the size of lions, and of such strength that a tame one led by four persons seized a mule by its hinder leg, overpowered it, and dragged it to him. The monkeys are larger than the largest dogs; they are of a white colour, except the face, which is black. The contrary is observed in other places. Their tails are more than two cubits in length. They are very tame, and not of a mischievous disposition. They neither attack people, nor steal.

Stones are found there of the colour of frankincense, and sweeter than figs or honey.

In some places there are serpents of two cubits in length, with membraneous wings like bats. They fly at night, and let fall drops of urine or sweat, which occasions the skin of persons[Pg 98] [CAS. 703] who are not on their guard to putrefy. There are also winged scorpions of great size.

Ebony grows there. There are also dogs of great courage, which do not loose their hold till water is poured into their nostrils: some of them destroy their sight, and the eyes of others even fall out, by the eagerness of their bite. Both a lion and a bull were held fast by one of these dogs. The bull was caught by the muzzle, and died before the dog could be loosened.

38. In the mountainous country is a river, the Silas, on the surface of which nothing will float. Democritus, who had travelled over a large part of Asia, disbelieves this, and Aristotle does not credit it, although atmospheres exist so rare that no bird can sustain its flight in them. Vapours also, which ascend (from some substances), attract and absorb, as it were, whatever is flying over them; as amber attracts straw, and the magnet iron, and perhaps there may be in water a similar power.

As these matters belong to physics and to the question of floating bodies, these must be referred to them. At present we must proceed to what follows, and to the subjects more nearly relating to geography.

39. It is said that the Indians are divided into seven castes. The first in rank, but the smallest in number, are the philosophers. Persons who intend to offer sacrifice, or to perform any sacred rite, have the services of these persons on their private account; but the kings employ them in a public capacity at the time of the Great Assembly, as it is called, where at the beginning of the new year all the philosophers repair to the king at the gate, and anything useful which they have committed to writing, or observed, tending to improve the productions of the earth or animals, or of advantage to the government of the state, is then publicly declared.

Whoever has been detected in giving false information thrice is enjoined silence by law during the rest of his life; but he who has made correct observations is exempted from all contributions and tribute.

40. The second caste is that of husbandmen, who constitute the majority of natives, and are a most mild and gentle people, as they are exempted from military service, and cultivate[Pg 99] their land free from alarm; they do not resort to cities, either to transact private business, or take part in public tumults. It therefore frequently happens that at the same time, and in the same part of the country, one body of men are in battle array, and engaged in contests with the enemy, while others are ploughing or digging in security, having these soldiers to protect them. The whole of the territory belongs to the king. They cultivate it on the terms of receiving as wages a fourth part of the produce.

41. The third caste consists of shepherds and hunters, who alone are permitted to hunt, to breed cattle, to sell and to let out for hire beasts of burden. In return for freeing the country from wild beasts and birds, which infest sown fields, they receive an allowance of corn from the king. They lead a wandering life, and dwell in tents. No private person is allowed to keep a horse or an elephant. The possession of either one or the other is a royal privilege, and persons are appointed to take care of them.

42. The manner of hunting the elephant is as follows: Round a bare spot a ditch is dug, of about four or five stadia in extent, and at the place of entrance a very narrow bridge is constructed. Into the enclosure three or four of the tamest female elephants are driven. The men themselves lie in wait under cover of concealed huts. The wild elephants do not approach the females by day, but at night they enter the enclosure one by one; when they have passed the entrance, the men secretly close it. They then introduce the strongest of the tame combatants, the drivers of which engage with the wild animals, and also wear them out by famine; when the latter are exhausted by fatigue, the boldest of the drivers gets down

unobserved, and creeps under the belly of his own elephant. From this position he creeps beneath the belly of the wild elephant, and ties his legs together; when this is done, a signal is given to the tame elephants to beat those which are tied by the legs, till they fall to the ground. After they have fallen down, they fasten the wild and tame elephants together by the neck with thongs of raw cow-hide, and, in order that they may not be able to shake off those who are attempting to mount them, cuts are made round the neck, and thongs of leather are put into these incisions, so that they submit to their bonds through pain, and so remain quiet. Among the elephants[Pg 100] [CAS. 705] which are taken, those are rejected which are too old or too young for service; the remainder are led away to the stables. They tie their feet one to another, and their necks to a pillar firmly fastened in the ground, and tame them by hunger. They recruit their strength afterwards with green cane and grass. They then teach them to obey; some by words; others they pacify by tunes, accompanied with the beating of a drum. Few are difficult to be tamed; for they are naturally of a mild and gentle disposition, so as to approximate to the character of a rational animal. Some have taken up their drivers, who have fallen on the ground lifeless, and carried them safe out of battle. Others have fought, and protected their drivers, who have crept between their fore-legs. If they have killed any of their feeders or masters in anger, they feel their loss so much that they refuse their food through grief, and sometimes die of hunger.

43. They copulate like horses, and produce young chiefly in the spring. It is the season for the male, when he is in heat and is ferocious. At this period he discharges some fatty matter through an opening in the temples. It is the season also for the females, when this same passage is open. Eighteen months is the longest, and sixteen the shortest period that they go with young. The dam suckles her young six years. Many of them live as long as men who attain to the greatest longevity, some even to the protracted age of two hundred years.

They are subject to many diseases, which are difficult to be cured. A remedy for diseases of the eye is to bathe them with cow's milk. For complaints in general, they drink dark wine. In cases of wounds, they drink butter; for it draws out iron instruments. Their sores are fomented with swine's flesh.

Onesicritus says, that they live three hundred years, and rarely five hundred; and that they go with young ten years. He and other writers say, that they are larger and stronger than the African elephants. They will pull down with their trunks battlements, and uproot trees, standing erect upon their hind feet.

According to Nearchus, traps are laid in the hunting grounds, at certain places where roads meet; the wild elephants are forced into the toils by the tame elephants, which[Pg 101] are stronger, and guided by a driver. They become so tame and docile, that they learn even to throw a stone at a mark, to use military weapons, and to be excellent swimmers. A chariot drawn by elephants is esteemed a most important possession, and they are driven without bridles.[384]

A woman is greatly honoured who receives from her lover a present of an elephant, but this does not agree with what he said before, that a horse and an elephant are the property of kings alone.

44. This writer says that he saw skins of the myrmeces (or ants), which dig up gold, as large as the skins of leopards. Megasthenes, however, speaking of the myrmeces, says, among the Derdæ a populous nation of the Indians, living towards the east, and among the mountains, there was a mountain plain of about 3000 stadia in circumference; that below this plain were mines containing gold, which the myrmeces, in size not less than foxes, dig up. They are excessively fleet, and subsist on what they catch. In winter they dig holes, and pile up the earth in heaps, like moles, at the mouths of the openings.

The gold-dust which they obtain requires little preparation by fire. The neighbouring people go after it by stealth, with beasts of burden; for if it is done openly, the myrmeces fight furiously, pursuing those that run away, and if they seize them, kill them and the beasts. In order to prevent discovery, they place in various parts pieces of the flesh of wild beasts, and when the myrmeces are dispersed in various directions, they take away the gold-dust, and, not being acquainted with the mode of smelting it, dispose of it in its rude state at any price to merchants.

45. Having mentioned what Megasthenes and other writers relate of the hunters and the beasts of prey, we must add the following particulars.

Nearchus is surprised at the multitude and the noxious nature of the tribe of reptiles. They retreat from the plains to the settlements, which are not covered with water at the period of inundations, and fill the houses. For this reason the inhabitants raise their beds at some   height   from   the [Pg   102] [CAS. 706]ground, and are sometimes compelled to abandon their dwellings, when they are infested by great multitudes of these animals; and, if a great proportion of these multitudes were not destroyed by the waters, the country would be a desert. Both the minuteness of some animals and the excessive magnitude of others are causes of danger; the former, because it is difficult to guard against their attacks; the latter, on account of their strength, for snakes are to be seen of sixteen cubits in length. Charmers go about the country, and are supposed to cure wounds made by serpents. This seems to comprise nearly their whole art of medicine, for disease is not frequent among them, which is owing to their frugal manner of life, and to the absence of wine; whenever diseases do occur, they are treated by the Sophistæ (or wise men).

Aristobulus says, that he saw no animals of these pretended magnitudes, except a snake, which was nine cubits and a span in length. And I myself saw one in Egypt, nearly of the same size, which was brought from India. He says also, that he saw many serpents of a much inferior size, and asps and large scorpions. None of these, however, are so noxious as the slender small serpents, a span long, which are found concealed in tents, in vessels, and in hedges. Persons wounded by them bleed from every pore, suffering great pain, and die, unless they have immediate assistance; but this assistance is easily obtained, by means of the virtues of the Indian roots and drugs.

Few crocodiles, he says, are found in the Indus, and these are harmless, but most of the other animals, except the hippopotamus, are the same as those found in the Nile; but Onesicritus says that this animal also is found there.

According to Aristobulus, none of the sea fish ascend the Nile from the sea, except the shad,[385] the grey mullet,[386] and dolphin, on account of the crocodiles; but great numbers ascend the Indus. Small craw-fish[387] go up as far as the mountains,[388] and the larger as far as the confluence of the Indus and the Acesines.

[Pg 103]So much then on the subject of the wild animals of India. We shall return to Megasthenes, and resume our account where we digressed.

46. After the hunters and the shepherds, follows the fourth caste, which consists, he says, of those who work at trades, retail wares, and who are employed in bodily labour. Some of these pay taxes, and perform certain stated services. But the armour-makers and ship-builders receive wages and provisions from the king, for whom only they work. The general-in-chief furnishes the soldiers with arms, and the admiral lets out ships for hire to those who undertake voyages and traffic as merchants.

47. The fifth caste consists of fighting men, who pass the time not employed in the field in idleness and drinking, and are maintained at the charge of the king. They are ready whenever they are wanted to march on an expedition, for they bring nothing of their own with them, except their bodies.

48. The sixth caste is that of the Ephori, or inspectors. They are intrusted with the superintendence of all that is going on, and it is their duty to report privately to the king. The city inspectors employ as their coadjutors the city courtesans; and the inspectors of the camp, the women who follow it. The best and the most faithful persons are appointed to the office of inspector.

49. The seventh caste consists of counsellors and assessors of the king. To these persons belong the offices of state, tribunals of justice, and the whole administration of affairs.

It is not permitted to contract marriage with a person of another caste, nor to change from one profession or trade to another, nor for the same person to undertake several, except he is of the caste of philosophers, when permission is given, on account of his superior qualifications.

50. Of the magistrates, some have the charge of the market, others of the city, others of the soldiery. Some have the care of the rivers, measure the land, as in Egypt, and inspect the closed reservoirs, from which water is distributed by canals, so that all may have an equal use of it. These persons have charge also of the hunters, and have the power of rewarding or punishing those who merit either. They collect the taxes, and superintend the occupations connected with land, as wood-cutters, carpenters, workers in brass, and miners. They[Pg 104] [CAS. 708] superintend the public roads, and place a pillar at every ten stadia, to indicate the by-ways and distances.

51. Those who have charge of the city are divided into six bodies of five each. The first has the inspection of everything relating to the mechanical arts; the second entertain strangers, assign lodgings, observe their mode of life, by means of attendants whom they attach to them, escort them out of the country on their departure; if they die, take charge of their property, have the care of them when sick, and when they die, bury them.

The third class consists of those who inquire at what time and in what manner births and deaths take place, which is done with a view to tax (on these occasions), and in order that the deaths and births of persons both of good and bad character should not be concealed.

The fourth division consists of those who are occupied in sales and exchanges; they have the charge of measures, and of the sale of the products in season, by a signal. The same person is not allowed to exchange various kinds of articles, except he pays a double tax.

The fifth division presides over works of artisans, and disposes of articles by public notice. The new are sold apart from the old, and there is a fine imposed for mixing them together. The sixth and last comprises those who collect the tenth of the price of the articles sold. Death is the punishment for committing a fraud with regard to the tax.

These are the peculiar duties performed by each class, but in their collective capacity they have the charge both of their own peculiar province and of civil affairs, the repairs of public works, prices[389] of articles, of markets, harbours, and temples.

52. Next to the magistrates of the city is a third body of governors, who have the care of military affairs. This class also consists of six divisions, each composed of five persons. One division is associated with the chief naval superintendent, another with the person who has the charge of the bullock-teams, by which military engines are transported, of provisions both for the men and beasts, and other requisites for the army. They furnish attendants, who beat a drum, and carry gongs;[390] [Pg 105]and besides these, grooms, mechanists, and their assistants. They despatch by the sound of the gong the foragers for grass, and insure expedition and security by rewards and punishments. The third division has the care of the infantry; the fourth, of the horses; the fifth, of the chariots; the sixth, of the elephants. There are royal stables for the horses and elephants. There is also a royal magazine of arms; for the soldier returns his arms to the armoury, and the horse and elephant to the stables. They use the elephants without bridles. The chariots are drawn on the march by oxen. The horses are led by a halter, in order that their legs may not be chafed and inflamed, nor their spirit damped, by drawing chariots. Besides the charioteer, there are two persons who fight by his side in the chariot. With the elephant are four persons, the driver and three bowmen, who discharge arrows from his back.

53. All the Indians are frugal in their mode of life, and especially in camp. They do not tolerate useless and undisciplined multitudes, and consequently observe good order. Theft is very rare among them. Megasthenes, who was in the camp of Sandrocottus, which consisted of 400,000 men, did not witness on any day thefts reported, which exceeded the sum of two hundred drachmæ, and this among a people who have no written laws, who are ignorant even of writing, and regulate everything by memory. They are, however, happy on account of their simple manners and frugal way of life. They never drink wine, but at sacrifices. Their beverage is made from rice instead of barley, and their food consists for the most part of rice pottage. The simplicity of their laws and contracts appears from their not having many law-suits. They have no suits respecting pledges and deposits, nor do they

require witnesses or seals, but make their deposits, and confide in one another. Their houses and property are unguarded. These things denote temperance and sobriety; others no one would approve, as their eating always alone, and their not having all of them one common hour for their meals, but each taking it as he likes. The contrary custom is more agreeable to the habits of social and civil life.

54. As an exercise of the body they prefer friction in various ways, but particularly by making use of smooth sticks of ebony, which they pass over the surface of the body.

[Pg 106]
[CAS. 709]

Their sepulchres are plain, and the tumuli of earth low.

In contrast to their parsimony in other things, they indulge in ornament. They wear dresses worked with gold and precious stones, and flowered (variegated) robes, and are attended by persons following them with umbrellas; for as they highly esteem beauty, everything is attended to, which can improve their looks.

They respect alike truth and virtue; therefore they do not assign any privilege to the old, unless they possess superior wisdom.

They marry many wives, who are purchased from their parents, and give in exchange for them a yoke of oxen. Some marry wives to possess obedient attendants, others with a view to pleasure and numerous offspring, and the wives prostitute themselves, unless chastity is enforced by compulsion.

No one wears a garland when sacrificing, or burning incense, or pouring out a libation. They do not stab, but strangle the victim, that nothing mutilated, but that which is entire, may be offered to the Deity.

A person convicted of bearing false testimony suffers a mutilation of his extremities. He who has maimed another not only undergoes in return the loss of the same limb, but his hand also is cut off. If he has caused a workman to lose his hand or his eye, he is put to death.

Megasthenes says, that none of the Indians employ slaves. But, according to Onesicritus, this is peculiar to the people in the territory of Musicanus. He speaks of this as an excellent rule, and mentions many others to be found in that country, as the effects of a government by good laws.

55. The care of the king's person is committed to women, who are also purchased of their parents. The body-guard, and the rest of the military, are stationed without the gates. A woman, who puts to death a king when drunk, is rewarded by becoming the wife of his successor. The sons succeed the father. The king may not sleep during the day-time, and at night he is obliged from time to time to change his bed, from dread of treachery.

The king leaves his palace in time of war; he leaves it also when he goes to sit in his court as a judge. He remains there all day thus occupied, not suffering himself to be interrupted even though the time arrives for attending to his person.[Pg 107] This attention to his person consists of friction with pieces of wood, and he continues to listen to the cause, while the friction is performed by four attendants who surround him.

Another occasion of leaving his palace is to offer sacrifice.

The third is a sort of Bacchanalian departure to the chace. Crowds of women surround him, and on the outside (of these) are spear-men. The road is set off with ropes; a man, or even a woman, who passes within the ropes is put to death.

The king is preceded by drums and gongs. He hunts in the enclosures, and discharges his arrows from a high seat. Near him stand two or three armed women. When hunting in the open ground, he shoots his arrows from an elephant; of the women some are in chariots, some on horses, and others on elephants; they are provided with all kinds of weapons, as if they were going on a military expedition.

56. These customs when compared with ours are very strange, but the following are still more extraordinary. According to Megasthenes, the nations who inhabit the Caucasus have commerce with women in public; and eat the bodies of their relatives; the monkeys climb precipices, and roll down large stones upon their pursuers; most of the animals which are tame in our country are wild in theirs; the horses have a single horn, with heads like those of deer; reeds which grow to the height of thirty orguiæ,[321] others which grow on the

ground, fifty orguiæ in length, and in thickness some are three and others six cubits in diameter.

57. He then deviates into fables, and says that there are men of five, and even three spans in height, some of whom are without nostrils, with only two breathing orifices above the mouth. Those of three spans in height wage war with the cranes (described by Homer) and with the partridges, which are as large as geese; these people collect and destroy the eggs of the cranes which lay their eggs there; and nowhere else are the eggs or the young cranes to be found; frequently a crane escapes from this country with a brazen point of a weapon in its body, wounded by these people.

Similar to this is the account of the Enotocoitæ,[322] of the wild men, and of other monsters. The wild men could not be brought to Sandrocottus, for they died by abstaining from [Pg 108]
[CAS. 711]food. Their heels are in front, the instep and toes are turned backwards. Some have been taken, which had no mouths, and were tame. They live near the sources of the Ganges, and are supported by the smell of dressed meat and the fragrance of fruits and flowers, having instead of mouths orifices through which they breathe. They are distressed by strong-smelling substances, and therefore their lives are sustained with difficulty, particularly in a camp.

With respect to the other singular animals, the philosophers informed him of a people called Ocypodæ, so swift of foot that they leave horses behind them; of Enotocoitæ, or persons having ears hanging down to their feet, so that they lie and sleep upon them, and so strong as to be able to pluck up trees and to break the sinew string of a bow; of others (Monommati) who have only one eye, and the ears of a dog, the eye placed in the middle of the forehead, the hair standing erect, and the breasts shaggy; of others (Amycteres) without nostrils, devouring everything, eaters of raw meat, short-lived, and dying before they arrive at old age; the upper part of their mouths projects far beyond the lower lip.

With respect to the Hyperboreans, who live to the age of a thousand years, his description is the same as that of Simonides, Pindar, and other mythological writers.

The story told by Timagenes of a shower of drops of brass, which were raked together, is a fable. The account of Megasthenes is more probable, namely, that the rivers bring down gold-dust, a part of which is paid as a tax to the king; and this is the case in Iberia (of Armenia).

58. Speaking of the philosophers, he says, that those who inhabit the mountains are worshippers of Bacchus, and show as a proof (of the god having come among them) the wild vine, which grows in their country only; the ivy, the laurel, the myrtle, the box-tree, and other evergreens, none of which are found beyond the Euphrates, except a few in parks, which are only preserved with great care. To wear robes and turbans, to use perfumes, and to be dressed in dyed and flowered garments, for their kings to be preceded when they leave their palaces, and appear abroad, by gongs and drums, are Bacchanalian customs. But the philosophers who live in the plains worship Hercules.

These are fabulous stories, contradicted by many writers,[Pg 109] particularly what is said of the vine and wine, for a great part of Armenia, the whole of Mesopotamia and Media, as far as Persia and Carmania, is beyond the Euphrates, the greater part of which countries is said to have excellent vines, and to produce good wine.

59. Megasthenes divides the philosophers again into two kinds, the Brachmanes[393]and the Garmanes.[394] The Brachmanes are held in greater repute, for they agree more exactly in their opinions. Even from the time of their conception in the womb they are under the care and guardianship of learned men, who go to the mother, and seem to perform some incantation for the happiness and welfare of the mother and the unborn child, but in reality they suggest prudent advice, and the mothers who listen to them most willingly are thought to be the most fortunate in their offspring. After the birth of the children, there is a succession of persons who have the care of them, and as they advance in years, masters more able and accomplished succeed.

The philosophers live in a grove in front of the city within a moderate-sized enclosure. Their diet is frugal, and they lie upon straw pallets and on skins. They abstain from animal food, and from sexual intercourse with women; their time is occupied in grave discourse, and they communicate with those who are inclined to listen to them; but the hearer is not permitted to speak or cough, or even to spit on the ground; otherwise, he is expelled that very day from their society, on the ground of having no control over himself. After living thirty-seven years in this manner, each individual retires to his own possessions, and lives with less restraint, wearing robes of fine linen, and rings of gold, but without profuseness, upon the hands and in the ears. They eat the flesh of animals, of those particularly which do not assist man in his labour, and abstain from hot and seasoned food. They have as many wives as they please with a view to numerous offspring, for from many wives greater advantages are derived.

As they have no slaves, they require more the services, which are at hand, of their children.

The Brachmanes do not communicate their philosophy to their wives, for fear they should   divulge   to   the   profane,   if [Pg   110] [CAS. 712]they became depraved, anything which ought to be concealed; or lest they should abandon their husbands in case they became good (philosophers) themselves. For no one who despises alike pleasure and pain, life and death, is willing to be subject to the authority of another; and such is the character of a virtuous man and a virtuous woman.

They discourse much on death, for it is their opinion that the present life is the state of one conceived in the womb, and that death to philosophers is birth to a real and a happy life. They therefore discipline themselves much to prepare for death, and maintain that nothing which happens to man is bad or good, for otherwise the same things would not be the occasion of sorrow to some and of joy to others, opinions being merely dreams, nor that the same persons could be affected with sorrow and joy by the same things, on different occasions.

With regard to opinions on physical phenomena, they display, says Megasthenes, great simplicity, their actions being better than their reasoning, for their belief is chiefly founded on fables. On many subjects their sentiments are the same as those of the Greeks. According to the Brachmanes, the world was created, and is liable to corruption; it is of a spheroïdal figure; the god who made and governs it pervades the whole of it; the principles of all things are different, but the principle of the world's formation was water; in addition to the four elements there is a fifth nature, of which the heavens and the stars are composed; the earth is situated in the centre of the universe. Many other peculiar things they say of the principle of generation and of the soul. They invent fables also, after the manner of Plato, on the immortality of the soul, and on the punishments in Hades, and other things of this kind. This is the account which Megasthenes gives of the Brachmanes.

60. Of the Garmanes, the most honourable, he says, are the Hylobii, who live in the forests, and subsist on leaves and wild fruits: they are clothed with garments made of the bark of trees,[395] and abstain from commerce with women and from wine. The kings hold communication with them by messengers, concerning the causes of things, and through them worship and supplicate the Divinity.

[Pg 111]

Second in honour to the Hylobii, are the physicians, for they apply philosophy to the study of the nature of man. They are of frugal habits, but do not live in the fields, and subsist upon rice and meal, which every one gives when asked, and receive them hospitably. They are able to cause persons to have a numerous offspring, and to have either male or female children, by means of charms. They cure diseases by diet, rather than by medicinal remedies. Among the latter, the most in repute are unguents and cataplasms. All others they suppose partake greatly of a noxious nature.

Both this and the other class of persons practise fortitude, as well in supporting active toil as in enduring suffering, so that they will continue a whole day in the same posture, without motion.

There are enchanters and diviners, versed in the rites and customs relative to the dead, who go about villages and towns begging. There are others who are more civilized and better informed than these, who inculcate the vulgar opinions concerning Hades, which, according to their ideas, tend to piety and sanctity. Women study philosophy with some of them, but abstain from sexual intercourse.

61. Aristobulus says, that he saw at Taxila two sophists (wise men), both Brachmanes, the elder had his head shaved, but the younger wore his hair; both were attended by disciples. When not otherwise engaged, they spent their time in the market-place. They are honoured as public counsellors, and have the liberty of taking away, without payment, whatever article they like which is exposed for sale; when any one accosts them, he pours over them oil of jessamine, in such profusion that it runs down from their eyes. Of honey and sesamum, which is exposed for sale in large quantity, they take enough to make cakes, and are fed without expense.

They came up to Alexander's table and took their meal standing, and they gave an example of their fortitude by retiring to a neighbouring spot, where the elder, falling on the ground supine, endured the sun and the rain, which had now set in, it being the commencement of spring. The other stood on one leg, with a piece of wood three cubits in length raised in both hands; when one leg was fatigued he changed the support to the other, and thus continued the whole day. The younger appeared to possess much more self-command;   for,[Pg   112]
[CAS. 714] after following the king a short distance, he soon returned to his home. The king sent after him, but he bade the king to come to him, if he wanted anything of him. The other accompanied the king to the last: during his stay he changed his dress, and altered his mode of life, and when reproached for his conduct, answered, that he had completed the forty years of discipline which he had promised to observe: Alexander made presents to his children.

62. Aristobulus relates also some strange and unusual customs of the people of Taxila. Those, who through poverty are unable to marry their daughters, expose them for sale in the market-place, in the flower of their age, to the sound of shell trumpets and drums, with which the war-note is given. A crowd is thus assembled. First her back, as far as the shoulders, is uncovered, then the parts in front, for the examination of any man who comes for this purpose. If she pleases him, he marries her on such conditions as may be determined upon.

The dead are thrown out to be devoured by vultures. To have many wives is a custom common to these and to other nations. He says, that he had heard, from some persons, of wives burning themselves voluntarily with their deceased husbands; and that those women who refused to submit to this custom were disgraced. The same things have been told by other writers.[396]

63. Onesicritus says, that he himself was sent to converse with these wise men. For Alexander heard that they went about naked, practised constancy and fortitude, and were held in the highest honour; that, when invited, they did not go to other persons, but commanded others to come to them, if they wished to participate in their exercises or their conversation. Such being their character, Alexander did not consider it to be consistent with propriety to go to them, nor to compel them to do anything contrary to their inclination or against the custom of their country; he therefore despatched Onesicritus to them.

Onesicritus found, at the distance of 20 stadia from the city, fifteen men standing in different postures, sitting or [Pg 113]lying down naked, who continued in these positions until the evening, and then returned to the city. The most difficult thing to endure was the heat of the sun, which was so powerful, that no one else could endure without pain to walk on the ground at mid-day with bare feet.

64. He conversed with Calanus, one of these sophists, who accompanied the king to Persia, and died after the custom of his country, being placed on a pile of [burning] wood. When Onesicritus came, he was lying upon stones. Onesicritus approached, accosted him, and told him that he had been sent by the king, who had heard the fame of his wisdom, and that he was to give an account of his interview, if there were no objection, he was ready to listen to his discourse. When Calanus saw his mantle, head-covering, and shoes, he laughed, and said, "Formerly, there was abundance everywhere of corn and barley, as there is now of dust; fountains then flowed with water, milk, honey, wine, and oil, but mankind by repletion and luxury became proud and insolent. Jupiter, indignant at this state of things, destroyed all, and appointed for man a life of toil. On the reäppearance of temperance and other virtues, there was again an abundance of good things. But at present the condition of mankind

63

approaches satiety and insolence, and there is danger lest the things which now exist should disappear."

When he had finished, he proposed to Onesicritus, if he wished to hear his discourse, to strip off his clothes, to lie down naked by him on the same stones, and in that manner to listen to him; while he was hesitating what to do, Mandanis,327who was the oldest and wisest of the sophists, reproached Calanus for his insolence, although he censured such insolence himself. Mandanis called Onesicritus to him, and said, I commend the king, because, although he governs so large an empire, he is yet desirous of acquiring wisdom, for he is the only philosopher in arms that I ever saw; it would be of the greatest advantage, if those were philosophers who have the power of persuading the willing and of compelling the unwilling to learn temperance; but I am entitled to indulgence, if, when conversing by means of three interpreters, who, except the language, know no more than the vulgar, I am not [Pg 114] [CAS. 716]able to demonstrate the utility of philosophy. To attempt it is to expect water to flow pure through mud.

65. "The tendency of his discourse," he said, "was this, that the best philosophy was that which liberated the mind from pleasure and grief; that grief differed from labour, in that the former was inimical, the latter friendly to men; for that men exercised their bodies with labour in order to strengthen the mental powers, by which means they would be able to put an end to dissensions, and give good counsel to all, to the public and to individuals; that he certainly should at present advise Taxiles to receive Alexander as a friend; for if he entertained a person better than himself, he might be improved; but if a worse person, he might dispose him to good."

After this Mandanis inquired, whether such doctrines were taught among the Greeks. Onesicritus answered, that Pythagoras taught a similar doctrine, and enjoined his disciples to abstain from whatever has life; that Socrates and Diogenes, whose discourses he had heard, held the same opinions. Mandanis replied, "that in other respects he thought them wise, but that in one thing they were mistaken, namely, in preferring custom to nature, for otherwise they would not be ashamed of going naked, like himself, and of subsisting on frugal fare; for the best house was that which required least repairs." He says also that they employ themselves much on natural subjects, as prognostics, rain, drought, and diseases. When they repair to the city, they disperse themselves in the market-places; if they meet any one carrying figs or bunches of grapes, they take what is offered gratuitously; if it is oil, it is poured over them, and they are anointed with it. Every wealthy house, even to the women's apartment, is open to them; when they enter it, they engage in conversation, and partake of the repast. Disease of the body they regard as most disgraceful, and he who apprehends it, after preparing a pyre, destroys himself by fire; he (previously) anoints himself, and sitting down upon it orders it to be lighted, remaining motionless while he is burning.

66. Nearchus gives the following account of the Sophists. The Brachmanes engage in public affairs, and attend the kings as counsellors; the rest are occupied in the study of nature.[Pg 115] Calanus belonged to the latter class. Women study philosophy with them, and all lead an austere life.

Of the customs of the other Indians, he says, that their laws, whether relating to the community or to individuals, are not committed to writing, and differ altogether from those of other people. For example, it is the practice among some tribes, to propose virgins as prizes to the conquerors in a trial of skill in boxing; wherefore they marry without portions; among other tribes the ground is cultivated by families and in common; when the produce is collected, each takes a load sufficient for his subsistence during the year; the remainder is burnt, in order to have a reason for renewing their labour, and not remaining inactive. Their weapons consist of a bow and arrows, which are three cubits in length, or a javelin, and a shield, and a sword three cubits long. Instead of bridles, they use muzzles,398 which differ little from a halter, and the lips are perforated with spikes.

67. Nearchus, producing proofs of their skill in works of art, says, that when they saw sponges in use among the Macedonians, they imitated them by sewing hairs, thin threads, and strings in wool; after the wool was felted, they drew out the hairs, threads, and strings, and dyed it with colours. There quickly appeared also manufactures of brushes for the body, and of vessels for oil (lecythi). They write, he says, letters upon cloth, smoothed by being

well beaten, although other authors affirm that they have no knowledge of writing. They use brass, which is cast, and not wrought; he does not give the reason of this, although he mentions the strange effect, namely, if that vessels of this description fall to the ground, they break like those made of clay.

This following custom also is mentioned in accounts of India, that, instead of prostrating themselves before their kings, it is usual to address them, and all persons in authority and high station, with a prayer.

The country produces precious stones, as crystal, carbuncles of all kinds, and pearls.

68. As an instance of the disagreement among historians, [Pg 116] [CAS. 717]we may adduce their (different) accounts of Calanus. They all agree that he accompanied Alexander, and underwent a voluntary death by fire in his presence, but they differ as to the manner and cause of his death. Some give the following account. Calanus accompanied the king, as the rehearser of his praises, beyond the boundaries of India, contrary to the common Indian custom; for the philosophers attend upon their kings, and act as instructors in the worship of the gods, in the same manner as the Magi attend the Persian kings. When he fell sick at Pasargadæ, being then attacked with disease for the first time in his life, he put himself to death at the age of seventy-three years, regardless of the entreaties of the king. A pyre was raised, and a golden couch placed upon it. He laid down upon it, and covering himself up, was burnt to death.

Others say, that a chamber was constructed of wood, which was filled with the leaves of trees, and a pyre being raised upon the roof, he was shut up in it, according to his directions, after the procession, with which he had been accompanied, had arrived at the spot. He threw himself upon the pyre, and was consumed like a log of wood, together with the chamber.

Megasthenes says, that self-destruction is not a dogma of the philosophers, and that those who commit this act are accounted fool-hardy; that some, who are by nature harsh, inflict wounds upon their bodies, or cast themselves down precipices; those who are impatient of pain drown themselves; those who can endure pain strangle themselves; and those of ardent tempers throw themselves into the fire. Of this last description was Calanus, who had no control over himself, and was a slave to the table of Alexander. Calanus is censured, while Mandanis is applauded. When Alexander's messengers invited the latter to come to the son of Jove, promising a reward if he would comply, and threatening punishment if he refused, he answered, "Alexander was not the son of Jove, for he did not govern even the smallest portion of the earth; nor did he himself desire a gift of one who[399] was satisfied with nothing. Neither did he fear his threats, for as long as he lived India would supply him with food enough; and when he died, he should be delivered from the flesh [Pg 117]wasted by old age, and be translated to a better and purer state of existence." Alexander commended and pardoned him.

69. Historians also relate that the Indians worship Jupiter Ombrius (or, the Rainy), the river Ganges, and the indigenous deities of the country; that when the king washes his hair,[400] a great feast is celebrated, and large presents are sent, each person displaying his wealth in competition with his neighbour.

They say, that some of the gold-digging myrmeces (ants) have wings; and that the rivers, like those of Iberia,[401] bring down gold-dust.

In processions at their festivals, many elephants are in the train, adorned with gold and silver, numerous carriages drawn by four horses and by several pairs of oxen; then follows a body of attendants in full dress, (bearing) vessels of gold, large basins and goblets, an orguia[402] in breadth, tables, chairs of state, drinking-cups, and lavers of Indian copper, most of which were set with precious stones, as emeralds, beryls, and Indian carbuncles; garments embroidered and interwoven with gold; wild beasts, as buffaloes,[403] panthers, tame lions, and a multitude of birds of variegated plumage and of fine song.

Cleitarchus speaks of four-wheeled carriages bearing trees with large leaves, from which were suspended (in cages) different kinds of tame birds, among which the orion[404] was said to possess the sweetest note, but the catreus[405] was the most beautiful in appearance, and had the most variegated plumage. In shape it approached nearest to the peacock, but the rest of the description must be taken from Cleitarchus.

70. Opposed to the Brachmanes there are philosophers, called Pramnæ, contentious people, and fond of argument. They ridicule the Brachmanes as boasters and fools for occupying themselves with physiology and astronomy. Some of the Pramnæ are called Pramnæ of the mountains, others Gymnetæ, others again are called Townsmen and Countrymen. [Pg 118]

[CAS. 719]The Pramnæ of the mountains wear deer-skins, and carry scrips filled with roots and drugs; they profess to practise medicine by means of incantations, charms, and amulets.

The Gymnetæ, as their name imports, are naked and live chiefly in the open air, practising fortitude for the space of thirty-seven years; this I have before mentioned; women live in their society, but without cohabitation. The Gymnetæ are held in singular estimation.

71. The (Pramnæ) Townsmen are occupied in civil affairs, dwell in cities, and wear fine linen, or (as Countrymen they live) in the fields, clothed in the skins of fawns or antelopes. In short, the Indians wear white garments, white linen and muslin, contrary to the accounts of those who say that they wear garments of a bright colour; all of them wear long hair and long beards, plait their hair, and bind it with a fillet.

72. Artemidorus says that the Ganges descends from the Emoda mountains and proceeds towards the south; when it arrives at the city Ganges,⁴⁰⁶ it turns to the east, and keeps this direction as far as Palibothra,⁴⁰⁷ and the mouth by which it discharges itself into the sea. He calls one of the rivers which flow into it Œdanes,⁴⁰⁸ which breeds crocodiles and dolphins. Some other circumstances besides are mentioned by him, but in so confused and negligent a manner that they are not to be regarded. To these accounts may be added that of Nicolaus Damascenus.

73. This writer states that at Antioch, near Daphne,⁴⁰⁹ he met with ambassadors from the Indians, who were sent to Augustus Cæsar. It appeared from the letter that several persons were mentioned in it, but three only survived, whom he says he saw. The rest had died chiefly in consequence of [Pg 119]the length of the journey. The letter was written in Greek upon a skin; the import of it was, that Porus was the writer, that although he was sovereign of six hundred kings, yet that he highly esteemed the friendship of Cæsar; that he was willing to allow him a passage through his country, in whatever part he pleased, and to assist him in any undertaking that was just.

Eight naked servants, with girdles round their waists, and fragrant with perfumes, presented the gifts which were brought. The presents were a Hermes (i. e. a man) born without arms, whom I have seen, large snakes, a serpent ten cubits in length, a river tortoise of three cubits in length, and a partridge (?) larger than a vulture. They were accompanied by the person, it is said, who burnt himself to death at Athens. This is the practice with persons in distress, who seek escape from existing calamities, and with others in prosperous circumstances, as was the case with this man. For as everything hitherto had succeeded with him, he thought it necessary to depart, lest some unexpected calamity should happen to him by continuing to live; with a smile, therefore, naked, anointed, and with the girdle round his waist, he leaped upon the pyre. On his tomb was this inscription,— ZARMANOCHEGAS,⁴¹⁰ AN INDIAN, A NATIVE OF BARGOSA,⁴¹¹ HAVING IMMORTALIZED HIMSELF ACCORDING TO THE CUSTOM OF HIS COUNTRY, HERE LIES.

CHAPTER II.
ARIANA.

1. NEXT to India is Ariana, the first portion of the country subject to the Persians, lying beyond⁴¹² the Indus, and the first [Pg 120] [CAS. 720]of the higher satrapies without the Taurus.⁴¹³ On the north it is bounded by the same mountains as India, on the south by the same sea, and by the same river Indus, which separates it from India. It stretches thence towards the west as far as the line drawn from the Caspian Gates⁴¹⁴ to Carmania,⁴¹⁵ whence its figure is quadrilateral.

The southern side begins from the mouths of the Indus, and from Patalene, and terminates at Carmania and the mouth of the Persian Gulf, by a promontory projecting a considerable distance to the south. It then makes a bend towards the gulf in the direction of Persia.

The Arbies, who have the same name as the river Arbis,[416] are the first inhabitants we meet with in this country. They are separated by the Arbis from the next tribe, the Oritæ, and according to Nearchus, occupy a tract of sea-coast of about 1000 stadia in length; this country also is a part of India. Next are the Oritæ, a people governed by their own laws. The voyage along the coast belonging to this people extends 1800 stadia, that along the country of the Ichthyophagi, who follow next, extends 7400 stadia; that along the country of the Carmani as far as Persia, 3700 stadia. The whole number of stadia is 13,900.

2. The greater part of the country inhabited by the Ichthyophagi is on a level with the sea. No trees, except palms and a kind of thorn, and the tamarisk, grow there. There is also a scarcity of water, and of food produced by cultivation. Both they and their cattle subsist upon fish, and are supplied by rain water and wells. The[Pg 121]flesh of the animals has the smell of fish. Their dwellings are built with the bones of large whales and shells, the ribs furnishing beams and supports, and the jaw-bones, door-ways. The vertebral bones serve as mortars in which fish, which have been previously dried in the sun, are pounded. Of this, with the addition of flour, cakes are made; for they have grinding mills (for corn), although they have no iron. This however is not so surprising, because it is possible for them to import it from other parts. But how do they hollow out the mills again, when worn away? with the same stones, they say, with which their arrows and javelins, which are hardened in the fire, are sharpened. Some fish are dressed in ovens, but the greater part is eaten raw. The fish are taken in nets made of the bark of the palm.

3. Above the Ichthyophagi is situated Gedrosia,[417] a country less exposed to the heat of the sun than India, but more so than the rest of Asia. As it is without fruits and water, except in summer, it is not much better than the country of the Ichthyophagi. But it produces aromatics, particularly nard and myrrh, in such quantity, that the army of Alexander used them on the march for tent coverings and beds; they thus breathed an air full of odours, and at the same time more salubrious.

The summer was purposely chosen for leaving India, for at that season it rains in Gedrosia, and the rivers and wells are filled, but in winter they fail. The rain falls in the higher parts to the north, and near the mountains: when the rivers swell, the plains near the sea are watered, and the wells are also filled. Alexander sent persons before him into the desert country to dig wells and to prepare stations for himself and his fleet.

4. Having separated his forces into three divisions, he set out with one division through Gedrosia, keeping at the utmost from the sea not more than 500 stadia, in order to secure the coast for his fleet; but he frequently approached the sea-side, although the beach was impracticable and rugged. The second division he sent forward under the command of Craterus through the interior, with a view of reducing Ariana, and of proceeding to the same places to which he himself was directing his march. (The third division), the fleet he intrusted to Nearchus and Onesicritus, his master pilot, giving them orders to [Pg 122] [CAS. 721]take up convenient positions in following him, and to sail along the coast parallel to his line of march.

5. Nearchus says, that while Alexander was on his march, he himself commenced his voyage, in the autumn, about the achronical rising of the Pleiades,[418] the wind not being before favourable. The Barbarians however, taking courage at the departure of the king, became daring, and attempted to throw off their subjection, attacked them, and endeavoured to drive them out of the country. But Craterus set out from the Hydaspes, and proceeded through the country of the Arachoti and of the Drangæ into Carmania.

Alexander was greatly distressed throughout the whole march, as his road lay through a barren country. The supplies of provisions which he obtained came from a distance, and were scanty and unfrequent, so much so that the army suffered greatly from hunger, the beasts of burden dropped down, and the baggage was abandoned, both on the march and in the camp. The army was saved by eating dates and the marrow of the palm-tree.[419]

Alexander however (says Nearchus), although acquainted with the hardships of the enterprise, was ambitious of conducting this large army in safety, as a conqueror, through the same country where, according to the prevailing report, Semiramis escaped by flight from India with about twenty, and Cyrus with about seven men.

6. Besides the want of provisions, the scorching heat was distressing, as also the deep and burning sand. In some places there were sand-hills, so that in addition to the difficulty of lifting the legs, as out of a pit, there were ascents and descents. It was necessary also, on account of the watering places, to make long marches of two, four, and sometimes even of six hundred stadia, for the most part during the night. Frequently the encampment was at a distance of 30 stadia from the watering places, in order that the soldiers might [Pg 123]not be induced by thirst to drink to excess. For many of them plunged into the water in their armour, and continued drinking until they were drowned; when swollen after death they floated, and corrupted the shallow water of the cisterns. Others, exhausted by thirst, lay exposed to the sun, in the middle of the road. They then became tremulous, their hands and their feet shook, and they died like persons seized with cold and shivering. Some turned out of the road to indulge in sleep, overcome with drowsiness and fatigue; some were left behind, and perished, being ignorant of the road, destitute of everything, and overpowered by heat. Others escaped after great sufferings. A torrent of water, which fell in the night time, overwhelmed and destroyed many persons, and much baggage; a great part even of the royal equipage was swept away.

The guides, through ignorance, deviated so far into the interior, that the sea was no longer in sight. The king, perceiving the danger, immediately set out in search of the coast; when he had discovered it, and by sinking wells had found water fit for drinking, he sent for the army: afterwards he continued his march for seven days near the shore, with a good supply of water. He then again returned into the interior.

7. There was a plant resembling the laurel, which if eaten by the beasts of burden caused them to die of epilepsy, accompanied with foaming at the mouth. A thorn also, the fruit of which, like gourds, strewed the ground, and was full of a juice; if drops of it fell into the eyes of any kind of animal it became completely blind. Many persons were suffocated by eating unripe dates. Danger also was to be apprehended from serpents; for on the sand-hills there grew a plant, underneath which they crept and hid themselves. The persons wounded by them died.

The Oritæ, it was said, smeared the points of their arrows, which were of wood hardened in the fire, with deadly poisons. When Ptolemy was wounded and in danger of his life, a person appeared in a dream to Alexander, and showed him a root with leaves and branches, which he told him to bruise and place upon the wound. Alexander awoke from his dream, and remembering the vision, searched and found the root growing in abundance, of which   both   he   and   others   made   use;[Pg   124] [CAS. 723]when the Barbarians perceived that the antidote for the poison was discovered, they surrendered to the king. It is probable, however, that some one acquainted with the plant informed the king of its virtues, and that the fabulous part of the story was invented for the purpose of flattery.

Having arrived at the palace[420] of the Gedrosii on the sixtieth day after leaving the Ori,[421] and allowed his army a short period of rest, he set out for Carmania.

8. The position of the southern side of Ariana is thus situated, with reference to the sea-coast, the country of the Gedrosii and the Oritæ lying near and above it. A great part of Gedrosia extends into the interior until it touches upon the Drangæ, Arachoti, and Paropamisadæ, of whom Eratosthenes speaks in the following manner: we cannot give a better description. "Ariana," he says, "is bounded on the east by the Indus, on the south by the Great Sea, on the north by the Paropamisus and the succeeding chain of mountains as far as the Caspian Gates, on the west by the same limits[422] by which the territory of the Parthians is separated from Media, and Carmania from Parætacene and Persia.

The breadth of the country is the length of the Indus, reckoned from the Paropamisus as far as the mouths of that river, and amounts to 12,000, or according to others to 13,000, stadia. The length, beginning from the Caspian Gates, as it is laid down in Asiatic Stathmi,[423] is estimated in two different ways. From the Caspian Gates to Alexandreia among the Arii[424] through Parthia is one and the same road. Then a road leads in a straight line through Bactriana, and over the pass of the mountain to Ortospana,[425] to the meeting of the three roads from Bactra, which is among the Paropamisadæ. The other branch turns off a little from Aria towards the south to Prophthasia in Drangiana; then the remainder leads as far as the confines of India and of the Indus; so that the road through the Drangæ and the Arachoti is longer, the whole amounting to 15,300 stadia. But if we deduct 1300 stadia, we shall have the remainder as the length of the country in a straight line, namely, 14,000 stadia; for the length of the coast is not much less, although some persons increase this sum by [Pg 125]adding to the 10,000 stadia Carmania, which is reckoned at 6000 stadia. For they seem to reckon it either together with the gulfs, or together with the Carmanian coast within the Persian Gulf. The name also of Ariana is extended so as to include some part of Persia, Media, and the north of Bactria and Sogdiana; for these nations speak nearly the same language.[426]

9. The order in which these nations are disposed is as follows. Along the Indus are the Paropamisadæ, above whom lies the mountain Paropamisus; then towards the south are the Arachoti; then next to these towards the south, the Gedroseni, together with other tribes who occupy the sea-coast; the Indus runs parallel along the breadth of these tracts. The Indians occupy [in part][427] some of the countries situated along the Indus, which formerly belonged to the Persians: Alexander deprived the Ariani of them, and established there settlements of his own. But Seleucus Nicator gave them to Sandrocottus in consequence of a marriage contract, and received in return five hundred elephants.

The Arii are situated on the west, by the side of the Paropamisadæ and the Drangæ[428] by the Arachoti and Gedrosii. The Arii are situated by the side of the Drangæ both on the north and west, and nearly encompass them. Bactriana adjoins Aria on the north, and the Paropamisadæ, through whose territory Alexander passed when he crossed the Caucasus on his way to Bactra. Towards the west, next to the Arii, are the Parthians, and the parts about the Caspian Gates. Towards the south of Parthia is the desert of Carmania; then follows the remainder of Carmania and Gedrosia.

10. We shall better understand the position of the places about the above-mentioned mountainous tract, if we further examine the route which Alexander took from the Parthian territory to Bactriana, when he was in pursuit of Bessus. He came first to Ariana, next to the Drangæ, where he put to death Philotas, the son of Parmenio, having detected his traitorous intentions. He despatched persons to Ecbatana[429] also [Pg 126] [CAS. 725]to put the father to death as an accomplice in the conspiracy. It is said that these persons performed in eleven days, upon dromedaries, a journey of 30 or 40 days, and executed their business.

The Drangæ resemble the Persians in all other respects in their mode of life, except that they have little wine. Tin is found in the country.[430]

Alexander next went from the Drangæ to the Euergetæ,[431] (to whom Cyrus gave this name,) and to the Arachoti; then through the territory of the Paropamisadæ at the setting of the Pleiad.[432] It is a mountainous country, and at that time was covered with snow, so that the march was performed with difficulty. The numerous villages, however, on their march, which were well provided with everything except oil, afforded relief in their distress. On their left hand were the summits of the mountains.

The southern parts of the Paropamisus belong to India and Ariana; the northern parts towards the west belong to Bactriana [towards the east to Sogdiana * *[433]Bactrian barbarians]. Having wintered there, with India above to the right hand, and having founded a city, he crossed the summits of the mountains into Bactriana. The road was bare of everything except a few trees of the bushy terminthus;[434] the army was driven from want of food to eat the flesh of the beasts of burthen, and that in a raw state for want of firewood; but silphium grew in great abundance, which promoted the digestion of this raw food. Fifteen days after founding the city and leaving winter quarters, he came to Adrapsa[435] (Darapsa?), a city of Bactriana.

11. Chaarene is situated somewhere about this part of the country bordering upon India. This, of all the places subject to the Parthians, lies nearest to India. It is distant 10,000 or [Pg 127]9000 stadia[436] from Bactriana,[437] through the country of the Arachoti, and the above-mentioned mountainous tract. Craterus traversed this country, subjugating those who refused to submit, and hastened with the greatest expedition to form a junction with the king. Nearly about the same time both armies, consisting of infantry, entered Carmania together, and at a short interval afterwards Nearchus sailed with his fleet into the Persian

Gulf, having undergone great danger and distress from wandering in his course, and among other causes, from great whales.

12. It is probable that those who sailed in the expedition greatly exaggerated many circumstances; yet their statements prove the sufferings to which they were exposed, and that their apprehensions were greater than the real danger. That which alarmed them the most was the magnitude of the whales, which occasioned great commotion in the sea from their numbers; their blowing was attended with so great a darkness, that the sailors could not see where they stood. But when the pilots informed the sailors, who were terrified at the sight and ignorant of the cause, that they were animals which might easily be driven away by the sound of a trumpet, and by loud noises, Nearchus impelled the vessels with violence in the direction of the impediment, and at the same time frightened the animals with the sound of trumpets. The whales dived, and again rose at the prow of the vessels, so as to give the appearance of a naval combat; but they soon made off.

13. Those who now sail to India speak of the size of these animals and their mode of appearance, but as coming neither in bodies nor frequently, yet as repulsed by shouts and by the sound of trumpets. They affirm that they do not approach the land, but that the bones of those which die, bared of flesh, are readily thrown up by the waves, and supply the Ichthyophagi with the above-mentioned material for the construction of their cabins. According to Nearchus, the size of these animals is three and twenty orguiæ in length.[438]

[Pg 128]

[CAS. 726]Nearchus says that he proved the confident belief of the sailors in the existence of an island situated in the passage, and destructive to those who anchored near it, to be false.

A bark in its course, when it came opposite to this island, was never afterwards seen, and some men who were sent in search did not venture to disembark upon the island, but shouted and called to the crew, when, receiving no answer, they returned. But as all imputed this disappearance to the island, Nearchus said that he himself sailed to it, went ashore, disembarked with a part of his crew, and went round it. But not discovering any trace of those of whom he was in search, he abandoned the attempt, and informed his men that no fault was to be imputed to the island (for otherwise destruction would have come upon himself and those who disembarked with him), but that some other cause (and innumerable others were possible) might have occasioned the loss of the vessel.

14. Carmania is the last portion of the sea-coast which begins from the Indus. Its first promontory projects towards the south into the Great Sea.[439] After it has formed the mouth of the Persian Gulf towards the promontory, which is in sight, of Arabia Felix, it bends towards the Persian Gulf, and is continued till it touches Persia.

Carmania is large, situated in the interior, and extending itself between Gedrosia and Persia, but stretches more to the north than Gedrosia. This is indicated by its fertility, for it not only produces everything, but the trees are of a large size, excepting however the olive; it is also watered by rivers. Gedrosia also differs little from the country of the Ichthyophagi, so that frequently there is no produce from the ground. They therefore keep the annual produce in store for several years.

Onesicritus says, that a river in Carmania brings down gold-dust; that there are mines of silver, copper, and minium; and that there are two mountains, one of which contains arsenic, the other salt.

There belongs to it a desert tract, which is contiguous to Parthia and Parætacene. The produce of the ground is like that of Persia; and among other productions the vine. The[Pg 129] Carmanian vine, as we call it, often bears bunches of grapes of two cubits in size; the seeds are very numerous and very large; probably the plant grows in its native soil with great luxuriance.

Asses, on account of the scarcity of horses, are generally made use of even in war. They sacrifice an ass to Mars, who is the only deity worshipped by them, for they are a warlike people. No one marries before he has cut off the head of an enemy and presented it to the king, who deposits the scull in the royal treasury. The tongue is minced and mixed with flour, which the king, after tasting it, gives to the person who brought it, to be eaten by himself and his family. That king is the most highly respected, to whom the greatest number of heads are presented.

According to Nearchus, most of the customs and the language of the inhabitants of Carmania resemble those of the Persians and Medes.

The passage across the mouth of the Persian Gulf does not occupy more than one day.

CHAPTER III.

1. NEXT to Carmania is Persis. A great part of it extends along the coast of the Gulf, which has its name from the country, but a much larger portion stretches into the interior, and particularly in its length, reckoned from the south, and Carmania to the north, and to the nations of Media.

It is of a threefold character, as we regard its natural condition and the quality of the air. First, the coast, extending for about 4400 or 4300 stadia, is burnt up with heat; it is sandy, producing little except palm trees, and terminates at the greatest river in those parts, the name of which is Oroatis.[440] Secondly, the country above the coast produces everything, and is a plain; it is excellently adapted for the rearing of cattle, and abounds with rivers and lakes.

The third portion lies towards the north, and is bleak and mountainous. On its borders live the camel-breeders.

[Pg 130]
[CAS. 727]

Its length, according to Eratosthenes, towards the north and Media,[441] is about 8000, or, including some projecting promontories, 9000 stadia; the remainder (from Media) to the Caspian Gates is not more than 3000 stadia. The breadth in the interior of the country from Susa to Persepolis is 4200 stadia, and thence to the borders of Carmania 1600 stadia more.

The tribes inhabiting this country are those called the Pateischoreis, the Achæmenidæ, and Magi; these last affect a sedate mode of life; the Curtii and Mardi are robbers, the rest are husbandmen.

2. Susis also is almost a part of Persis. It lies between Persis and Babylonia, and has a very considerable city, Susa. For the Persians and Cyrus, after the conquest of the Medes, perceiving that their own country was situated towards the extremities, but Susis more towards the interior, nearer also to Babylon and the other nations, there placed the royal seat of the empire. They were pleased with its situation on the confines of Persis, and with the importance of the city; besides the consideration that it had never of itself undertaken any great enterprise, had always been in subjection to other people, and constituted a part of a greater body, except, perhaps, anciently in the heroic times.

It is said to have been founded by Tithonus, the father of Memnon. Its compass was 120 stadia. Its shape was oblong. The Acropolis was called Memnonium. The Susians have the name also of Cissii. Æschylus[442] calls the mother of Memnon, Cissia. Memnon is said to be buried near Paltus in Syria, by the river Badas, as Simonides says in his Memnon, a dithyrambic poem among the Deliaca. The wall of the city, the temples and palaces, were constructed in the same manner as those of the Babylonians, of baked brick and asphaltus, as some writers relate. Polycletus however says, that its circumference was 200 stadia, and that it was without walls.

3. They embellished the palace at Susa more than the rest, but they did not hold in less veneration and honour the [Pg 131]palaces at Persepolis and Pasargadæ.[443] For in these stronger and hereditary places were the treasure-house, the riches, and tombs of the Persians. There was another palace at Gabæ, in the upper parts of Persia, and another on the sea-coast, near a place called Taoce.[444]

This was the state of things during the empire of the Persians. But afterwards different princes occupied different palaces; some, as was natural, less sumptuous, after the power of Persis had been reduced first by the Macedonians, and secondly still more by the Parthians. For although the Persians have still a kingly government, and a king of their own, yet their power is very much diminished, and they are subject to the king of Parthia.

4. Susa is situated in the interior, upon the river Choaspes, beyond the bridge; but the territory extends to the sea: and the sea-coast of this territory, from the borders of the Persian coast nearly as far as the mouths of the Tigris, is a distance of about 3000 stadia.

The Choaspes flows through Susis, terminating on the same coast, and has its source in the territory of the Uxii.[445] For a rugged and precipitous range of mountains lies between the Susians and Persis, with narrow defiles, difficult to pass; they were inhabited by robbers, who constantly exacted payment even from the kings themselves, at their entrance into Persis from Susis.

Polycletus says, that the Choaspes, and the Eulæus,[446] and the Tigris also enter a lake, and thence discharge themselves into the sea; that on the side of the lake is a mart, as the rivers do not receive the merchandise from the sea, nor convey it down to the sea, on account of dams in the river, purposely constructed, and that the goods are transported by land a distance of 800 stadia[447] to Susa; according to others, the rivers which flow through Susis discharge themselves by the intermediate canals of the Euphrates into the single stream of the Tigris, which on this account has at its mouth the name of Pasitigris.

[Pg 132]

[CAS. 729]5. According to Nearchus, the sea-coast of Susis is swampy, and terminates at the river Euphrates; at its mouth is a village, which receives the merchandise from Arabia; for the coast of Arabia approaches close to the mouths of the Euphrates and the Pasitigris; the whole intermediate space is occupied by a lake which receives the Tigris; on sailing up the Pasitigris 150 stadia is the bridge of rafts leading to Susa from Persis, and is distant from Susa 60 (600?) stadia; the Pasitigris is distant from the Oroatis about 2000 stadia; the ascent through the lake to the mouth of the Tigris is 600 (6000?) stadia;[448] near the mouth stands the Susian village (Aginis), distant from Susa 500 stadia; the journey by water from the mouth of the Euphrates, up to Babylon, through a well-inhabited tract of country, is a distance of more than 3000 stadia.

Onesicritus says that all the rivers discharge themselves into the lake, both the Euphrates and the Tigris; and that the Euphrates, again issuing from the lake, discharges itself into the sea by a separate mouth.

6. There are many other narrow defiles in passing out through the territory of the Uxii, and entering Persis. These Alexander forced in his march through the country at the Persian Gates, and at other places, when he was hastening to see the principal parts of Persis, and the treasure-holds, in which wealth had been accumulated during the long period that Asia was tributary to Persis.

He crossed many rivers, which flow through the country and discharge themselves into the Persian Gulf.

Next to the Choaspes are the Copratas[449] and the Pasitigris, which has its source in the country of the Uxii. There is also the river Cyrus, which flows through Cœle Persis,[450] as it is called, near Pasargadæ. The king changed his name, which was formerly Agradatus, to that of this river. Alexander crossed the Araxes[451] close to Persepolis. Persepolis was distinguished for the magnificence of the treasures which it contained. The Araxes flows out of the Parætacene,[452] and receives the Medus,[453] which has its source in Media. These rivers run through a very fruitful valley, which, like Persepolis,[Pg 133] lies close to Carmania and to the eastern parts of the country. Alexander burnt the palace at Persepolis, to avenge the Greeks, whose temples and cities the Persians had destroyed by fire and sword.

7. He next came to Pasargadæ,[454] which also was an ancient royal residence. Here he saw in a park the tomb of Cyrus. It was a small tower, concealed within a thick plantation of trees solid below, but above consisting of one story and a shrine which had a very narrow opening; Aristobulus says, he entered through this opening, by order of Alexander, and decorated the tomb. He saw there a golden couch, a table with cups, a golden coffin, and a large quantity of garments and dresses ornamented with precious stones. These objects he saw at his first visit, but on a subsequent visit the place had been robbed, and everything had been removed except the couch and the coffin which were only broken. The dead body had been removed from its place; whence it was evident that it was the act not of the Satrap,[455] but of robbers, who had left behind what they could not easily carry off. And this occurred although there was a guard of Magi stationed about the place, who received for

72

their daily subsistence a sheep, and every month a horse.[456] The remote distance to which the army of Alexander had advanced, to Bactra and India, gave occasion to the introduction of many disorderly acts, and to this among others.

Such is the account of Aristobulus, who records the following inscription on the tomb. "O MAN, I AM CYRUS,[457] I ESTABLISHED THE PERSIAN EMPIRE AND AS KING OF ASIA, GRUDGE ME NOT THEREFORE THIS MONUMENT."

Onesicritus however says that the tower had ten stories, that Cyrus lay in the uppermost, and that there was an inscription in Greek, cut in Persian letters, "I CYRUS, KING OF KINGS, LIE HERE." And another inscription to the same effect in the Persian language.

8. Onesicritus mentions also this inscription on the tomb of Darius: "I WAS A FRIEND TO MY FRIENDS, I WAS THE FIRST OF HORSEMEN AND ARCHERS, I EXCELLED AS HUNTER, I COULD DO EVERYTHING."

[Pg 134]

[CAS. 730]Aristus of Salamis, a writer of a much later age than these, says, that the tower consisted of two stories, and was large; that it was built at the time the Persians succeeded to the kingdom (of the Medes); that the tomb was preserved; that the above-mentioned inscription was in the Greek, and that there was another to the same purport in the Persian language.

Cyrus held in honour Pasargadæ, because he there conquered, in his last battle, Astyages the Mede, and transferred to himself the empire of Asia; he raised it to the rank of a city, and built a palace in memory of his victory.

9. Alexander transferred everything that was precious in Persis to Susa, which was itself full of treasures and costly materials; he did not, however, consider this place, but Babylon, as the royal residence, and intended to embellish it. There too his treasure was deposited.

They say that, besides the treasures in Babylon and in the camp of Alexander, which were not included in the sum, the treasure found at Susa and in Persis was reckoned to amount to 40,000, and according to some writers to 50,000, talents. But others say, that the whole treasure, collected from all quarters, and transported to Ecbatana, amounted to 180,000 talents, and that the 8,000 talents which Darius carried away with him in his flight from Media became the booty of those who put him to death.

10. Alexander preferred Babylon, because he saw that it far surpassed the other cities in magnitude, and had other advantages. Although Susis is fertile, it has a glowing and scorching atmosphere, particularly near the city, as he (Aristobulus?) says. Lizards and serpents at mid-day in the summer, when the sun is at its greatest height, cannot cross the streets of the city quick enough to prevent their being burnt to death midway by the heat. This happens nowhere in Persis, although it lies more towards the south.

Cold water for baths is suddenly heated by exposure to the sun. Barley spread out in the sun is roasted[458] like barley prepared in ovens. For this reason earth is laid to the depth of two cubits upon the roofs of the houses. They are obliged to construct their houses narrow, on account of the weight placed upon them, and from want of long beams, but, as large dwellings[Pg 135] are required to obviate the suffocating heat, the houses are long.

The beam made of the palm tree has a peculiar property, for although it retains its solidity, it does not as it grows old give way downwards, but curves upwards with the weight, and is a better support to the roof.

The cause of the scorching heat is said to be high, overhanging mountains on the north, which intercept the northern winds. These, blowing from the tops of the mountains at a great height, fly over without touching the plains, to the more southern parts of Susis. There the air is still, particularly when the Etesian winds cool the other parts of the country which are burnt up by heat.

11. Susis is so fertile in grain, that barley and wheat produce, generally, one hundred, and sometimes two hundred fold. Hence the furrows are not ploughed close together, for the roots when crowded impede the sprouting of the plant.

The vine did not grow there before the Macedonians planted it, both there and at Babylon. They do not dig trenches, but thrust down into the ground iron-headed stakes, which when drawn out are immediately replaced by the plants.

Such is the character of the inland parts. The sea-coast is marshy and without harbours; hence Nearchus says, that he met with no native guides, when coasting with his fleet from India to Babylonia, for nowhere could his vessels put in, nor was he able to procure persons who could direct him by their knowledge and experience.

12. The part of Babylonia formerly called Sitacene, and afterwards Apolloniatis,[459]is situated near Susis.

Above both, on the north and towards the east, are the Elymæi[460] and the Parætaceni, predatory people relying for security on their situation in a rugged and mountainous country. The Parætaceni lie more immediately above the Apolloniatæ, and therefore annoy them the more. The Elymæi are at war with this people and with the Susians, and the Uxii with the Elymæi, but not so constantly at present as might be expected, on account of the power of the Parthians, to whom all the inhabitants[Pg 136] [CAS. 732] of those regions are under subjection. When therefore the Parthians are quiet, all are tranquil, and their subject nations. But when, as frequently happens, there is an insurrection, which has occurred even in our own times, the event is not the same to all, but different to different people. For the disturbance has benefited some, but disappointed the expectation of others.

Such is the nature of the countries of Persis and Susiana.

13. The manners and customs of the Persians are the same as those of the Susians and the Medes, and many other people; and they have been described by several writers, yet I must mention what is suitable to my purpose.

The Persians do not erect statues nor altars, but, considering the heaven as Jupiter, sacrifice on a high place.[461] They worship the sun also, whom they call Mithras, the moon, Venus, fire, earth, winds, and water. They sacrifice, having offered up prayers, in a place free from impurities, and present the victim crowned.[462]

After the Magus, who directs the sacrifice, has divided the flesh, each goes away with his share, without setting apart any portion to the gods; for the god, they say, requires the soul of the victim, and nothing more. Nevertheless, according to some writers, they lay a small piece of the caul upon the fire.

14. But it is to fire and water especially that they offer sacrifice. They throw upon the fire dry wood without the bark, and place fat over it; they then pour oil upon it, and light it below; they do not blow the flame with their breath, but fan it; those who have blown the flame with their breath, or thrown any dead thing or dirt upon the fire, are put to death.

They sacrifice to water by going to a lake, river, or fountain; having dug a pit, they slaughter the victim over it, taking care that none of the pure water near be sprinkled with blood, and thus be polluted. They then lay the flesh in order upon myrtle or laurel branches; the Magi touch it with slender twigs,[463] and make incantations, pouring oil mixed with [Pg 137]milk and honey, not into the fire, nor into the water, but upon the earth. They continue their incantations for a long time, holding in the hands a bundle of slender myrtle rods.

15. In Cappadocia (for in this country there is a great body of Magi, called Pyræthi,[464] and there are many temples dedicated to the Persian deities) the sacrifice is not performed with a knife, but the victim is beaten to death with a log of wood, as with a mallet.

The Persians have also certain large shrines, called Pyrætheia.[465] In the middle of these is an altar, on which is a great quantity of ashes, where the Magi maintain an unextinguished fire. They enter daily, and continue their incantation for nearly an hour, holding before the fire a bundle of rods, and wear round their heads high turbans of felt, reaching down on each side so as to cover the lips and the sides of the cheeks. The same customs are observed in the temples of Anaïtis and of Omanus. Belonging to these temples are shrines, and a wooden statue of Omanus is carried in procession. These we have seen ourselves.[466] Other usages, and such as follow, are related by historians.

16. The Persians never pollute a river with urine, nor wash nor bathe in it; they never throw a dead body, nor anything unclean, into it. To whatever god they intend to sacrifice, they first address a prayer to fire.

17. They are governed by hereditary kings. Disobedience is punished by the head and arms being cut off, and the body cast forth. They marry many women, and maintain at the same time a great number of concubines, with a view to a numerous offspring.

The kings propose annual prizes for a numerous family of children. Children are not brought into the presence of their parents until they are four years old.

Marriages are celebrated at the beginning of the vernal equinox. The bridegroom passes into the bride-chamber, having previously eaten some fruit, or camel's marrow, but nothing else during the day.

18. From the age of five to twenty-four years they are taught to use the bow, to throw the javelin, to ride, and to speak the truth. They have the most virtuous preceptors, [Pg 138] [CAS. 733]who interweave useful fables in their discourses, and rehearse, sometimes with, sometimes without, music, the actions of the gods and of illustrious men.

The youths are called to rise before day-break, at the sound of brazen instruments, and assemble in one spot, as if for arming themselves or for the chase. They are arranged in companies of fifty, to each of which one of the king's or a satrap's son is appointed as leader, who runs, followed at command by the others, an appointed distance of thirty or forty stadia.

They require them to give an account of each lesson, when they practise loud speaking, and exercise the breath and lungs. They are taught to endure heat, cold, and rains; to cross torrents, and keep their armour and clothes dry; to pasture animals, to watch all night in the open air, and to eat wild fruits, as the terminthus,[467] acorns, and wild pears.

[These persons are called Cardaces, who live upon plunder, for "carda" means a manly and warlike spirit.][468]

The daily food after the exercise of the gymnasium is bread, a cake, cardamum,[469]a piece of salt, and dressed meat either roasted or boiled, and their drink is water.

Their mode of hunting is by throwing spears from horseback, or with the bow or the sling.

In the evening they are employed in planting trees, cutting roots, fabricating armour, and making lines and nets. The youth do not eat the game, but carry it home. The king gives rewards for running, and to the victors in the other contests of the pentathla (or five games). The youths are adorned with gold, esteeming it for its fiery appearance. They do not ornament the dead with gold, nor apply fire to them, on account of its being an object of veneration.

19. They serve as soldiers in subordinate stations, and in [Pg 139]those of command from twenty to fifty years of age, both on foot and on horseback. They do not concern themselves with the public markets, for they neither buy nor sell. They are armed with a romb-shaped shield. Besides quivers, they have battle-axes and short swords. On their heads they wear a cap rising like a tower. The breastplate is composed of scales of iron.

The dress of the chiefs consists of triple drawers, a double tunic with sleeves reaching to the knees; the under garment is white, the upper of a variegated colour. The cloak for summer is of a purple or violet colour, but for winter of a variegated colour. The turbans are similar to those of the Magi; and a deep double shoe. The generality of people wear a double tunic reaching to the half of the leg. A piece of fine linen is wrapped round the head. Each person has a bow and a sling.

The entertainments of the Persians are expensive. They set upon their table entire animals in great number, and of various kinds. Their couches, drinking-cups, and other articles are so brilliantly ornamented that they gleam with gold and silver.

20. Their consultations on the most important affairs are carried on while they are drinking, and they consider the resolutions made at that time more to be depended upon than those made when sober.

On meeting persons of their acquaintance, and of equal rank with themselves, on the road, they approach and kiss them, but to persons of an inferior station they offer the cheek, and in that manner receive the kiss. But to persons of still lower condition they only bend the body.

Their mode of burial is to smear the bodies over with wax, and then to inter them. The Magi are not buried, but the birds are allowed to devour them. These persons, according to the usage of the country, espouse even their mothers.

Such are the customs of the Persians.

21. The following, mentioned by Polycletus, are perhaps customary practices:

At Susa each king builds in the citadel, as memorials of the administration of his government, a dwelling for himself, treasure-houses, and magazines for tribute collected (in kind).

From the sea-coast they obtain silver, from the interior the[Pg 140] [CAS. 735] produce of each province, as dyes, drugs, hair, wool, or anything else of this sort, and cattle. The apportionment of the tribute was settled by Darius [Longimanus, who was a very handsome person with the exception of the length of his arms, which reached to his knees].[470] The greater part both of gold and silver is wrought up, and there is not much in coined money. The former they consider as best adapted for presents, and for depositing in store-houses. So much coined money as suffices for their wants they think enough; but, on the other hand, money is coined in proportion to what is required for expenditure.[471]

22. Their habits are in general temperate. But their kings, from the great wealth which they possessed, degenerated into a luxurious way of life. They sent for wheat from Assos in Æolia, for Chalybonian[472] wine from Syria, and water from the Eulæus, which is the lightest of all, for an Attic cotylus measure of it weighs less by a drachm (than the same quantity of any other water).

23. Of the barbarians the Persians were the best known to the Greeks, for none of the other barbarians who governed Asia governed Greece. The barbarians were not acquainted with the Greeks, and the Greeks were but slightly acquainted, and by distant report only, with the barbarians. As an instance, Homer was not acquainted with the empire of the Syrians nor of the Medes, for otherwise as he mentions the wealth of Egyptian Thebes and of Phœnicia, he would not have passed over in silence the wealth of Babylon, of Ninus, and of Ecbatana.

The Persians were the first people that brought Greeks under their dominion; the Lydians (before them) did the [Pg 141]same, they were not however masters of the whole, but of a small portion only of Asia, that within the river Halys; their empire lasted for a short time, during the reigns of Crœsus and Alyattes; and they were deprived of what little glory they had acquired, when conquered by the Persians.

The Persians, (on the contrary, increased in power and,) as soon as they had destroyed the Median empire, subdued the Lydians and brought the Greeks of Asia under their dominion. At a later period they even passed over into Greece and were worsted in many great battles, but still they continued to keep possession of Asia, as far as the places on the sea-coast, until they were completely subdued by the Macedonians.

24. The founder of their empire was Cyrus. He was succeeded by his son Cambyses, who was put to death by the Magi. The seven Persians who killed the Magi delivered the kingdom into the hands of Darius, the son of Hystaspes. The succession terminated with Arses, whom Bagous the eunuch having killed set up Darius, who was not of the royal family. Alexander overthrew Darius, and reigned himself twelve years.[473] The empire of Asia was partitioned out among his successors, and transmitted to their descendants, but was dissolved after it had lasted about two hundred and fifty years.[474]

At present the Persians are a separate people, governed by kings, who are subject to other kings; to the kings of Macedon in former times, but now to those of Parthia.

[Pg   142]
[CAS. 736]

BOOK XVI.
SUMMARY

The sixteenth Book contains Assyria, in which are the great cities Babylon and Nisibis; Adiabene, Mesopotamia, all Syria; Phœnicia, Palestine; the whole of Arabia; all that part of India which touches upon Arabia; the territory of the Saracens, called by our author Scenitis; and the whole country bordering the Dead and Red Seas.

CHAPTER I.

1. ASSYRIA is contiguous to Persia and Susiana. This name is given to Babylonia, and to a large tract of country around; this tract contains Aturia,[475] in which is Nineveh, the Apolloniatis, the Elymæi, the Parætacæ, and the Chalonitis about Mount Zagrum,[476]—the plains about Nineveh, namely, Dolomene, Calachene, Chazene, and Adiabene,—the nations of Mesopotamia, bordering upon the Gordyæi;[477] the Mygdones about Nisibis, extending to the Zeugma[478] of the Euphrates, and to the great range of country on the other side that river, occupied by Arabians, and by those people who are properly called Syrians in the present age. This last people extend as far as the Cilicians, Phœnicians, and Jews, to the sea opposite the Sea of Egypt, and to the Bay of Issus.

2. The name of Syrians seems to extend from Babylonia as far as the Bay of Issus, and, anciently, from this bay to the Euxine.

Both tribes of the Cappadocians, those near the Taurus and those near the Pontus, are called to this time Leuco-Syrians (or White Syrians),[479] as though there existed a nation[Pg 143] of Black Syrians. These are the people situated beyond the Taurus, and I extend the name of Taurus as far as the Amanus.[480]

When the historians of the Syrian empire say that the Medes were overthrown by the Persians, and the Syrians by the Medes, they mean no other Syrians than those who built the royal palaces at Babylon and Nineveh; and Ninus, who built Nineveh in Aturia, was one of these Syrians. His wife, who succeeded her husband, and founded Babylon, was Semiramis. These sovereigns were masters of Asia. Many other works of Semiramis, besides those at Babylon, are extant in almost every part of this continent, as, for example, artificial mounds, which are called mounds of Semiramis, and walls[481] and fortresses, with subterraneous passages; cisterns for water; roads[482] to facilitate the ascent of mountains; canals communicating with rivers and lakes; roads and bridges.

The empire they left continued with their successors to the time of [the contest between] Sardanapalus and Arbaces.[483] It was afterwards transferred to the Medes.

3. The city Nineveh was destroyed immediately upon the overthrow of the Syrians.[484] It was much larger than Babylon, and situated in the plain of Aturia. Aturia borders upon the places about Arbela; between these is the river Lycus.[485] Arbela and the parts about it[486] belong to Babylonia. In the country on the other side of the Lycus are the plains of Aturia, which surround Nineveh.[487]

[Pg   144]
[CAS. 737]In Aturia is situated Gaugamela, a village where Darius was defeated and lost his kingdom. This place is remarkable for its name, which, when interpreted, signifies the Camel's House. Darius, the son of Hystaspes, gave it this name, and assigned (the revenues of) the place for the maintenance of a camel, which had undergone the greatest possible labour and fatigue in the journey through the deserts of Scythia, when carrying baggage and provision for the king. The Macedonians, observing that this was a mean village, but Arbela a considerable settlement (founded, as it is said, by Arbelus, son of Athmoneus), reported that the battle was fought and the victory obtained near Arbela, which account was transmitted to historians.

4. After Arbela and the mountain Nicatorium[488] (a name which Alexander, after the victory at Arbela, superadded), is the river Caprus,[489] situated at the same distance from Arbela as the Lycus. The country is called Artacene.[490] Near Arbela is the city Demetrias; next is the spring of naphtha, the fires, the temple of the goddess Anæa,[491] Sadracæ, the palace of Darius, son of Hystaspes, the Cyparisson, or plantation of Cypresses, and the passage across the Caprus, which is close to Seleucia and Babylon.

5. Babylon itself also is situated in a plain. The wall is 385[492] stadia in circumference, and 32 feet in thickness. The height of the space between the towers is 50, and of the towers 60 cubits. The roadway upon the walls will allow chariots with four horses when they meet to pass each other with ease. Whence, among the seven wonders of the world, are reckoned

this wall and the hanging garden: the shape of the garden [Pg 145]is a square, and each side of it measures four plethra. It consists of vaulted terraces, raised one above another, and resting upon cube-shaped pillars. These are hollow and filled with earth to allow trees of the largest size to be planted. The pillars, the vaults, and the terraces are constructed of baked brick and asphalt.

The ascent to the highest story is by stairs, and at their side are water engines, by means of which persons, appointed expressly for the purpose, are continually employed in raising water from the Euphrates into the garden. For the river, which is a stadium in breadth, flows through the middle of the city, and the garden is on the side of the river. The tomb also of Belus is there. At present it is in ruins, having been demolished, as it is said, by Xerxes. It was a quadrangular pyramid of baked brick, a stadium in height, and each of the sides a stadium in length. Alexander intended to repair it. It was a great undertaking, and required a long time for its completion (for ten thousand men were occupied two months in clearing away the mound of earth), so that he was not able to execute what he had attempted, before disease hurried him rapidly to his end. None of the persons who succeeded him attended to this undertaking; other works also were neglected, and the city was dilapidated, partly by the Persians, partly by time, and, through the indifference of the Macedonians to things of this kind, particularly after Seleucus Nicator had fortified Seleucia on the Tigris near Babylon, at the distance of about 300 stadia.

Both this prince and all his successors directed their care to that city, and transferred to it the seat of empire. At present it is larger than Babylon; the other is in great part deserted, so that no one would hesitate to apply to it what one of the comic writers said of Megalopolitæ in Arcadia,

"The great city is a great desert."

On account of the scarcity of timber, the beams and pillars of the houses were made of palm wood. They wind ropes of twisted reed round the pillars, paint them over with colours, and draw designs upon them; they cover the doors with a coat of asphaltus. These are lofty, and all the houses are vaulted on account of the want of timber. For the country is bare, a great part of it is covered with shrubs, and produces [Pg 146] [CAS. 739]nothing but the palm. This tree grows in the greatest abundance in Babylonia. It is found in Susiana also in great quantity, on the Persian coast, and in Carmania.

They do not use tiles for their houses, because there are no great rains. The case is the same in Susiana and in Sitacene.

6. In Babylon a residence was set apart for the native philosophers called Chaldæans, who are chiefly devoted to the study of astronomy. Some, who are not approved of by the rest, profess to understand genethlialogy, or the casting of nativities. There is also a tribe of Chaldæans, who inhabit a district of Babylonia, in the neighbourhood of the Arabians, and of the sea called the Persian Sea.[493] There are several classes of the Chaldæan astronomers. Some have the name of Orcheni, some Borsippeni, and many others, as if divided into sects, who disseminate different tenets on the same subjects. The mathematicians make mention of some individuals among them, as Cidenas, Naburianus, and Sudinus. Seleucus also of Seleuceia is a Chaldæan, and many other remarkable men.

7. Borsippa is a city sacred to Diana and Apollo. Here is a large linen manufactory. Bats of much larger size than those in other parts abound in it. They are caught and salted for food.

8. The country of the Babylonians is surrounded on the east by the Susans, Elymæi, and Parætaceni; on the south by the Persian Gulf, and the Chaldæans as far as the Arabian Meseni; on the west by the Arabian Scenitæ as far as Adiabene and Gordyæa; on the north by the Armenians and Medes as far as the Zagrus, and the nations about that river.

9. The country is intersected by many rivers, the largest of which are the Euphrates and the Tigris: next to the Indian rivers, the rivers in the southern parts of Asia are said to hold the second place. The Tigris is navigable upwards from its mouth to Opis,[494] and to the present Seleuceia. Opis is a village and a mart for the surrounding places. The [Pg 147]Euphrates also is navigable up to Babylon, a distance of more than 3000 stadia. The Persians, through fear of incursions from without, and for the purpose of preventing vessels

from ascending these rivers, constructed artificial cataracts. Alexander, on arriving there, destroyed as many of them as he could, those particularly [on the Tigris from the sea] to Opis. But he bestowed great care upon the canals; for the Euphrates, at the commencement of summer, overflows. It begins fill in the spring, when the snow in Armenia melts: the ploughed land, therefore, would be covered with water and be submerged, unless the overflow of the superabundant water were diverted by trenches and canals, as in Egypt the water of the Nile is diverted. Hence the origin of canals. Great labour is requisite for their maintenance, for the soil is deep, soft, and yielding, so that it would easily be swept away by the stream; the fields would be laid bare, the canals filled, and the accumulation of mud would soon obstruct their mouths. Then, again, the excess of water discharging itself into the plains near the sea forms lakes, and marshes, and reed-grounds, supplying the reeds with which all kinds of platted vessels are woven; some of these vessels are capable of holding water, when covered over with asphaltus; others are used with the material in its natural state. Sails are also made of reeds; these resemble mats or hurdles.

10. It is not, perhaps, possible to prevent inundations of this kind altogether, but it is the duty of good princes to afford all possible assistance. The assistance required is to prevent excessive overflow by the construction of dams, and to obviate the filling of rivers, produced by the accumulation of mud, by cleansing the canals, and removing stoppages at their mouths. The cleansing of the canals is easily performed, but the construction of dams requires the labour of numerous workmen. For the earth being soft and yielding, does not support the superincumbent mass, which sinks, and is itself carried away, and thus a difficulty arises in making dams at the mouth. Expedition is necessary in closing the canals to prevent all the water flowing out. When the canals dry up in the summer time, they cause the river to dry up also; and if the river is low (before the canals are closed), it cannot supply the canals in time with water, of which the country, burnt up and scorched, requires a very large quantity;[Pg 148]
[CAS. 740] for there is no difference, whether the crops are flooded by an excess or perish by drought and a failure of water. The navigation up the rivers (a source of many advantages) is continually obstructed by both the above-mentioned causes, and it is not possible to remedy this unless the mouths of the canals were quickly opened and quickly closed, and the canals were made to contain and preserve a mean between excess and deficiency of water.

11. Aristobulus relates that Alexander himself, when he was sailing up the river, and directing the course of the boat, inspected the canals, and ordered them to be cleared by his multitude of followers; he likewise stopped up some of the mouths, and opened others. He observed that one of these canals, which took a direction more immediately to the marshes, and to the lakes in front of Arabia, had a mouth very difficult to be dealt with, and which could not be easily closed on account of the soft and yielding nature of the soil; he (therefore) opened a new mouth at the distance of 30 stadia, selecting a place with a rocky bottom, and to this the current was diverted. But in doing this he was taking precautions that Arabia should not become entirely inaccessible in consequence of the lakes and marshes, as it was already almost an island from the quantity of water (which surrounded it). For he contemplated making himself master of this country; and he had already provided a fleet and places of rendezvous; and had built vessels in Phœnicia and at Cyprus, some of which were in separate pieces, others were in parts, fastened together by bolts. These, after being conveyed to Thapsacus in seven distances of a day's march, were then to be transported down the river to Babylon. He constructed other boats in Babylonia, from cypress trees in the groves and parks, for there is a scarcity of timber in Babylonia. Among the Cossæi, and some other tribes, the supply of timber is not great.

The pretext for the war, says Aristobulus, was that the Arabians were the only people who did not send their ambassadors to Alexander; but the true reason was his ambition to be lord of all.

When he was informed that they worshipped two deities only, Jupiter and Bacchus, who supply what is most requisite for the subsistence of mankind, he supposed that, after his conquests, they would worship him as a third, if he permitted[Pg 149] them to enjoy their former national independence. Thus was Alexander employed in clearing the canals, and in examining minutely the sepulchres of the kings, most of which are situated among the lakes.

12. Eratosthenes, when he is speaking of the lakes near Arabia, says, that the water, when it cannot find an outlet, opens passages under-ground, and is conveyed through these as far as the Cœle-Syrians,[495] it is also compressed and forced into the parts near Rhinocolura[496] and Mount Casius,[497] and there forms lakes and deep pits.[498] But I know not whether this is probable. For the overflowings of the water of the Euphrates, which form the lakes and marshes near Arabia, are near the Persian Sea. But the isthmus which separates them is neither large nor rocky, so that it was more probable that the water forced its way in this direction into the sea, either under the ground, or across the surface, than that it traversed so dry and parched a soil for more than 6000 stadia; particularly, when we observe, situated midway in this course, Libanus, Antilibanus, and Mount Casius.[499]

[Pg 150]

[CAS. 742]Such, then, are the accounts of Eratosthenes and Aristobulus.

13. But Polycleitus says, that the Euphrates does not overflow its banks, because its course is through large plains; that of the mountains (from which it is supplied), some are distant 2000, and the Cossæan mountains scarcely 1000 stadia, that they are not very high, nor covered with snow to a great depth, and therefore do not occasion the snow to melt in great masses, for the most elevated mountains are in the northern parts above Ecbatana; towards the south they are divided, spread out, and are much lower; the Tigris also receives the greater part of the water [which comes down from them], and thus overflows its banks.[500]

The last assertion is evidently absurd, because the Tigris descends into the same plains (as the Euphrates); and the above-mentioned mountains are not of the same height, the northern being more elevated, the southern extending in breadth, but are of a lower altitude. The quantity of snow is not, however, to be estimated by altitude only, but by aspect. The same mountain has more snow on the northern than on the southern side, and the snow continues longer on the former than on the latter. As the Tigris therefore receives from the most southern parts of Armenia, which are near Babylon, the water of the melted snow, of which there is no great quantity, since it comes from the southern side, it should overflow in a less degree than the Euphrates, which receives the water from both parts (northern and southern); and not from a single mountain only, but from many, as I have mentioned in the description of Armenia. To this we must add the length of the river, the large tract of country which it traverses in the Greater and in the Lesser Armenia, the large space it takes in its course in passing out of the Lesser Armenia and Cappadocia, after issuing out of the Taurus in its way to Thapsacus (forming the boundary between Syria below and Mesopotamia), and the large remaining portion of country as far as Babylon and to its mouth, a course in all of 36,000 stadia.

This, then, on the subject of the canals (of Babylonia).

[Pg 151]

14. Babylonia produces barley in larger quantity than any other[501] country, for a produce of three hundred-fold is spoken of. The palm tree furnishes everything else, bread, wine, vinegar, and meal; all kinds of woven articles are also procured from it. Braziers use the stones of the fruit instead of charcoal. When softened by being soaked in water, they are food for fattening oxen and sheep.

It is said that there is a Persian song in which are reckoned up 360 useful properties of the palm.

They employ for the most part the oil of sesamum, a plant which is rare in other places.

15. Asphaltus is found in great abundance in Babylonia. Eratosthenes describes it as follows.

The liquid asphaltus, which is called naphtha, is found in Susiana; the dry kind, which can be made solid, in Babylonia. There is a spring of it near the Euphrates. When this river overflows at the time of the melting of the snow, the spring also of asphaltus is filled, and overflows into the river, where large clods are consolidated, fit for buildings constructed of baked bricks. Others say that the liquid kind also is found in Babylonia. With respect to the

80

solid kind, I have described its great utility in the construction of buildings. They say that boats (of reeds) are woven,[502]which, when besmeared with asphaltus, are firmly compacted. The liquid kind, called naphtha, is of a singular nature. When it is brought near the fire, the fire catches it; and if a body smeared over with it is brought near the fire, it burns with a flame, which it is impossible to extinguish, except with a large quantity of water; with a small quantity it burns more violently, but it may be smothered and extinguished by mud, vinegar, alum, and glue. It is said that Alexander, as an experiment, ordered naphtha to be poured over a boy in a bath, and a lamp to be brought near his body. The boy became enveloped in flames, and would have perished if the bystanders had not mastered the fire by pouring upon him a great quantity of water, and thus saved his life.

Poseidonius says that there are springs of naphtha in Babylonia, some of which produce white, others black, naphtha; the first of these, I mean the white naphtha, which attracts flame, [Pg 152] [CAS. 743]is liquid sulphur; the second, or black naphtha, is liquid asphaltus, and is burnt in lamps instead of oil.

16. In former times the capital of Assyria was Babylon; it is now called Seleuceia upon the Tigris. Near it is a large village called Ctesiphon. This the Parthian kings usually made their winter residence, with a view to spare the Seleucians the burden of furnishing quarters for the Scythian soldiery. In consequence of the power of Parthia, Ctesiphon[503] may be considered as a city rather than a village; from its size it is capable of lodging a great multitude of people; it has been adorned with public buildings by the Parthians, and has furnished merchandise, and given rise to arts profitable to its masters.

The kings usually passed the winter there, on account of the salubrity of the air, and the summer at Ecbatana and in Hyrcania,[504] induced by the ancient renown of these places.

As we call the country Babylonia, so we call the people Babylonians, not from the name of the city, but of the country; the case is not precisely the same, however, as regards even natives of Seleuceia, as, for instance, Diogenes, the stoic philosopher [who had the appellation of the Babylonian, and not the Seleucian].[505]

17. At the distance of 500 stadia from Seleuceia is Artemita, a considerable city, situated nearly directly to the east, which is the position also of Sitacene.[506] This extensive and fertile tract of country lies between Babylon and Susiana, so that the whole road in travelling from Babylon to Susa passes through Sitacene. The road from Susa[507] into the interior of Persis, through the territory of the Uxii,[508] and from Persis into the middle of Carmania,[509] leads also towards the east.

Persis, which is a large country, encompasses Carmania on the [west][510] and north. Close to it adjoin Paraetacene,[511] and [Pg 153]the Cossæan territory as far as the Caspian Gates, inhabited by mountainous and predatory tribes. Contiguous to Susiana is Elymaïs, a great part of which is rugged, and inhabited by robbers. To Elymaïs adjoin the country about the Zagrus[512] and Media.[513]

18. The Cossæi, like the neighbouring mountaineers, are for the most part archers, and are always out on foraging parties. For as they occupy a country of small extent, and barren, they are compelled by necessity to live at the expense of others. They are also necessarily powerful, for they are all fighting men. When the Elymæi were at war with the Babylonians and Susians, they supplied the Elymæi with thirteen thousand auxiliaries.

The Paraetaceni attend to the cultivation of the ground more than the Cossæi, but even these people do not abstain from robbery.

The Elymæi occupy a country larger in extent, and more varied, than that of the Paraetaceni. The fertile part of it is inhabited by husbandmen. The mountainous tract is a nursery for soldiers, the greatest part of whom are archers. As it is of considerable extent, it can furnish a great military force; their king, who possesses great power, refuses to be subject, like others, to the king of Parthia. The country was similarly independent in the time of the Persians, and afterwards[514] in the time of the Macedonians, who governed Syria. When Antiochus the Great attempted to plunder the temple of Belus, the neighbouring barbarians, unassisted, attacked and put him to death. In after-times the king of Parthia[515] heard that the temples in their country contained great wealth, but knowing that the people would not submit, and admonished by the fate of Antiochus, he invaded their country with a large army; he took the temple of Minerva, and that of Diana, called Azara, and carried away treasure to the amount of 10,000 [Pg 154] [CAS. 744]talents. Seleuceia also, a large city on the river Hedyphon,[516] was taken. It was formerly called Soloce.

There are three convenient entrances into this country; one from Media and the places about the Zagrus, through Massabatice; a second from Susis, through the district Gabiane. Both Gabiane and Massabatice are provinces of Elymæa. A third passage is that from Persis. Corbiane also is a province of Elymaïs.

Sagapeni and Silaceni, small principalities, border upon Elymaïs.

Such, then, is the number and the character of the nations situated above Babylonia towards the east.

We have said that Media and Armenia lie to the north, and Adiabene and Mesopotamia to the west of Babylonia.

19. The greatest part of Adiabene consists of plains, and, although it is a portion of Babylon, has its own prince. In some places it is contiguous to Armenia.[517] For the Medes, Armenians, and Babylonians, the three greatest nations in these parts, were from the first in the practice, on convenient opportunities, of waging continual war with each other, and then making peace, which state of things continued till the establishment of the Parthian empire.

The Parthians subdued the Medes and Babylonians, but never at any time conquered the Armenians. They made frequent inroads into their country, but the people were not subdued, and Tigranes, as I have mentioned in the description of Armenia,[518] opposed them with great vigour and success.

Such is the nature of Adiabene. The Adiabeni are also called Saccopodes.[519]

We shall describe Mesopotamia and the nations towards [Pg 155]the south, after premising a short account of the customs of the Assyrians.

20. Their other customs are like those of the Persians, but this is peculiar to themselves: three discreet persons, chiefs of each tribe, are appointed, who present publicly young women who are marriageable, and give notice by the crier, beginning with those most in estimation, of a sale of them to men intending to become husbands. In this manner marriages are contracted.

As often as the parties have sexual intercourse with one another, they rise, each apart from the other, to burn perfumes. In the morning they wash, before touching any household vessel. For as ablution is customary after touching a dead body, so it is practised after sexual intercourse.[520] There is a custom prescribed by an oracle for all the Babylonian women to have intercourse with strangers. The women repair to a temple of Venus, accompanied by numerous attendants and a crowd of people. Each woman has a cord round her head. The man approaches a woman, and places on her lap as much money as he thinks proper; he then leads her away to a distance from the sacred grove, and has intercourse with her. The money is considered as consecrated to Venus.

There are three tribunals, one consisting of persons who are past military service, another of nobles, and a third of old men, besides another appointed by the king. It is the business of the latter[521] to dispose of the virgins in marriage, and to determine causes respecting adultery; of another to decide those relative to theft; and of the third, those of assault and violence.

The sick are brought out of their houses into the highways, and inquiry is made of passengers whether any of them can give information of a remedy for the disease. There is no one so ill-disposed as not to accost the sick person, and acquaint[Pg 156] [CAS. 745] him with anything that he considers may conduce to his recovery.

Their dress is a tunic reaching to the feet, an upper garment of wool, [and] a white cloak. The hair is long. They wear a shoe resembling a buskin. They wear also a seal, and carry a staff not plain, but with a figure upon the top of it, as an apple, a rose, a lily, or something of the kind. They anoint themselves with oil of sesamum. They bewail the dead, like the Egyptians and many other nations. They bury the body in honey, first besmearing it with wax.

There are three communities which have no corn. They live in the marshes, and subsist on fish. Their mode of life is like that of the inhabitants of Gedrosia.[522]

21. Mesopotamia has its name from an accidental circumstance. We have said that it is situated between the Euphrates and the Tigris, that the Tigris washes its eastern side only, and the Euphrates its western and southern sides. To the north is the Taurus, which separates Armenia from Mesopotamia. The greatest distance by which they are separated from each other is that towards the mountains. This distance may be the same which Eratosthenes mentions, and is reckoned from Thapsacus,⁵²³ where there was the (Zeugma) old bridge of the Euphrates, to the (Zeugma) passage over the Tigris, where Alexander crossed it, a distance, that is, of 2400 stadia. The least distance between them is somewhere about Seleuceia and Babylon, and is a little more than 200 stadia.

The Tigris flows through the middle of the lake called Thopitis⁵²⁴ in the direction of its breadth, and after traversing it to the opposite bank, sinks under ground with a loud noise and rushing of air. Its course is for a long space invisible, but it rises again to the surface not far from Gordyæa. According to Eratosthenes, it traverses the lake with such rapidity, that although the lake is saline and without fish,⁵²⁵ yet in this part it is fresh, has a current, and abounds with fish.

22. The contracted shape of Mesopotamia extends far in length, and somewhat resembles a ship. The Euphrates forms the larger part of its boundary. The distance from Thapsacus to Babylon, according to Eratosthenes, is 4800 [Pg 157]stadia, and from the (Zeugma)⁵²⁶ bridge in Commagene, where Mesopotamia begins, to Thapsacus, is not less than 2000 stadia.

23. The country lying at the foot of the mountains is very fertile. The people, called by the Macedonians Mygdones, occupy the parts towards the Euphrates, and both Zeugmata, that is, the Zeugma in Commagene, and the ancient Zeugma at Thapsacus. In their territory is Nisibis,⁵²⁷ which they called also Antioch in Mygdonia, situated below Mount Masius,⁵²⁸ and Tigranocerta,⁵²⁹ and the places about Carrhæ, Nicephorium,⁵³⁰ Chordiraza,⁵³¹ and Sinnaca, where Crassus was taken prisoner by stratagem, and put to death by Surena, the Parthian general.⁵³²

24. Near the Tigris are the places belonging to the Gordyæi,⁵³³ whom the ancients called Carduchi; their cities are Sareisa, Satalca, and Pinaca, a very strong fortress with three citadels, each enclosed by its own wall, so that it is as it were a triple city. It was, however, subject to the king of Armenia; the Romans also took it by storm, although the Gordyæi had the reputation of excelling in the art of building, and to be skilful in the construction of siege engines. It was for this reason Tigranes took them into his service. The rest of Mesopotamia (Gordyæa?) was subject to the Romans. Pompey assigned to Tigranes the largest and best portion of the country; for it has fine pastures, is rich in plants, and produces evergreens and an aromatic, the amomum. It breeds lions also. It furnishes naphtha, and the stone called Gangitis,⁵³⁴ which drives away reptiles.

25. Gordys, the son of Triptolemus, is related to have colonized Gordyene. The Eretrians⁵³⁵ afterwards, who were carried away by force by the Persians, settled here. We shall soon speak of Triptolemus in our description of Syria.

26. The parts of Mesopotamia inclining to the south, and [Pg 158] [CAS. 747]at a distance from the mountains, are an arid and barren district, occupied by the Arabian Scenitæ, a tribe of robbers and shepherds, who readily move from place to place, whenever pasture or booty begin to be exhausted. The country lying at the foot of the mountains is harassed both by these people and by the Armenians. They are situated above, and keep them in subjection by force. It is at last subject for the most part to these people, or to the Parthians, who are situated at their side, and possess both Media and Babylonia.

27. Between the Tigris and the Euphrates flows a river called Basileios (or the Royal river), and about Anthemusia another called the Aborrhas.⁵³⁶ The road for merchants going from Syria to Seleuceia and Babylon lies through the country of the (Arabian) Scenitæ, [now called Malii,]⁵³⁷ and through the desert belonging to their territory. The Euphrates is crossed in the latitude of Anthemusia, a place in Mesopotamia.⁵³⁸ Above the river, at the distance of four schœni, is Bambyce, which is called by the names of Edessa and Hierapolis,⁵³⁹ where the Syrian goddess Atargatis is worshipped. After crossing the river, the road lies through a desert country on the borders of Babylonia to Scenæ, a considerable city, situated on the banks of a canal. From the passage across the river to [Pg 159]Scenæ is a journey of five and

twenty days. There are (on the road) owners of camels, who keep resting-places, which are well supplied with water from cisterns, or transported from a distance.

The Scenitæ exact a moderate tribute from merchants, but [otherwise] do not molest them: the merchants, therefore, avoid the country on the banks of the river, and risk a journey through the desert, leaving the river on the right hand at a distance of nearly three days' march. For the chiefs of the tribes living on both banks of the river, who occupy not indeed a fertile territory, yet one less sterile than the rest (of the country), are settled in the midst of their own peculiar domains, and each exacts a tribute of no moderate amount for himself. And it is difficult among so large a body of people, and of such daring habits, to establish any common standard of tribute advantageous to the merchant.

Scenæ is distant from Seleuceia 18 schœni.

28. The Euphrates and its eastern banks are the boundaries of the Parthian empire. The Romans and the chiefs of the Arabian tribes occupy the parts on this side the Euphrates as far as Babylonia. Some of the chiefs attach themselves in preference to the Parthians, others to the Romans, to whom they adjoin. The Scenitæ nomades, who live near the river, are less friendly to the Romans than those tribes who are situated at a distance near Arabia Felix. The Parthians were once solicitous of conciliating the friendship of the Romans, but having repulsed Crassus,[540] who began the war with them, they suffered reprisals, when they themselves commenced hostilities, and sent Pacorus into Asia.[541] But Antony, following the advice of the Armenian,[542] was betrayed, and was unsuccessful (against them). Phraates, his[543] successor, was so anxious to obtain the friendship of Augustus Cæsar, that he even sent the trophies, which the Parthians had set up as memorials of[Pg 160] [CAS. 748]the defeat of the Romans. He also invited Titius to a conference, who was at that time præfect of Syria, and delivered into his hands, as hostages, four of his legitimate sons, Seraspadanes, Rhodaspes, Phraates, and Bonones, with two of their wives and four of their sons; for he was apprehensive of conspiracy and attempts on his life.[544] He knew that no one could prevail against him, unless he was opposed by one of the Arsacian family, to which race the Parthians were strongly attached. He therefore removed the sons out of his way, with a view of annihilating the hopes of the disaffected.

The surviving sons, who live at Rome, are entertained as princes at the public expense. The other kings (his successors) have continued to send ambassadors (to Rome), and to hold conferences (with the Roman præfects).

CHAPTER II.

1. SYRIA is bounded on the north by Cilicia and the mountain Amanus; from the sea to the bridge on the Euphrates (that is, from the Issic Bay to the Zeugma in Commagene) is a distance of 1400 stadia, and forms the above-mentioned (northern) boundary; on the east it is bounded by the Euphrates and the Arabian Scenitæ, who live on this side the Euphrates; on the south, by Arabia Felix and Egypt; on the west, by the Egyptian and Syrian Seas as far as Issus.

2. Beginning from Cilicia and Mount Amanus, we set down as parts of Syria, Commagene, and the Seleucis of Syria, as it is called, then Cœle-Syria, lastly, on the coast, Phœnicia, and in the interior, Judæa. Some writers divide the whole of Syria into Cœlo-Syrians, Syrians, and Phœnicians, and say that there are intermixed with these four other nations, Jews, Idumæans, Gazæans, and Azotii, some of whom are husbandmen, as the Syrians and Cœlo-Syrians, and others merchants, as the Phœnicians.

[Pg 161]

3. This is the general description [of Syria].[545]

In describing it in detail, we say that Commagene is rather a small district. It contains a strong city, Samosata, in which was the seat of the kings. At present it is a (Roman) province. A very fertile but small territory lies around it. Here is now the Zeugma, or bridge, of the Euphrates, and near it is situated Seleuceia, a fortress of Mesopotamia, assigned by Pompey to the Commageneans. Here Tigranes confined in prison for some time and put to death Selene, surnamed Cleopatra, after she was dispossessed of Syria.[546]

4. Seleucis is the best of the above-mentioned portions of Syria. It is called and is a Tetrapolis, and derives its name from the four distinguished cities which it contains; for there are more than four cities, but the four largest are Antioch Epidaphne,[547]Seleuceia in Pieria,[548] Apameia,[549] and Laodiceia.[550] They were called Sisters from the concord which existed between them. They were founded by Seleucus Nicator. The largest bore the name of his father, and the strongest his own. Of the others, Apameia had its name from his wife Apama, and Laodiceia from his mother.

In conformity with its character of Tetrapolis, Seleucis, according to Poseidonius, was divided into four satrapies; Cœle-Syria into the same number, but [Commagene, like] Mesopotamia, consisted of one.[551]

Antioch also is a Tetrapolis, consisting (as the name implies)[Pg 162] [CAS. 750] of four portions, each of which has its own, and all of them a common wall.[552]

[Seleucus] Nicator founded the first of these portions, transferring thither settlers from Antigonia, which a short time before Antigonus, son of Philip, had built near it. The second was built by the general body of settlers; the third by Seleucus, the son of Callinicus; the fourth by Antiochus, the son of Epiphanes.

5. Antioch is the metropolis of Syria. A palace was constructed there for the princes of the country. It is not much inferior in riches and magnitude to Seleuceia on the Tigris and Alexandreia in Egypt.

[Seleucus] Nicator settled here the descendants of Triptolemus, whom we have mentioned a little before.[553] On this account the people of Antioch regard him as a hero, and celebrate a festival to his honour on Mount Casius[554] near Seleuceia. They say that when he was sent by the Argives in search of Io, who first disappeared at Tyre, he wandered through Cilicia; that some of his Argive companions separated from him and founded Tarsus; that the rest attended him along the sea-coast, and, relinquishing their search, settled with him on the banks of the Orontes;[555] that Gordys the son of Triptolemus, with some of those who had accompanied his father, founded a colony in Gordyæa, and that the descendants of the rest became settlers among the inhabitants of Antioch.

6. Daphne,[556] a town of moderate size, is situated above Antioch at the distance of 40 stadia. Here is a large forest, with a thick covert of shade and springs of water flowing through it. In the midst of the forest is a sacred grove, which is a sanctuary, and a temple of Apollo and Diana. It is the custom for the inhabitants of Antioch and the neighbouring people to assemble here to celebrate public festivals. The forest is 80 stadia in circumference.

[Pg 163]7. The river Orontes flows near the city. Its source is in Cœle-Syria. Having taken its course under-ground, it reäppears, traverses the territory of Apameia to Antioch, approaching the latter city, and then descends to the sea at Seleuceia. The name of the river was formerly Typhon, but was changed to Orontes, from the name of the person who constructed the bridge over it.

According to the fable, it was somewhere here that Typhon was struck with lightning, and here also was the scene of the fable of the Arimi, whom we have before mentioned.[557] Typhon was a serpent, it is said, and being struck by lightning, endeavoured to make its escape, and sought refuge in the ground; it deeply furrowed the earth, and (as it moved along) formed the bed of the river; having descended under-ground, it caused a spring to break out, and from Typhon the river had its name.

On the west the sea, into which the Orontes discharges itself, is situated below Antioch in Seleuceia, which is distant from the mouth of the river 40, and from Antioch 120 stadia. The ascent by the river to Antioch is performed in one day.

To the east of Antioch are the Euphrates, Bambyce,[558] Berœa,[559] and Heracleia, small towns formerly under the government of Dionysius, the son of Heracleon. Heracleia is distant 20 stadia from the temple of Diana Cyrrhestis.

8. Then follows the district of Cyrrhestica,[560] which extends as far as that of Antioch. On the north near it are Mount Amanus and Commagene. Cyrrhestica extends as far as these places, and touches them. Here is situated a city, Gindarus, the acropolis of Cyrrhestica, and a convenient resort for robbers, and near it a place called Heracleium. It was near these places that Pacorus, the eldest of the sons of the Parthian king, who had invaded Syria, was defeated by Ventidius, and killed.

Pagræ,[561] in the district of Antioch, is close to Gindarus. It [Pg 164] [CAS. 751]is a strong fortress situated on the pass over the Amanus, which leads from the gates of the Amanus into Syria. Below Pagræ lies the plain of Antioch, through which flow the rivers Arceuthus, Orontes, and Labotas.[562] In this plain is also the trench of Meleagrus, and the river Œnoparas,[563] on the banks of which Ptolemy Philometor, after having defeated Alexander Balas, died of his wounds.[564]

Above these places is a hill called Trapezon from its form,[565] and upon it Ventidius engaged Phranicates[566] the Parthian general.

After these places, near the sea, are Seleuceia[567] and Pieria, a mountain continuous with the Amanus and Rhosus, situated between Issus and Seleuceia.

Seleuceia formerly had the name of Hydatopotami (rivers of water). It is a considerable fortress, and may defy all attacks; wherefore Pompey, having excluded from it Tigranes, declared it a free city.

To the south of Antioch is Apameia, situated in the interior, and to the south of Seleuceia, the mountains Casius and Anti-Casius.

Still further on from Seleuceia are the mouths of the Orontes, then the Nymphæum, a kind of sacred cave, next Casium, then follows Poseidium[568] a small city, and Heracleia.[569]

9. Then follows Laodiceia, situated on the sea; it is a very well-built city, with a good harbour; the territory, besides its fertility in other respects, abounds with wine, of which the greatest part is exported to Alexandreia. The whole mountain overhanging the city is planted almost to its summit with vines. The summit of the mountain is at a great distance from Laodiceia, sloping gently and by degrees upwards from the city; but it rises perpendicularly over Apameia.

Laodiceia suffered severely when Dolabella took refuge there. Being besieged by Cassius, he defended it until his death, but he involved in his own ruin the destruction of many parts of the city.[570]

[Pg 165]10. In the district of Apameia is a city well fortified in almost every part. For it consists of a well-fortified hill, situated in a hollow plain, and almost surrounded by the Orontes, which, passing by a large lake in the neighbourhood, flows through wide-spread marshes and meadows of vast extent, affording pasture for cattle and horses.[571] The city is thus securely situated, and received the name Cherrhonesus (or the peninsula) from the nature of its position. It is well supplied from a very large fertile tract of country, through which the Orontes flows with numerous windings. Seleucus Nicator, and succeeding kings, kept there five hundred elephants, and the greater part of their army.

It was formerly called Pella by the first Macedonians, because most of the soldiers of the Macedonian army had settled there; for Pella, the native place of Philip and Alexander, was held to be the metropolis of the Macedonians. Here also the soldiers were mustered, and the breed of horses kept up. There were in the royal stud more than thirty thousand brood mares and three hundred stallions. Here were employed colt-breakers, instructors in the method of fighting in heavy armour, and all who were paid to teach the arts of war.

The power Trypho, surnamed Diodotus, acquired is a proof of the influence of this place; for when he aimed at the empire of Syria, he made Apameia the centre of his operations. He was born at Casiana, a strong fortress in the Apameian district, and educated in Apameia; he was a favourite of the king and the persons about the court. When he attempted to effect a revolution in the state, he obtained his supplies from Apameia and from the neighbouring cities, Larisa,[572] Casiana, Megara, Apollonia, and others like them, all of which were reckoned to belong to the district of Apameia. He was proclaimed king of this country, and maintained his sovereignty for a long time. Cæcilius Bassus, at the head of two legions, caused Apameia to revolt, and was besieged by two large Roman armies, but his resistance was so vigorous and long that he only surrendered voluntarily and on his own conditions.[573] For the country supplied his army with provisions, [Pg 166] [CAS. 753]and a great many of the chiefs of the neighbouring tribes were his allies, who possessed strongholds, among which was Lysias, situated above the lake, near Apameia, Arethusa,[574]belonging to Sampsiceramus and Iamblichus his son, chiefs of the tribe of the Emeseni.[575] At no great distance were Heliopolis and Chalcis,[576] which were subject to Ptolemy, son of Mennæus,[577] who possessed the Massyas[578] and the mountainous country of the Ituræans. Among the auxiliaries of Bassus was Alchædamnus,[579] king of the Rhambæi, a tribe of the Nomades on this side of the Euphrates. He was a friend of the Romans, but, considering himself as having been unjustly treated by their governors, he retired to Mesopotamia, and then became a tributary of Bassus. Poseidonius the Stoic was a native of this place, a man of the most extensive learning among the philosophers of our times.

11. The tract called Parapotamia, belonging to the Arab chiefs, and Chalcidica, extending from the Massyas, border upon the district of Apameia on the east; and nearly all the country further to the south of Apameia belongs to the Scenitæ, who resemble the Nomades of Mesopotamia. In proportion as the nations approach the Syrians they become more civilized, while the Arabians and Scenitæ are less so. Their [Pg 167]governments are better constituted [as that of Arethusa under Sampsiceramus, that of Themella under Gambarus, and other states of this kind].[580]

12. Such is the nature of the interior parts of the district of Seleuceia.

The remainder of the navigation along the coast from Laodiceia is such as I shall now describe.

Near Laodiceia are the small cities, Poseidium, Heracleium, and Gabala. Then follows the maritime tract[581] of the Aradii, where are Paltus,[582] Balanæa, and Carnus,[583] the arsenal of Aradus, which has a small harbour; then Enydra,[584] and Marathus, an ancient city of the Phœnicians in ruins. The Aradii[585] divided the territory by lot. Then follows the district Simyra.[586] Continuous with these places is Orthosia,[587] then the river Eleutherus, which some make the boundary of Seleucis towards Phœnicia and Cœle-Syria.

13. Aradus is in front of a rocky coast without harbours, and situated nearly between its arsenal[588] and Marathus. It is distant from the land 20 stadia. It is a rock, surrounded by the sea, of about seven stadia in circuit, and covered with dwellings. The population even at present is so large that the houses have many stories. It was colonized, it is said, by fugitives from Sidon. The inhabitants are supplied with water partly from cisterns containing rain water,   and   partly [Pg   168] [CAS. 754]from the opposite coast. In war time they obtain water a little in front of the city, from the channel (between the island and the mainland), in which there is an abundant spring. The water is obtained by letting down from a boat, which serves for the purpose, and inverting over the spring (at the bottom of the sea), a wide-mouthed funnel of lead, the end of which is contracted to a moderate-sized opening; round this is fastened a (long) leathern pipe, which we may call the neck, and which receives the water, forced up from the spring through the funnel. The water first forced up is sea water, but the boatmen wait for the flow of pure and potable water, which is received into vessels ready for the purpose, in as large a quantity as may be required, and carry it to the city.[589]

14. The Aradii were anciently governed by their own kings in the same manner as all the other Phœnician cities. Afterwards the Persians, Macedonians, and now the Romans have changed the government to its present state.

The Aradii, together with the other Phœnicians, consented to become allies of the Syrian kings; but upon the dissension of the two brothers, Callinicus Seleucus and Antiochus Hierax, as he was called, they espoused the party of Callinicus; they entered into a treaty, by which they were allowed to receive persons who quitted the king's dominions, and took refuge among them, and were not obliged to deliver them up against their will. They were not, however, to suffer them to embark and quit the island without the king's permission. From this they derived great advantages; for those who took refuge there were not ordinary people, but persons who had held the highest trusts, and apprehended the worst consequences (when they fled). They regarded those who received them with hospitality as their benefactors; they acknowledged their preservers, and remembered with gratitude the kindness which they had received, particularly after their return to their own country. It was thus that the Aradii acquired possession of a large part of the opposite continent, most of which they possess even at present, and were otherwise successful.[Pg 169] To this good fortune they added prudence and industry in the conduct of their maritime affairs; when they saw their neighbours, the Cilicians, engaged in piratical adventures, they never on any occasion took part with them in such (a disgraceful) occupation.[590]

15. After Orthosia and the river Eleutherus is Tripolis, which has its designation from the fact of its consisting of three cities, Tyre, Sidon, and Aradus. Contiguous to Tripolis is Theoprosopon,[591] where the mountain Libanus terminates. Between them lies a small place called Trieres.

16. There are two mountains, which form Cœle-Syria, as it is called, lying nearly parallel to each other; the commencement of the ascent of both these mountains, Libanus and Antilibanus, is a little way from the sea; Libanus rises above the sea near Tripolis and Theoprosopon, and Antilibanus, above the sea near Sidon. They terminate somewhere near the Arabian mountains, which are above the district of Damascus and the Trachones as they are there called, where they form fruitful hills. A hollow plain lies between them, the breadth of which towards the sea is 200 stadia, and the length from the sea to the interior is about twice that number of stadia. Rivers flow through it, the largest of which is the Jordan, which water a country fertile and productive of all things. It contains also a lake, which produces the aromatic rush and reed. In it are also marshes. The name of the lake is Gennesaritis. It produces also balsamum.[592]

Among the rivers is the Chrysorrhoas, which commences [Pg 170] [CAS. 755] from the city and territory of Damascus, and is almost entirely drained by water-courses; for it supplies with water a large tract of country, with a very deep soil.

The Lycus[593] and the Jordan are navigated upwards chiefly by the Aradii, with vessels of burden.

17. Of the plains, the first reckoning from the sea is called Macras and Macra-pedium. Here Poseidonius says there was seen a serpent lying dead, which was nearly a plethrum in length, and of such a bulk and thickness that men on horseback standing on each side of its body could not see one another; the jaws when opened could take in a man on horseback, and the scales of the skin were larger than a shield.

18. Next to the plain of Macras is that of Massyas, which also contains some mountainous parts, among which is Chalcis, the acropolis, as it were, of the Massyas. The commencement of this plain is at Laodiceia,[594] near Libanus. The Ituræans and Arabians, all of whom are freebooters, occupy the whole of the mountainous tracts. The husbandmen live in the plains, and when harassed by the freebooters, they require protection of various kinds. The robbers have strongholds from which they issue forth; those, for example, who occupy Libanus have high up on the mountain the fortresses Sinna, Borrhama, and some others like them; lower down, Botrys and Gigartus, caves also near the sea, and the castle on the promontory Theoprosopon. Pompey destroyed these fastnesses, from whence the robbers overran Byblus,[595] and Berytus[596] situated next to it, and which lie between Sidon and Theoprosopon.

Byblus, the royal seat of Cinyrus, is sacred to Adonis. Pompey delivered this place from the tyranny of Cinyrus, by striking off his head. It is situated upon an eminence at a little distance from the sea.

19. After Byblus is the river Adonis,[597] and the mountain Climax, and Palæ-Byblus, then the river Lycus, and Berytus. This latter place was razed by Tryphon, but now the Romans have restored it, and two legions were stationed there by Agrippa, who also added to it a large portion of the territory of Massyas, as far as the sources of the Orontes. These sources are near Libanus, the Paradeisus, and the Egyptian [Pg 171] Fort near the district of Apameia. These places lie near the sea.

20. Above the Massyas is the Royal Valley, as it is called, and the territory of Damascus, so highly extolled. Damascus is a considerable city, and in the time of the Persian empire was nearly the most distinguished place in that country.

Above Damascus are the two (hills) called Trachones; then, towards the parts occupied by Arabians and Ituræans promiscuously, are mountains of difficult access, in which were caves extending to a great depth. One of these caves was capable of containing four thousand robbers, when the territory of Damascus was subject to incursions from various quarters. The Barbarians used to rob the merchants most generally on the side of Arabia Felix,[598] but this happens less frequently since the destruction of the bands of the robbers under Zenodorus, by the good government of the Romans, and in consequence of the security afforded by the soldiers stationed and maintained in Syria.

21. The whole country[599] above Seleucis, extending towards Egypt and Arabia, is called Cœle-Syria, but peculiarly the tract bounded by Libanus and Antilibanus, of the remainder one part is the coast extending from Orthosia[600] as far as Pelusium,[601] and is called Phœnicia, a narrow strip of land along the sea; the other, situated above Phœnicia in the interior between Gaza and Antilibanus, and extending to the Arabians, called Judæa.

22. Having described Cœle-Syria properly so called, we pass on to Phœnicia, of which we have already described[602] the part extending from Orthosia to Berytus.

Next to Berytus is Sidon, at the distance of 400 stadia. Between these places is the river Tamyras,[603] and the grove of Asclepius and Leontopolis.

Next to Sidon is Tyre,[604] the largest and most ancient city of the Phœnicians. This city is the rival of Sidon in magnitude, fame, and antiquity, as recorded in many fables. For although poets have celebrated Sidon more than Tyre (Homer, however, does not even mention Tyre), yet the colonies sent into Africa and Spain, as far as, and beyond the Pillars, extol [Pg 172]
[CAS. 756]much more the glory of Tyre. Both however were formerly, and are at present, distinguished and illustrious cities, but which of the two should be called the capital of Phœnicia is a subject of dispute among the inhabitants.[605] Sidon is situated upon a fine naturally-formed harbour on the mainland.

23. Tyre is wholly an island, built nearly in the same manner as Aradus. It is joined to the continent by a mound, which Alexander raised, when he was besieging it. It has two harbours, one close, the other open, which is called the Egyptian harbour. The houses here, it is said, consist of many stories, of more even than at Rome; on the occurrence, therefore, of an earthquake, the city was nearly demolished.[606] It sustained great injury when it was taken by siege by Alexander, but it rose above these misfortunes, and recovered itself both by the skill of the people in the art of navigation, in which the Phœnicians in general have always excelled all nations, and by (the export of) purple-dyed manufactures, the Tyrian purple being in the highest estimation. The shell-fish from which it is procured is caught near the coast, and the Tyrians have in great abundance other requisites for dyeing. The great number of dyeing works renders the city unpleasant as a place of residence, but the superior skill of the people in the practice of this art is the source of its wealth. Their independence was secured to them at a small expense to themselves, not only by the kings of Syria, but also by the Romans, who confirmed what the former had conceded.[607] They pay extravagant honours to Hercules.

[Pg 173]The great number and magnitude of their colonies and cities are proofs of their maritime skill and power.

Such then are the Tyrians.

24. The Sidonians are said by historians to excel in various kinds of art, as the words of Homer also imply.[608] Besides, they cultivate science and study astronomy and arithmetic, to which they were led by the application of numbers (in accounts) and night sailing, each of which (branches of knowledge) concerns the merchant and seaman; in the same manner the Egyptians were led to the invention of geometry by the mensuration of ground, which was required in consequence of the Nile confounding, by its overflow, the respective boundaries of the country. It is thought that geometry was introduced into Greece from Egypt, and astronomy and arithmetic from Phœnicia. At present the best opportunities are afforded in these cities of acquiring a knowledge of these, and of all other branches of philosophy.

If we are to believe Poseidonius, the ancient opinion about atoms originated with Mochus, a native of Sidon, who lived before the Trojan times. Let us, however, dismiss subjects relating to antiquity. In my time there were distinguished philosophers, natives of Sidon, as Boethus, with whom I studied the philosophy of Aristotle,[609] and Diodotus his brother. Antipater was of Tyre, and a little before my time Apollonius, who published a table of the philosophers of the school of Zeno, and of their writings.

Tyre is distant from Sidon not more than 200 stadia. Between the two is situated a small town, called Ornithopolis, (the city of birds); next a river[610] which empties itself near Tyre into the sea. Next after Tyre is Palæ-tyrus (ancient Tyre), at the distance of 30 stadia.[611]
[Pg 174]
[CAS. 758]25. Then follows Ptolemaïs, a large city, formerly called Ace.[612] It was the place of

rendezvous for the Persians in their expeditions against Egypt. Between Ace and Tyre is a sandy beach, the sand of which is used in making glass. The sand, it is said, is not fused there, but carried to Sidon to undergo that process. Some say that the Sidonians have, in their own country, the vitrifiable sand; according to others, the sand of every place can be fused. I heard at Alexandria from the glass-workers, that there is in Egypt a kind of vitrifiable earth, without which expensive works in glass of various colours could not be executed, but in other countries other mixtures are required; and at Rome, it is reported, there have been many inventions both for producing various colours, and for facilitating the manufacture, as for example in glass wares, where a glass bowl may be purchased for a copper coin,[613] and glass is ordinarily used for drinking.

26. A phenomenon[614] of the rarest kind is said to have occurred on the shore between Tyre and Ptolemaïs. The people of Ptolemaïs had engaged in battle with Sarpedon the general, and after a signal defeat were left in this place, when a wave from the sea, like the rising tide, overwhelmed the fugitives; some were carried out to sea and drowned, others perished in hollow places; then again the ebb succeeding, uncovered and displayed to sight the bodies lying in confusion among dead fish.

A similar phenomenon took place at Mount Casium in Egypt. The ground, to a considerable distance, after a violent and single shock fell in parts, at once exchanging places; the elevated parts opposed the access of the sea, and parts which had subsided admitted it. Another shock occurred, and the place recovered its ancient position, except that there was an alteration (in the surface of the ground) in some places, and none in others. Perhaps such occurrences are connected with periodical returns the nature of which is unknown to us. This is said to be the case with the rise of the waters of the Nile, which exhibits a variety in its effects, but observes (in general) a certain order, which we do not comprehend.

27. Next to Ace is the Tower of Strato, with a station for [Pg 175]vessels.[615] Between these places is Mount Carmel, and cities of which nothing but the names remain, as Sycaminopolis, Bucolopolis, Crocodeilopolis, and others of this kind; next is a large forest.[616]

28. Then Joppa,[617] where the coast of Egypt, which at first stretches towards the east, makes a remarkable bend towards the north. In this place, according to some writers, Andromeda was exposed to the sea-monster. It is sufficiently elevated; it is said to command a view of Jerusalem, the capital of the Jews,[618] who, when they descended to the sea, used this place as a naval arsenal. But the arsenals of robbers are the haunts of robbers. Carmel, and the forest, belonged to the Jews. The district was so populous that the neighbouring village Iamneia,[619] and the settlements around, could furnish forty thousand soldiers.

Thence to Casium,[620] near Pelusium, are little more than 1000 stadia, and 1300 to Pelusium itself.

29. In the interval is Gadaris,[621] which the Jews have appropriated to themselves, then Azotus and Ascalon.[622] From Iamneia to Azotus and Ascalon are about 200 stadia. The country of the Ascalonitæ produces excellent onions; the town is small. Antiochus the philosopher, who lived a little before our time, was a native of this place. Philodemus the Epicurean was a native of Gadara, as also Meleagrus, Menippus the satirist, and Theodorus the rhetorician, my contemporary.

[Pg   176]
[CAS. 759]30. Next and near Ascalon is the harbour of the Gazæi. The city is situated inland at the distance of seven stadia. It was once famous, but was razed by Alexander, and remains uninhabited. There is said to be a passage thence across, of 1260 stadia, to the city Aila[623] (Aelana), situated on the innermost recess of the Arabian Gulf. This recess has two branches, one, in the direction of Arabia and Gaza, is called Ailanites, from the city upon it; the other is in the direction of Egypt, towards Heroopolis,[624] to which from Pelusium is the shortest road (between the two seas). Travelling is performed on camels, through a desert and sandy country, in the course of which snakes are found in great numbers.

31. Next to Gaza is Raphia,[625] where a battle was fought between Ptolemy the Fourth and Antiochus the Great.[626] Then Rhinocolura,[627] so called from the colonists, whose noses had been mutilated. Some Ethiopian invaded Egypt, and, instead of putting the malefactors to death, cut off their noses, and settled them at Rhinocolura, supposing that they would not

venture to return to their own country, on account of the disgraceful condition of their faces.

32. The whole country from Gaza is barren and sandy, and still more so is that district next to it, which contains the lake Sirbonis,[628] lying above it in a direction almost parallel to the sea, and leaving a narrow pass between, as far as what is called the Ecregma.[629] The length of the pass is about 200, and the greatest breadth 50 stadia. The Ecregma is filled up with earth. Then follows another continuous tract of the same kind to Casium,[630] and thence to Pelusium.

33. The Casium is a sandy hill without water, and forms a promontory: the body of Pompey the Great is buried there, and on it is a temple of Jupiter Casius.[631] Near this place Pompey the Great was betrayed by the Egyptians, and put to death. Next is the road to Pelusium, on which is situated[Pg 177]Gerrha;[632] and the rampart, as it is called, of Chabrias, and the pits near Pelusium, formed by the overflowing of the Nile in places naturally hollow and marshy.

Such is the nature of Phœnicia. Artemidorus says, that from Orthosia to Pelusium is 3650 stadia, including the winding of the bays, and from Melænæ or Melania in Cilicia to Celenderis,[633] on the confines of Cilicia and Syria, are 1900 stadia; thence to the Orontes 520 stadia, and from Orontes to Orthosia 1130 stadia.

34. The western extremities of Judæa towards Casius are occupied by Idumæans, and by the lake [Sirbonis]. The Idumæans are Nabatæans. When driven from their country[634] by sedition, they passed over to the Jews, and adopted their customs.[635]The greater part of the country along the coast to Jerusalem is occupied by the Lake Sirbonis, and by the tract contiguous to it; for Jerusalem is near the sea, which, as we have said,[636] may be seen from the arsenal of Joppa.[637] These districts (of Jerusalem and Joppa) lie towards the north; they are inhabited generally, and each place in particular, by mixed tribes of Egyptians, Arabians, and Phœnicians. Of this description are the inhabitants of Galilee, of the plain of Jericho, and of the territories of Philadelphia and Samaria,[638] surnamed Sebaste by Herod;[639] but although there is such a mixture of inhabitants, the report most credited, [one] among many things believed respecting the temple [and the inhabitants] of Jerusalem, is, that the Egyptians were the ancestors of the present Jews.[640]

35. An Egyptian priest named Moses, who possessed a portion of the country called the Lower [Egypt] * * * *, being dissatisfied with the established institutions there, left it and came to Judæa with a large body of people who worshipped the Divinity. He declared and taught that the Egyptians and Africans entertained erroneous sentiments, [Pg 178] [CAS. 761]in representing the Divinity under the likeness of wild beasts and cattle of the field; that the Greeks also were in error in making images of their gods after the human form. For God [said he] may be this one thing which encompasses us all, land and sea, which we call heaven, or the universe, or the nature of things.[641] Who then of any understanding would venture to form an image of this Deity, resembling anything with which we are conversant? on the contrary, we ought not to carve any images, but to set apart some sacred ground and a shrine worthy of the Deity, and to worship Him without any similitude.[642]He taught that those who made fortunate dreams were to be permitted to sleep in the temple, where they might dream both for themselves and others; that those who practised temperance and justice, and none else, might expect good, or some gift or sign from the God, from time to time.

36. By such doctrine Moses[643] persuaded a large body of right-minded persons to accompany him to the place where Jerusalem now stands. He easily obtained possession of it, as the spot was not such as to excite jealousy, nor for which there could be any fierce contention; for it is rocky, and, although well supplied with water, it is surrounded by a barren and waterless territory.[644] The space within [the city] is 60 stadia [in circumference], with rock underneath the surface.

Instead of arms, he taught that their defence was in their sacred things and the Divinity, for whom he was desirous of finding a settled place, promising to the people to deliver such a kind of worship and religion as should not burthen those who adopted it with great expense, nor molest them with [so-called] divine possessions, nor other absurd practices.

Moses thus obtained their good opinion, and established no ordinary kind of government. All the nations around willingly united themselves to him, allured by his discourses and promises.

[Pg 179]37. His successors continued for some time to observe the same conduct, doing justly, and worshipping God with sincerity. Afterwards superstitious persons were appointed to the priesthood, and then tyrants. From superstition arose abstinence from flesh, from the eating of which it is now the custom to refrain, circumcision, excision,⁶⁴⁵ and other practices which the people observe. The tyrannical government produced robbery; for the rebels plundered both their own and the neighbouring countries. Those also who shared in the government seized upon the property of others, and ravaged a large part of Syria and of Phœnicia.

Respect, however, was paid to the Acropolis; it was not abhorred as the seat of tyranny, but honoured and venerated as a temple.

38. This is according to nature, and common both to Greeks and barbarians. For, as members of a civil community, they live according to a common law; otherwise it would be impossible for the mass to execute any one thing in concert (in which consists a civil state), or to live in a social state at all. Law is twofold, divine and human. The ancients regarded and respected divine, in preference to human, law; in those times, therefore, the number of persons was very great who consulted oracles, and, being desirous of obtaining the advice of Jupiter, hurried to Dodona,

"to hear the answer of Jove from the lofty oak."

The parent went to Delphi,

"anxious to learn whether the child which had been exposed (to die) was still living;"

while the child itself

"was gone to the temple of Apollo, with the hope of discovering its parents."

And Minos among the Cretans,

"the king who in the ninth year enjoyed converse with Great Jupiter,"

every nine years, as Plato says, ascended to the cave of Jupiter, received ordinances from him, and conveyed them to men. Lycurgus, his imitator, acted in a similar manner; for he was often accustomed, as it seemed, to leave his own country to inquire of the Pythian goddess what ordinances he was to promulgate to the Lacedæmonians.

[Pg   180]
[CAS. 762]

39. What truth there may be in these things I cannot say; they have at least been regarded and believed as true by mankind. Hence prophets received so much honour as to be thought worthy even of thrones, because they were supposed to communicate ordinances and precepts from the gods, both during their lifetime and after their death; as for example Teiresias,

"to whom alone Proserpine gave wisdom and understanding after death: the others flit about as shadows."⁶⁴⁶

Such were Amphiaraus, Trophonius, Orpheus, and Musæus: in former times there was Zamolxis, a Pythagorean, who was accounted a god among the Getæ; and in our time, Decæneus, the diviner of Byrebistas. Among the Bosporani, there was Achaïcarus; among the Indians, were the Gymnosophists; among the Persians, the Magi and Necyomanteis,⁶⁴⁷ and besides these the Lecanomanteis⁶⁴⁸ and Hydromanteis;⁶⁴⁹ among the Assyrians, were the Chaldæans; and among the Romans, the Tyrrhenian diviners of dreams.⁶⁵⁰

Such was Moses and his successors; their beginning was good, but they degenerated.

40. When Judæa openly became subject to a tyrannical government, the first person who exchanged the title of priest for that of king was Alexander.⁶⁵¹ His sons were Hyrcanus and Aristobulus. While they were disputing the succession to the kingdom, Pompey came upon them by surprise, deprived them of their power, and destroyed their fortresses, first taking Jerusalem itself by storm.⁶⁵² It was a stronghold, situated on a rock, well fortified and well supplied with water⁶⁵³[Pg 181]within, but externally entirely parched with drought. A ditch was cut in the rock, 60 feet in depth, and in width 250 feet. On the wall of the temple

were built towers, constructed of the materials procured when the ditch was excavated. The city was taken, it is said, by waiting for the day of fast, on which the Jews were in the habit of abstaining from all work. Pompey [availing himself of this], filled up the ditch, and threw bridges over it. He gave orders to raze all the walls, and he destroyed, as far as was in his power, the haunts of the robbers and the treasure-holds of the tyrants. Two of these forts, Thrax and Taurus, were situated in the passes leading to Jericho. Others were Alexandrium, Hyrcanium, Machærus, Lysias, and those about Philadelphia, and Scythopolis near Galilee.

41. Jericho is a plain encompassed by a mountainous district, which slopes towards it somewhat in the manner of a theatre. Here is the Phœnicon (or palm plantation), which contains various other trees of the cultivated kind, and producing excellent fruit; but its chief production is the palm tree. It is 100 stadia in length; the whole is watered with streams, and filled with dwellings. Here also is a palace and the garden of the balsamum.[654] The latter is a shrub with an aromatic smell, resembling the cytisus[655] and the terminthus.[656] Incisions are made in the bark, and vessels are placed beneath to receive the sap, which is like oily milk. After it is collected in vessels, it becomes solid. It is an excellent remedy for headache, incipient suffusion of the eyes, and dimness of sight. It bears therefore a high price, especially as it is produced in no other place.[657] This is the case also with the Phœnicon, which alone contains the caryotes[658] palm, if we except the Babylonian plain, and the country above it towards the east: a large revenue is derived from the palms and balsamum; xylobalsamum[659] is also used as a perfume.

[CAS. 764]42. The Lake Sirbonis[660] is of great extent. Some say that it is 1000 stadia in circumference. It stretches along the coast, to the distance of a little more than 200 stadia. It is deep, and the water is exceedingly heavy, so that no person can dive into it; if any one wades into it up to the waist, and attempts to move forward, he is immediately lifted out of the water.[661] It abounds with asphaltus, which rises, not however at any regular seasons, in bubbles, like boiling water, from the middle of the deepest part. The surface is convex, and presents the appearance of a hillock. Together with the asphaltus, there ascends a great quantity of sooty vapour, not perceptible to the eye, which tarnishes copper, silver, and everything bright—even gold. The neighbouring people know by the tarnishing of their vessels that the asphaltus is beginning to rise, and they prepare to collect it by means of rafts composed of reeds. The asphaltus is a clod of earth, liquefied by heat; the air forces it to the surface, where it spreads itself. It is again changed into so firm and solid a mass by cold water, such as the water of the lake, that it requires cutting or chopping (for use). It floats upon the water, which, as I have described, does not admit of diving or immersion, but lifts up the person who goes into it. Those who go on rafts for the asphaltus cut it in pieces, and take away as much as they are able to carry.

43. Such are the phenomena. But Posidonius says, that the people being addicted to magic, and practising incantations, (by these means) consolidate the asphaltus, pouring upon it urine and other fetid fluids, and then cut it into pieces. (Incantations cannot be the cause), but perhaps urine may have some peculiar power (in effecting the consolidation) in the [Pg 183]same manner that chrysocolla[662] is formed in the bladders of persons who labour under the disease of the stone, and in the urine of children.

It is natural for these phenomena to take place in the middle of the lake, because the source of the fire is in the centre, and the greater part of the asphaltus comes from thence. The bubbling up, however, of the asphaltus is irregular, because the motion of fire, like that of many other vapours, has no order perceptible to observers. There are also phenomena of this kind at Apollonia in Epirus.

44. Many other proofs are produced to show that this country is full of fire. Near Moasada[663] are to be seen rugged rocks, bearing the marks of fire; fissures in many places; a soil like ashes; pitch falling in drops from the rocks; rivers boiling up, and emitting a fetid odour to a great distance; dwellings in every direction overthrown; whence we are inclined to believe the common tradition of the natives, that thirteen cities[664] once existed there, the capital of which was Sodom, but that a circuit of about 60 stadia around it escaped uninjured; shocks of earthquakes, however, eruptions of flames and hot springs, containing asphaltus and sulphur, caused the lake to burst its bounds, and the rocks took fire; some of

the cities were swallowed up, others were abandoned by such of the inhabitants as were able to make their escape.

But Eratosthenes asserts, on the contrary, that the country was once a lake, and that the greater part of it was uncovered by the water discharging itself through a breach, as was the case in Thessaly.[665]

45. In the Gadaris, also, there is a lake of noxious water. If beasts drink it, they lose their hair, hoofs, and horns. At the place called Taricheæ,[666] the lake supplies the best fish for curing.[Pg 184]

[CAS. 765] On its banks grow trees which bear a fruit like the apple. The Egyptians use the asphaltus for embalming the bodies of the dead.

46. Pompey curtailed the territory which had been forcibly appropriated by the Jews, and assigned to Hyrcanus the priesthood. Some time afterwards, Herod, of the same family, and a native of the country,[667] having surreptitiously obtained the priesthood, distinguished himself so much above his predecessors, particularly in his intercourse, both civil and political, with the Romans, that he received the title and authority of king,[668] first from Antony, and afterwards from Augustus Cæsar. He put to death some of his sons, on the pretext of their having conspired against him;[669] other sons he left at his death, to succeed him, and assigned to each, portions of his kingdom. Cæsar bestowed upon the sons also of Herod marks of honour,[670] on his sister Salome,[671] and on her daughter Berenice. The sons were unfortunate, and were publicly accused. One[672] of them died in exile among the Galatæ Allobroges, whose country was assigned for his abode. The others, by great [Pg 185]interest and solicitation, but with difficulty, obtained leave to return[673] to their own country, each with his tetrarchy restored to him.

CHAPTER III.

1. ABOVE Judæa and Cœle-Syria, as far as Babylonia and the river tract, along the banks of the Euphrates towards the south, lies the whole of Arabia, except the Scenitæ in Mesopotamia. We have already spoken of Mesopotamia, and of the nations that inhabit it.[674]

The parts on the other (the eastern) side of the Euphrates, towards its mouth, are occupied by Babylonians and the nation of the Chaldæans. We have spoken of these people also.[675]

Of the rest of the country which follows after Mesopotamia, and extends as far as Cœle-Syria, the part approaching the river, as well as [a part of] Mesopotamia,[676]are occupied by Arabian Scenitæ, who are divided into small sovereignties, and inhabit tracts which are barren from want of water. They do not till the land at all, or only to a small extent, but they keep herds of cattle of all kinds, particularly of camels. Above these is a great desert; but the parts lying still more to the south are occupied by the nations inhabiting Arabia Felix, as it is called. The northern side of this tract is formed by the above-mentioned desert, the eastern by the Persian, the western by the Arabian Gulf, and the southern by [Pg 186] [CAS. 766]the great sea lying outside of both the gulfs, the whole of which is called the Erythræan Sea.[677]

2. The Persian Gulf has the name also of the Sea of Persia. Eratosthenes speaks of it in this manner: "They say that the mouth is so narrow, that from Harmozi,[678] the promontory of Carmania, may be seen the promontory at Macæ, in Arabia. From the mouth, the coast on the right hand is circular, and at first inclines a little from Carmania towards the east, then to the north, and afterwards to the west as far as Teredon and the mouth of the Euphrates.[679] In an extent of about 10,000 stadia, it comprises the coast of the Carmanians, Persians, and Susians, and in part of the Babylonians. (Of these we ourselves have before spoken.) Hence directly as far as the mouth are 10,000 stadia more, according, it is said, to the computation of Androsthenes of Thasos, who not only had accompanied Nearchus, but had also alone sailed along the sea-coast of Arabia.[680] It is hence evident that this sea is little inferior in size to the Euxine.

"He says that Androsthenes, who had navigated the gulf with a fleet, relates, that in sailing from Teredon with the continent on the right hand, an island Icaros[681] is met with, lying in front, which contained a temple sacred to Apollo, and an oracle of [Diana] Tauropolus.

3. "Having coasted the shore of Arabia to the distance of 2400 stadia, there lies, in a deep gulf, a city of the name of Gerrha,[682] belonging to Chaldæan exiles from Babylon, who [Pg 187]inhabit the district in which salt is found, and who have houses constructed of salt: as scales of salt separated by the burning heat of the sun are continually falling off, the houses are sprinkled with water, and the walls are thus kept firm together. The city is distant 200 stadia from the sea. The merchants of Gerrha generally carry the Arabian merchandise and aromatics by land; but Aristobulus says, on the contrary, that they frequently travel into Babylonia on rafts, and thence sail up the Euphrates to Thapsacus[683] with their cargoes, but afterwards carry them by land to all parts of the country.

4. "On sailing further, there are other islands, Tyre[684] and Aradus,[685] which have temples resembling those of the Phœnicians. The inhabitants of these islands (if we are to believe them) say that the islands and cities bearing the same name as those of the Phœnicians are their own colonies.[686] These islands are distant from Teredon ten days' sail, and from the promontory at the mouth of the gulf at Macæ one day's sail.

5. "Nearchus and Orthagoras relate, that an island Ogyris lies to the south, in the open sea, at the distance of 2000 stadia[687] from Carmania. In this island is shown the sepulchre of Erythras, a large mound, planted with wild palms. He [Pg 188] [CAS. 767]was king of the country, and the sea received its name from him. It is said that Mithropastes, the son of Arsites, satrap of Phrygia, pointed out these things to them. Mithropastes was banished by Darius, and resided in this island; he joined himself to those who had come down to the Persian Gulf, and hoped through their means to have an opportunity of returning to his own country.

6. "Along the whole coast of the Red Sea, in the deep part of the water grow trees resembling the laurel and the olive. When the tide ebbs, the whole trees are visible above the water, and at the full tide they are sometimes entirely covered. This is the more singular because the coast inland has no trees."

This is the description given by Eratosthenes of the Persian Sea, which forms, as we have said, the eastern side of Arabia Felix.

7. Nearchus says, that they were met by Mithropastes, in company with Mazenes, who was governor of one of the islands, called Doracta (Oaracta?)[688] in the Persian Gulf; that Mithropastes, after his retreat from Ogyris, took refuge there, and was hospitably received; that he had an interview with Mazenes, for the purpose of being recommended to the Macedonians, in the fleet of which Mazenes was the guide.

Nearchus also mentions an island, met with at the recommencement of the voyage along the coast of Persia, where are found pearls in large quantities and of great value; in other islands there are transparent and brilliant pebbles; in the islands in front of the Euphrates there are trees which send forth the odour of frankincense, and from their roots, when bruised, a (perfumed) juice flows out; the crabs and sea hedgehogs are of vast size, which is common in all the exterior seas, some being larger than Macedonian hats;[689] others of the capacity of two cotyli; he says also that he had seen driven on shore a whale fifty cubits in length.

[Pg 189]

CHAPTER IV.

1. ARABIA commences on the side of Babylonia with Mæcene.[690] In front of this district, on one side lies the desert of the Arabians, on the other are the marshes[691]opposite to the Chaldæans, formed by the overflowing of the Euphrates, and in another direction is the Sea of Persia. This country has an unhealthy and cloudy atmosphere; it is subject to showers, and also to scorching heat; still its products are excellent. The vine grows in the marshes; as much earth as the plant may require is laid upon hurdles of reeds;[692] the hurdle is frequently carried away by the water, and is then forced back again by poles to its proper situation.

2. I return to the opinions of Eratosthenes, which he next delivers respecting Arabia. He is speaking of the northern and desert part, lying between Arabia Felix, Cœle-Syria, and Judæa, to the recess of the Arabian Gulf.

From Heroopolis, situated in that recess of the Arabian Gulf which is on the side of the Nile, to Babylon, towards Petra of the Nabatæi, are 5600 stadia. The whole tract lies in the direction of the summer solstice (i. e. east and west), and passes through the adjacent Arabian tribes, namely Nabatæi, Chaulotæi, and Agræi. Above these people is Arabia Felix, stretching out 12,000 stadia towards the south to the Atlantic Sea.[693]

[Pg 190]

[CAS. 768]The first people, next after the Syrians and Jews, who occupy this country are husbandmen. These people are succeeded by a barren and sandy tract, producing a few palms, the acanthus,[694] and tamarisk; water is obtained by digging [wells] as in Gedrosia. It is inhabited by Arabian Scenitæ, who breed camels. The extreme parts towards the south, and opposite to Ethiopia, are watered by summer showers, and are sowed twice, like the land in India. Its rivers are exhausted in watering plains, and by running into lakes. The general fertility of the country is very great; among other products, there is in particular an abundant supply of honey; except horses,[695]there are numerous herds of animals, mules (asses?), and swine; birds also of every kind, except geese and the gallinaceous tribe.

Four of the most populous nations inhabit the extremity of the above-mentioned country; namely, the Minæi the part towards the Red Sea, whose largest city is Carna or Carnana.[696] Next to these are the Sabæans, whose chief city is Mariaba.[697] The third nation are the Cattabaneis,[698] extending to the straits and the passage across the Arabian Gulf. Their royal seat is called Tamna. The Chatramotitæ[699] are the furthest of these nations towards the east. Their city is Sabata.

3. All these cities are governed by one monarch, and are flourishing. They are adorned with beautiful temples and palaces. Their houses, in the mode of binding the timbers together, are like those in Egypt. The four countries comprise a greater territory than the Delta of Egypt.[700]

The son does not succeed the father in the throne, but the son who is born in a family of the nobles first after the accession of the king. As soon as any one is invested with the government, the pregnant wives of the nobles are registered, and guardians are appointed to watch which of them is first delivered of a son. The custom is to adopt and educate the [Pg 191]child in a princely manner as the future successor to the throne.

4. Cattabania produces frankincense, and Chatramotitis myrrh; these and other aromatics are the medium of exchange with the merchants. Merchants arrive in seventy days at Minæa from Ælana.[701] Ælana is a city on the other recess of the Arabian Gulf, which is called Ælanites, opposite to Gaza, as we have before described it.[702] The Gerrhæi arrive in Chatramotitis in forty days.

The part of the Arabian Gulf along the side of Arabia, if we reckon from the recess of the Ælanitic bay, is, according to the accounts of Alexander and Anaxicrates, 14,000 stadia in extent; but this computation is too great. The part opposite to Troglodytica, which is on the right hand of those who are sailing from Heroopolis[703] to Ptolemaïs, to the country where elephants are taken, extends 9000 stadia to the south, and inclines a little towards the east. Thence to the straits are about 4500 stadia, in a direction more towards the east. The straits at Ethiopia are formed by a promontory called Deire.[704] There is a small town upon it of the same name. The Ichthyophagi inhabit this country. Here it is said is a pillar of Sesostris the Egyptian, on which is inscribed, in hieroglyphics, an account of his passage (across the Arabian Gulf). For he appears to have subdued first Ethiopia and Troglodytica,[705] and afterwards to have passed over into Arabia. He then overran the whole of Asia. Hence in many places there are dykes called the dykes of Sesostris, and temples built in honour of Egyptian deities.

The straits at Deire are contracted to the width of 60 stadia; not indeed that these are now called the Straits, for ships proceed to a further distance, and find a passage of about 200 [Pg 192]

[CAS. 769]stadia between the two continents;[706] six islands contiguous to one another leave a very narrow passage through them for vessels, by filling up the interval between the continents. Through these goods are transported from one continent to the other on rafts; it is this passage which is called the Straits. After these islands, the subsequent navigation is among bays along the Myrrh country, in the direction of south and east, as far as the Cinnamon country, a distance of about 5000 stadia;[707] beyond this district no one to this time, it is said, has penetrated. There are not many cities upon the coast, but in the interior they are numerous and well inhabited. Such is the account of Arabia given by Eratosthenes. We must add what is related also by other writers.

5. Artemidorus[708] says, that the promontory of Arabia, opposite[Pg 193] to Deire, is called Acila,[709] and that the persons who live near Deire deprive themselves of the prepuce.

In sailing from Heroopolis along Troglodytica, a city is met with called Philotera,[710] after the sister of the second Ptolemy; it was founded by Satyrus, who was sent to explore the hunting-ground for the elephants, and Troglodytica itself. Next to this is another city, Arsinoë; and next to this, springs of hot water, which are salt and bitter; they are precipitated from a high rock, and discharge themselves into the sea. There is in a plain near (these springs) a mountain, which is of a red colour like minium. Next is Myus Hormus, which is also called Aphrodites Hormus;[711] it is a large harbour with an oblique entrance. In front are three islands; two are covered with olive trees, and one (the third) is less shaded with trees, and abounds with guinea-fowls.[712] Then follows Acathartus (or Foul Bay), which, like Myus Hormus, is in the latitude of the Thebaïs. The bay is really foul, for it is very dangerous from rocks (some of which are covered by the sea, others rise to the surface), as also from almost constant and furious tempests. At the bottom of the bay is situated the city Berenice.[713]

6. After the bay is the island Ophiodes,[714] so called from the accidental circumstance [of its having once been infested with serpents]. It was cleared of the serpents by the king,[715] on account of the destruction occasioned by those noxious animals to the persons who frequented the island, and on account of the topazes found there. The topaz is a transparent stone, sparkling with a golden lustre, which however is not easy to be distinguished in the day-time, on account of the brightness of the surrounding light, but at night the stones are visible to those who collect them. The collectors place a vessel over the spot [where the topazes are seen] as a mark, and dig them up in the day. A body of men was appointed and maintained by the kings of Egypt to guard the place [Pg 194] [CAS. 770]where these stones were found, and to superintend the collection of them.

7. Next after this island follow many tribes of Ichthyophagi and of Nomades; then succeeds the harbour of the goddess Soteira (the Preserver), which had its name from the circumstance of the escape and preservation of some masters [of vessels] from great dangers by sea.

After this the coast and the gulf seem to undergo a great change: for the voyage along the coast is no longer among rocks, and approaches almost close to Arabia; the sea is so shallow as to be scarcely of the depth of two orguiæ,[716] and has the appearance of a meadow, in consequence of the sea-weeds, which abound in the passage, being visible through and under the water. Even trees here grow from under the water, and the sea abounds with sea-dogs.

Next are two mountains,[717] the Tauri (or the Bulls), presenting at a distance a resemblance to these animals. Then follows another mountain, on which is a temple of Isis, built by Sesostris; then an island planted with olive trees, and at times overflowed. This is followed by the city Ptolemaïs, near the hunting-grounds of the elephants,[718] founded by Eumedes, who was sent by Philadelphus to the hunting-ground. He enclosed, without the knowledge of the inhabitants, a kind of peninsula with a ditch and wall, and by his courteous address gained over those who were inclined to obstruct the work, and instead of enemies made them his friends.

8. In the intervening space, a branch of the river Astaboras[719] discharges itself. It has its source in a lake, and empties part of its waters [into the bay], but the larger portion it contributes to the Nile. Then follow six islands, called Latomiæ,[720] after these the Sabaïtic mouth,[721] as it is called, and [Pg 195]in the inland parts a fortress built by Suchus.[722] Then a lake called Elæa, and the island of Strato;[723] next Saba[724] a port, and a hunting-ground for elephants of the same name. The country deep in the interior is called Tenessis. It is

occupied by those Egyptians who took refuge from the government of Psammitichus.[725] They are surnamed Sembritæ,[726] as being strangers. They are governed by a queen, to whom also Meroë, an island in the Nile near these places, is subject. Above this, at no great distance, is another island in the river, a settlement occupied by the same fugitives. From Meroë to this sea is a journey of fifteen days for an active person.

Near Meroë is the confluence of the Astaboras,[727] the Astapus,[728] and of the Astasobas with the Nile.

9. On the banks of these rivers live the Rhizophagi (or root-eaters) and Heleii (or marsh-men). They have their name from digging roots in the adjacent marsh, bruising them with stones, and forming them into cakes, which they dry in the sun for food. These countries are the haunts of lions. The wild beasts are driven out of these places, at the time of the rising of the dog-star, by large gnats.

Near these people live the Spermophagi (or seed-eaters), who, when seeds of plants fail, subsist upon seeds of trees,[729] [Pg 196] [CAS. 771]which they prepare in the same manner as the Rhizophagi prepare their roots.

Next to Elæa are the watch-towers of Demetrius, and the altars of Conon. In the interior Indian reeds grow in abundance. The country there is called the country of Coracius.

Far in the interior was a place called Endera, inhabited by a naked tribe,[730] who use bows and reed arrows, the points of which are hardened in the fire. They generally shoot the animals from trees, sometimes from the ground. They have numerous herds of wild cattle among them, on the flesh of which they subsist, and on that of other wild animals. When they have taken nothing in the chase, they dress dried skins upon hot coals, and are satisfied with food of this kind. It is their custom to propose trials of skill in archery for those who have not attained manhood.

Next to the altars of Conon is the port of Melinus, and above it is a fortress called that of Coraus and the chase of Coraus, also another fortress and more hunting-grounds. Then follows the harbour of Antiphilus, and above this a tribe, the Creophagi, deprived of the prepuce, and the women are excised after the Jewish custom.[731]

10. Further still towards the south are the Cynamolgi,[732] called by the natives Agrii, with long hair and long beards, who keep a breed of very large dogs for hunting the Indian cattle which come into their country from the neighbouring district, driven thither either by wild beasts or by scarcity of pasturage. The time of their incursion is from the summer solstice to the middle of winter.

Next to the harbour of Antiphilus is a port called the Grove of the Colobi (or the Mutilated), the city Berenice[733] of [Pg 197]Sabæ, and Sabæ[734] a considerable city; then the grove of Eumenes.[735]

Above is the city Darada, and a hunting-ground for elephants, called "At the Well." The district is inhabited by the Elephantophagi (or Elephant-eaters), who are occupied in hunting them. When they descry from the trees a herd of elephants directing their course through the forest, they do not [then] attack, but they approach by stealth and hamstring the hindmost stragglers from the herd. Some kill them with bows and arrows, the latter being dipped in the gall of serpents. The shooting with the bow is performed by three men, two, advancing in front, hold the bow, and one draws the string. Others remark the trees against which the elephant is accustomed to rest, and, approaching on the opposite side, cut the trunk of the tree low down. When the animal comes and leans against it, the tree and the elephant fall down together. The elephant is unable to rise, because its legs are formed of one piece of bone which is inflexible; the hunters leap down from the trees, kill it, and cut it in pieces. The Nomades call the hunters Acatharti, or impure.

11. Above this nation is situated a small tribe the Struthophagi[736] (or Bird-eaters), in whose country are birds of the size of deer, which are unable to fly, but run with the swiftness of the ostrich. Some hunt them with bows and arrows, others covered with the skins of birds. They hide the right hand in the neck of the skin, and move it as the birds move their necks. With the left hand they scatter grain from a bag suspended to the side; they thus entice the birds, till they drive them into pits, where the hunters despatch them with cudgels. The skins are used both as clothes and as coverings for beds. The Ethiopians called Simi are at war with these people, and use as weapons the horns of antelopes.

12. Bordering on this people is a nation blacker in complexion than the others,[737]shorter in stature, and very short-lived. They rarely live beyond forty years; for the flesh [Pg 198]

[CAS. 772]of their bodies is eaten up with worms.[738] Their food consists of locusts, which the south-west and west winds, when they blow violently in the spring-time, drive in bodies into the country. The inhabitants catch them by throwing into the ravines materials which cause a great deal of smoke, and light them gently. The locusts, as they fly across the smoke, are blinded and fall down. They are pounded with salt, made into cakes, and eaten as food.

Above these people is situated a desert tract with extensive pastures. It was abandoned in consequence of the multitudes of scorpions and tarantulas, called tetragnathi (or four-jawed), which formerly abounded to so great a degree as to occasion a complete desertion of the place long since by its inhabitants.

13. Next to the harbour of Eumenes, as far as Deire and the straits opposite the six islands,[739] live the Ichthyophagi, Creophagi, and Colobi, who extend into the interior.

Many hunting-grounds for elephants, and obscure cities and islands, lie in front of the coast.

The greater part are Nomades; husbandmen are few in number. In the country occupied by some of these nations styrax grows in large quantity. The Icthyophagi, on the ebbing of the tide, collect fish, which they cast upon the rocks and dry in the sun. When they have well broiled them, the bones are piled in heaps, and the flesh trodden with the feet is made into cakes, which are again exposed to the sun and used as food. In bad weather, when fish cannot be procured, the bones of which they have made heaps are pounded, made into cakes and eaten, but they suck the fresh bones. Some also live upon shell-fish, when they are fattened, which is done by throwing them into holes and standing pools of the sea, where they are supplied with small fish, and used as food when other fish are scarce. They have various kinds of places for preserving and feeding fish, from whence they derive their supply.

Some of the inhabitants of that part of the coast which is without water go inland every five days, accompanied by all [Pg 199]their families, with songs and rejoicings, to the watering-places, where, throwing themselves on their faces, they drink as beasts until their stomachs are distended like a drum. They then return again to the sea-coast. They dwell in caves or cabins, with roofs consisting of beams and rafters made of the bones and spines of whales, and covered with branches of the olive tree.

14. The Chelonophagi (or Turtle-eaters) live under the cover of shells (of turtles), which are large enough to be used as boats. Some make of the sea-weed, which is thrown up in large quantities, lofty and hill-like heaps, which are hollowed out, and underneath which they live. They cast out the dead, which are carried away by the tide, as food for fish.

There are three islands which follow in succession, the island of Tortoises, the island of Seals, and the island of Hawks. Along the whole coast there are plantations of palm trees, olive trees, and laurels, not only within, but in a great part also without the straits.

There is also an island [called the island] of Philip, opposite to it inland is situated the hunting-ground for elephants, called the chase of Pythangelus; then follows Arsinoë, a city with a harbour; after these places is Deire, and beyond them is a hunting-ground for elephants.

From Deire, the next country is that which bears aromatic plants. The first produces myrrh, and belongs to the Icthyophagi and the Creophagi. It bears also the persea, peach or Egyptian almond,[740] and the Egyptian fig. Beyond is Licha, a hunting-ground for elephants. There are also in many places standing pools of rain-water. When these are dried up, the elephants, with their trunks and tusks, dig holes and find water.

On this coast there are two very large lakes extending as far as the promontory Pytholaus.[741] One of them contains salt water, and is called a sea; the other, fresh water, and is the haunt of hippopotami and crocodiles. On the margin grows the papyrus. The ibis is seen in the neighbourhood of this place. The people who live near the promontory of Pytholaus (and beginning from this place) do not [Pg 200]

[CAS. 774]undergo any mutilation in any part of their body. Next is the country which produces frankincense; it has a promontory and a temple with a grove of poplars. In the inland parts is a tract along the banks of a river bearing the name of Isis, and another that of Nilus,[742] both of which produce myrrh and frankincense. Also a lagoon filled with water from the mountains; next the watch-post of the Lion, and the port of Pythangelus. The next tract bears the false cassia. There are many tracts in succession on the sides of rivers on which frankincense grows, and rivers extending to the cinnamon country. The river which bounds this tract produces (phlous) rushes[743] in great abundance. Then follows another river, and the port of Daphnus,[744] and a valley called Apollo's, which bears, besides frankincense, myrrh and cinnamon. The latter is more abundant in places far in the interior.

Next is the mountain Elephas,[745] a mountain projecting into the sea, and a creek; then follows the large harbour of Psygmus, a watering-place called that of Cynocephali, and the last promontory of this coast, Notu-ceras (or the Southern Horn).[746] After doubling this cape towards the south, we have [Pg 201]no more descriptions, he says, of harbours or places, because nothing is known of the sea-coast beyond this point.[747]

15. Along the coast there are both pillars and altars of Pytholaus, Lichas, Pythangelus, Leon, and Charimortus, that is, along the known coast from Deire as far as Notu-ceras; but the distance is not determined. The country abounds with elephants and lions called myrmeces (ants).[748] They have their genital organs reversed. Their skin is of a golden colour, but they are more bare than the lions of Arabia.

It produces also leopards of great strength and courage, and the rhinoceros. The rhinoceros is little inferior to the elephant; not, according to Artemidorus, in length to the crest,[749] although he says he had seen one at Alexandreia, but it is somewhat about [* * * less][750] in height, judging at least from the one I saw. Nor is the colour the pale yellow of box-wood, but like that of the elephant.[751] It was of the size of a bull. Its shape approached very nearly to that of the wild boar, and particularly the forehead; except the front, which is furnished with a hooked horn, harder than any bone. It uses it as a weapon, like the wild boar its tusks. It has also two hard welts, like folds of serpents, encircling the body from the chine to the belly, one on the withers, the other on the loins. This description is taken from one which I myself saw. Artemidorus adds to his account of this animal, that it is peculiarly inclined to dispute with the elephant for the place of pasture; thrusting its forehead under the belly [of the elephant] and ripping it up, unless prevented by the trunk and tusks of his adversary.

16. Camel-leopards are bred in these parts, but they do not in any respect resemble leopards, for their variegated skin is more like the streaked and spotted skin of fallow deer. The [Pg 202] [CAS. 775]hinder quarters are so very much lower than the fore quarters, that it seems as if the animal sat upon its rump, which is the height of an ox; the fore legs are as long as those of the camel. The neck rises high and straight up, but the head greatly exceeds in height that of the camel. From this want of proportion, the speed of the animal is not so great, I think, as it is described by Artemidorus, according to whom it is not to be surpassed. It is not however a wild animal, but rather like a domesticated beast; for it shows no signs of a savage disposition.

This country, continues Artemidorus, produces also sphinxes,[752] cynocephali,[753]and cebi,[754] which have the face of a lion, and the rest of the body like that of a panther; they are as large as deer. There are wild bulls also, which are carnivorous, and greatly exceed ours in size and swiftness. They are of a red colour. The crocuttas[755] is, according to this author, the mixed progeny of a wolf and a dog. What Metrodorus the Scepsian relates, in his book "on Custom," is like fable, and is to be disregarded.

Artemidorus mentions serpents also of thirty cubits in length, which can master elephants and bulls: in this he does not exaggerate.[756] But the Indian and African serpents are of a more fabulous size, and are said to have grass growing on their backs.

17. The mode of life among the Troglodytæ is nomadic. Each tribe is governed by tyrants. Their wives and children are common, except those of the tyrants. The offence of corrupting the wife of a tyrant is punished with the fine of a sheep.

The women carefully paint themselves with antimony. They wear about their necks shells, as a protection against fascination by witchcraft. In their quarrels, which are for pastures, they first push away each other with their hands, they then use stones, or, if wounds are inflicted, arrows and daggers. The women put an end to these disputes, by going into the midst of the combatants and using prayers and entreaties.

[Pg 203]Their food consists of flesh and bones pounded together, wrapped up in skins and then baked, or prepared after many other methods by the cooks, who are called Acatharti, or impure. In this way they eat not only the flesh, but the bones and skins also.

They use (as an ointment for the body?) a mixture of blood and milk; the drink of the people in general is an infusion of the paliurus (buckthorn);[757] that of the tyrants is mead; the honey being expressed from some kind of flower.

Their winter sets in when the Etesian winds begin to blow (for they have rain), and the remaining season is summer.

They go naked, or wear skins only, and carry clubs. They deprive themselves of the prepuce,[758] but some are circumcised like Egyptians. The Ethiopian Megabari have their clubs armed with iron knobs. They use spears and shields which are covered with raw hides. The other Ethiopians use bows and lances. Some of the Troglodytæ, when they bury their dead, bind the body from the neck to the legs with twigs of the buckthorn. They then immediately throw stones over the body, at the same time laughing and rejoicing, until they have covered the face. They then place over it a ram's horn, and go away.

They travel by night; the male cattle have bells fastened to them, in order to drive away wild beasts with the sound. They use torches also and arrows in repelling them. They watch during the night, on account of their flocks, and sing some peculiar song around their fires.

18. Having given this account of the Troglodytæ and of the neighbouring Ethiopians, Artemidorus returns to the Arabians. Beginning from Poseidium, he first describes those who border upon the Arabian Gulf, and are opposite to the Troglodytæ. He says that Poseidium is situated within the bay of [Heroopolis],[759]and that contiguous to Poseidium[760] is a grove of palm trees,[761] well supplied with water, which is [Pg 204] [CAS. 776]highly valued, because all the district around is burnt up and is without water or shade. But there the fertility of the palm is prodigious. A man and a woman are appointed by hereditary right to the guardianship of the grove. They wear skins, and live on dates. They sleep in huts built on trees, the place being infested with multitudes of wild beasts.

Next is the island of Phocæ (Seals),[762] which has its name from those animals, which abound there. Near it is a promontory,[763] which extends towards Petra, of the Arabians called Nabatæi, and to the country of Palestine, to this [island] the Minæi,[764] Gerrhæi, and all the neighbouring nations repair with loads of aromatics.

Next is another tract of sea-coast, formerly called the coast of the Maranitæ,[765]some of whom were husbandmen, others Scenitæ; but at present it is occupied by Garindæi, who destroyed the former possessors by treachery. They attacked those who were assembled to celebrate some quinquennial festival, and put them to death; they then attacked and exterminated the rest of the tribe.[766]

Next is the Ælanitic[767] Gulf and Nabatæa, a country well peopled, and abounding in cattle. The islands which lie near, and opposite, are inhabited by people who formerly lived without molesting others, but latterly carried on a piratical warfare in rafts[768] against vessels on their way from Egypt. But they suffered reprisals, when an armament was sent out against them, which devastated their country.

[Pg 205]Next is a plain, well wooded and well supplied with water; it abounds with cattle of all kinds, and, among other animals, mules, wild camels, harts, and hinds; lions also, leopards, and wolves are frequently to be found. In front lies an island called Dia. Then follows a bay of about 500 stadia in extent, closed in by mountains, the entrance into which is of difficult access. About it live people who are hunters of wild animals.

Next are three desert islands, abounding with olive trees, not like those in our own country, but an indigenous kind, which we call Ethiopic olives, the tears (or gum) of which have a medicinal virtue.

Then follows a stony beach, which is succeeded by a rugged coast,[769] not easily navigated by vessels, extending about 1000 stadia. It has few harbours and anchorages, for a rugged and lofty mountain stretches parallel to it; then the parts at its base, extending into the sea, form rocks under water, which, during the blowing of the Etesian winds and the storms of that period, present dangers, when no assistance can be afforded to vessels.

Next is a bay in which are some scattered islands,⁷⁷⁰ and continuous with the bay, are three very lofty mounds⁷⁷¹ of black sand. After these is Charmothas⁷⁷² a harbour, about 100 stadia in circumference, with a narrow entrance very dangerous for all kinds of vessels. A river empties itself into it. In the middle is a well-wooded island, adapted for cultivation.

Then follows a rugged coast, and after that are some bays and a country belonging to Nomades, who live by their camels. They fight from their backs; they travel upon them, and subsist on their milk and flesh. A river flows [Pg 206] [CAS. 777]through their country, which brings down gold-dust, but they are ignorant how to make any use of it. They are called Debæ;⁷⁷³ some of them are Nomades, others husbandmen.

I do not mention the greater part⁷⁷⁴ of the names of these nations, on account of the obscurity of the people, and because the pronunciation of them is strange⁷⁷⁵ [and uncouth].

Near these people is a nation more civilized, who inhabit a district with a more temperate climate; for it is well watered, and has frequent showers.⁷⁷⁶ Fossil gold is found there, not in the form of dust, but in lumps, which do not require much purification. The least pieces are of the size of a nut, the middle size of a medlar, the largest of a walnut. These are pierced and arranged alternately with transparent stones strung on threads and formed into collars. They are worn round the neck and wrists. They sell the gold to their neighbours at a cheap rate, exchanging it for three times the quantity of brass, and double the quantity of iron,⁷⁷⁷ through ignorance of the mode of working the gold, and the scarcity of the commodities received in exchange, which are more necessary for the purposes of life.

19. The country of the Sabæi,⁷⁷⁸ a very populous nation, is contiguous, and is the most fertile of all, producing myrrh, frank[Pg 207]incense, and cinnamon. On the coast is found balsamum and another kind of herb of a very fragrant smell, but which is soon dissipated. There are also sweet-smelling palms and the calamus. There are snakes also of a dark red colour, a span in length, which spring up as high as a man's waist, and whose bite is incurable.

On account of the abundance which the soil produces, the people are lazy and indolent in their mode of life. The lower class of people live on roots, and sleep on the trees.

The people who live near each other receive, in continued succession, the loads [of perfumes] and deliver them to others, who convey them as far as Syria and Mesopotamia. When the carriers become drowsy by the odour of the aromatics, the drowsiness is removed by the fumes of asphaltus and of goat's beard.

Mariaba,⁷⁷⁹ the capital of the Sabæans, is situated upon a mountain, well wooded. A king resides there, who determines absolutely all disputes and other matters; but he is forbidden to leave his palace, or if he does so, the rabble immediately assail him with stones, according to the direction of an oracle. He himself, and those about his person, pass their lives in effeminate voluptuousness.

The people cultivate the ground, or follow the trade of dealing in aromatics, both the indigenous sort and those brought from Ethiopia; in order to procure them, they sail through the straits in vessels covered with skins. There is such an abundance of these aromatics, that cinnamon, cassia, and other spices are used by them instead of sticks and firewood.

In the country of the Sabæans is found the larimnum, a most fragrant perfume.

By the trade [in these aromatics] both the Sabæans and the Gerrhæi have become the richest of all the tribes, and possess a great quantity of wrought articles in gold and silver, [Pg 208] [CAS. 778]as couches, tripods, basins, drinking-vessels, to which we must add the costly magnificence of their houses; for the doors, walls, and roofs are variegated with inlaid ivory, gold, silver, and precious stones.

This is the account of Artemidorus.⁷⁸⁰ The rest of the description is partly similar to that of Eratosthenes, and partly derived from other historians.

20. Some of these say, that the sea is red from the colour arising from reflection either from the sun, which is vertical, or from the mountains, which are red by being scorched with

intense heat; for the colour, it is supposed, may be produced by both these causes. Ctesias of Cnidus speaks of a spring which discharges into the sea a red and ochrous water. Agatharchides, his fellow-citizen, relates, on the authority of a person of the name of Boxus, of Persian descent, that when a troop of horses was driven by a lioness in heat as far as the sea, and had passed over to an island, a Persian of the name of Erythras constructed a raft, and was the first person who crossed the sea to it; perceiving the island to be well adapted for inhabitants, he drove the herd back to Persia, and sent out colonists both to this and the other islands and to the coast. He [thus] gave his own name to the sea. But according to others, it was Erythras the son of Perseus who was the king of this country.

According to some writers, from the straits in the Arabian Gulf to the extremity of the cinnamon country is a distance of 5000 stadia,[781] without distinguishing whether (the direction is) to the south or to the east.

It is said also that the emerald and the beryl are found in the gold mines. According to Poseidonius, an odoriferous salt is found in Arabia.

[Pg 209]21. The Nabatæans and Sabæans, situated above Syria, are the first people who occupy Arabia Felix. They were frequently in the habit of overrunning this country before the Romans became masters of it, but at present both they and the Syrians are subject to the Romans.

The capital of the Nabatæans is called Petra. It is situated on a spot which is surrounded and fortified by a smooth and level rock (petra), which externally is abrupt and precipitous, but within there are abundant springs of water both for domestic purposes and for watering gardens. Beyond the enclosure the country is for the most part a desert, particularly towards Judæa. Through this is the shortest road to Jericho, a journey of three or four days, and five days to the Phœnicon (or palm plantation). It is always governed by a king of the royal race. The king has a minister who is one of the Companions, and is called Brother. It has excellent laws for the administration of public affairs.

Athenodorus, a philosopher, and my friend, who had been at Petra, used to relate with surprise, that he found many Romans and also many other strangers residing there. He observed the strangers frequently engaged in litigation, both with one another and with the natives; but the natives had never any dispute amongst themselves, and lived together in perfect harmony.

22. The late expedition[782] of the Romans against the Arabians, under the command of Ælius Gallus, has made us acquainted with many peculiarities of the country. Augustus Cæsar despatched this general to explore the nature of these [Pg 210] [CAS. 779]places and their inhabitants, as well as those of Ethiopia; for he observed that Troglodytica, which is contiguous to Egypt, bordered upon Ethiopia; and that the Arabian Gulf was extremely narrow, where it separates the Arabians from the Troglodytæ. It was his intention either to conciliate or subdue the Arabians. He was also influenced by the report, which had prevailed from all time, that this people were very wealthy, and exchanged their aromatics and precious stones for silver and gold, but never expended with foreigners any part of what they received in exchange. He hoped to acquire either opulent friends, or to overcome opulent enemies. He was moreover encouraged to undertake this enterprise by the expectation of assistance from the Nabatæans, who promised to co-operate with him in everything.

23. Upon these inducements Gallus set out on the expedition. But he was deceived by Syllæus, the [king's] minister of the Nabatæans, who had promised to be his guide on the march, and to assist him in the execution of his design. Syllæus was however treacherous throughout; for he neither guided them by a safe course by sea along the coast, nor by a safe road for the army, as he promised, but exposed both the fleet and the army to danger, by directing them where there was no road, or the road was impracticable, where they were obliged to make long circuits, or to pass through tracts of country destitute of everything; he led the fleet along a rocky coast without harbours, or to places abounding with rocks concealed under water, or with shallows. In places of this description particularly, the flowing and ebbing of the tide did them the most harm.

The first mistake consisted in building long vessels [of war] at a time when there was no war, nor any likely to occur by sea. For the Arabians, being mostly engaged in traffic and commerce, are not a very warlike people even on land, much less so at sea. Gallus, notwithstanding, built not less than eighty biremes and triremes and galleys (phaseli) at Cleopatris,[783] near the old canal which leads from the Nile. When he discovered his mistake, he constructed a hundred and thirty vessels of burden, in which he embarked with about ten thousand infantry, collected from Egypt, consisting of Romans and allies, among whom were five hundred Jews and [Pg 211]a thousand Nabatæans, under the command of Syllæus. After enduring great hardships and distress, he arrived on the fifteenth day at Leuce-Come, a large mart in the territory of the Nabatæans, with the loss of many of his vessels, some with all their crews, in consequence of the difficulty of the navigation, but by no opposition from an enemy. These misfortunes were occasioned by the perfidy of Syllæus, who insisted that there was no road for an army by land to Leuce-Come, to which and from which place the camel-traders travel with ease and in safety from Petra, and back to Petra, with so large a body of men and camels as to differ in no respect from an army.

24. Another cause of the failure of the expedition was the fact of king Obodas not paying much attention to public affairs, and especially to those relative to war (as is the custom with all Arabian kings), but placed everything in the power of Syllæus the minister. His whole conduct in command of the army was perfidious, and his object was, as I suppose, to examine as a spy the state of the country, and to destroy, in concert with the Romans, certain cities and tribes: and when the Romans should be consumed by famine, fatigue, and disease, and by all the evils which he had treacherously contrived, to declare himself master of the whole country.

Gallus however arrived at Leuce-Come, with the army labouring under stomacacce and scelotyrbe, diseases of the country, the former affecting the mouth, the other the legs, with a kind of paralysis, caused by the water and the plants [which the soldiers had used in their food]. He was therefore compelled to pass the summer and the winter there, for the recovery of the sick.

Merchandise is conveyed from Leuce-Come to Petra, thence to Rhinocolura in Phœnicia, near Egypt, and thence to other nations. But at present the greater part is transported by the Nile to Alexandreia. It is brought down from Arabia and India to Myus Hormus, it is then conveyed on camels to Coptus[784] of the Thebaïs, situated on a canal of the Nile, and to Alexandreia. Gallus, setting out again from Leuce-Come on his return with his army, and through the treachery of his guide, traversed such tracts of country, that the army was obliged to carry water with them upon camels. After a [Pg 212] [CAS. 781]march of many days, therefore, he came to the territory of Aretas, who was related to Obodas. Aretas received him in a friendly manner, and offered presents. But by the treachery of Syllæus, Gallus was conducted by a difficult road through the country; for he occupied thirty days in passing through it. It afforded barley, a few palm trees, and butter instead of oil.

The next country to which he came belonged to Nomades, and was in great part a complete desert. It was called Ararene. The king of the country was Sabos. Gallus spent fifty days in passing through this territory, for want of roads, and came to a city of the Negrani, and to a fertile country peacefully disposed. The king had fled, and the city was taken at the first onset. After a march of six days from thence, he came to the river. Here the barbarians attacked the Romans, and lost about ten thousand men; the Romans lost only two men. For the barbarians were entirely inexperienced in war, and used their weapons unskilfully, which were bows, spears, swords, and slings; but the greater part of them wielded a double-edged axe. Immediately afterwards he took the city called Asca, which had been abandoned by the king. He thence came to a city Athrula, and took it without resistance; having placed a garrison there, and collected provisions for the march, consisting of corn and dates, he proceeded to a city Marsiaba, belonging to the nation of the Rhammanitæ, who were subjects of Ilasarus. He assaulted and besieged it for six days, but raised the siege in consequence of a scarcity of water. He was two days' march from the aromatic region, as he was informed by his prisoners. He occupied in his marches a period of six months, in consequence of the treachery of his guides. This he discovered when he was returning; and although he was late in discovering the design against him, he had time to take another road back; for he arrived in nine days at Negrana, where the battle was fought, and thence in eleven days he came to the "Seven Wells," as the place is called from the fact of their existing there. Thence he marched through a desert country, and came to Chaalla a village, and then to another called Malothas, situated on a river. His road then lay through a desert country, which had only a few watering-places, as far as Egra[785] [Pg 213]a village. It belongs to the territory of Obodas, and is situated upon the sea. He accomplished on his return the whole distance in sixty days, in which, on his first journey, he had consumed six months. From Negra he conducted his army in eleven days to Myus Hormus; thence across the country to Coptus, and arrived at Alexandreia with so much of his army could be saved. The remainder he lost, not by the enemy, but by disease, fatigue, famine, and marches through bad roads; for seven men only perished in battle. For these reasons this expedition contributed little in extending our knowledge of the country. It was however of some small service.

Syllæus, the author of these disasters, was punished for his treachery at Rome. He affected friendship, but he was convicted of other offences, besides perfidy in this instance, and was beheaded.

25. The aromatic country, as I have before said,[786] is divided into four parts. Of aromatics, the frankincense and myrrh are said to be the produce of trees, but cassia the growth of bushes; yet some writers say, that the greater part (of the cassia) is brought from India, and that the best frankincense is that from Persia.

According to another partition of the country, the whole of Arabia Felix is divided into five kingdoms (or portions), one of which comprises the fighting men, who fight for all the rest; another contains the husbandmen, by whom the rest are supplied with food; another includes those who work at mechanical trades. One division comprises the myrrh region; another the frankincense region, although the same tracts produce cassia, cinnamon, and nard. Trades are not changed from one family to another, but each workman continues to exercise that of his father.

The greater part of their wine is made from the palm.

A man's brothers are held in more respect than his children. The descendants of the royal family succeed as kings, and are invested with other governments, according to primogeniture. Property is common among all the relations. The eldest is the chief. There is one wife among them all. He who enters [Pg 214] [CAS. 783]the house before any of the rest, has intercourse with her, having placed his staff at the door; for it is a necessary custom, which every one is compelled to observe, to carry a staff. The woman however passes the night with the eldest. Hence the male children are all brothers. They have sexual intercourse also with their mothers. Adultery is punished with death, but an adulterer must belong to another family.

A daughter of one of the kings was of extraordinary beauty, and had fifteen brothers, who were all in love with her, and were her unceasing and successive visitors; she, being at last weary of their importunity, is said to have employed the following device. She procured staves to be made similar to those of her brothers; when one left the house, she placed before the door a staff similar to the first, and a little time afterwards another, and so on in succession, but making her calculation so that the person who intended to visit her might not have one similar to that at her door. On an occasion when the brothers were all of them together at the market-place, one left it, and came to the door of the house; seeing the staff there, and conjecturing some one to be in her apartment, and having left all the other brothers at the market-place, he suspected the person to be an adulterer; running therefore in haste to his father, he brought him with him to the house, but it was proved that he had falsely accused his sister.

26. The Nabatæans are prudent, and fond of accumulating property. The community fine a person who has diminished his substance, and confer honours on him who has increased it. They have few slaves, and are served for the most part by their relations, or by one another, or each person is his own servant; and this custom extends even to their kings. They eat their meals in companies consisting of thirteen persons. Each party is attended by two musicians. But the king gives many entertainments in great buildings. No one drinks more than eleven [appointed] cupfuls, from separate cups, each of gold.

The king courts popular favour so much, that he is not only his own servant, but sometimes he himself ministers to others. He frequently renders an account [of his administration] before the people, and sometimes an inquiry is made into his mode of life. [Pg 215]

The houses are sumptuous, and of stone. The cities are without walls, on account of the peace [which prevails among them]. A great part of the country is fertile, and produces everything except oil of olives; [instead of it], the oil of sesamum is used. The sheep have white fleeces, their oxen are large; but the country produces no horses.[787] Camels are the substitute for horses, and perform the [same kind of] labour. They wear no tunics, but have a girdle about the loins, and walk abroad in sandals.[788] The dress of the kings is the same, but the colour is purple.

Some merchandise is altogether imported into the country, others are not altogether imports, especially as some articles are native products, as gold and silver, and many of the aromatics; but brass and iron, purple garments, styrax, saffron, and costus (or white cinnamon), pieces of sculpture, paintings, statues, are not to be procured in the country.

They look upon the bodies of the dead as no better than dung, according to the words of Heracleitus, "dead bodies more fit to be cast out than dung;" wherefore they bury even their kings beside dung-heaps. They worship the sun, and construct the altar on the top of a house, pouring out libations and burning frankincense upon it every day.

27. When the poet says,

"I went to the country of the Ethiopians, Sidonians, and Erembi,"[789]

it is doubtful, what people he means by Sidonians, whether those who lived near the Persian Gulf, a colony from which nation are the Sidonians in our quarter (in the same manner as historians relate, that some Tyrian islanders are found there, and Aradii, from whom the Aradii in our country derive their origin), or whether the poet means actually the Sidonians themselves.

But there is more doubt about the Erembi, whether we are to suppose that he means the Troglodytæ, according to the opinion of those who, by a forced etymology, derive the word Erembi from ἔραν ἐμβαίνειν, that is, "entering into the earth," or whether he means the Arabians. Zeno the philosopher of our sect alters the reading in this manner,

[Pg   216]
[CAS. 784]

"And Sidoni, and Arabes;"

but Poseidonius alters it with a small variation,

"And Sidonii, and Arambi,"

as if the poet gave the name Arambi to the present Arabians, from their being so called by others in his time. He says also, that the situation of these three nations close to one another indicates a descent from some common stock, and that on this account they are called by names having a resemblance to one another, as Armenii, Aramæi, Arambi. For as we may suppose one nation to have been divided into three (according to the differences of latitude [in which they lived]), which successively became more marked [in proceeding from one to the other]), so in like manner we may suppose that several names were adopted in place of one. The proposed change of reading to Eremni is not probable, for that name is more applicable to the Ethiopians. The poet mentions also the Arimi, whom Poseidonius says are meant here, and not a place in Syria or Cilicia, or any other country, but Syria itself. For the Aramæi lived there. Perhaps these are the people whom the Greeks called Arimæi or Arimi. But the alterations of names, especially of barbarous nations, are frequent, Thus Darius was called Darieces; Parysatis, Pharziris; Athara, Atargata, whom Ctesias again calls Derceto.[790]

Alexander might be adduced to bear witness to the wealth of the Arabians, for he intended, it is said, after his return from India, to make Arabia the seat of empire. All his enterprises terminated with his death, which happened suddenly; but certainly one of his projects was to try whether the Arabians would receive him voluntarily, or resist him by force of arms; for having found that they did not send ambassadors to him, either before or after his expedition to India, he was beginning to make preparations for war, as we have said in a former part of this work.

BOOK XVII.
SUMMARY.
The Seventeenth Book contains the whole of Egypt and Africa.
CHAPTER I.

WHEN we were describing Arabia, we included in the description the gulfs which compress and make it a peninsula, namely the Gulfs of Arabia and of Persis. We described at the same time some parts of Egypt, and those of Ethiopia, inhabited by the Troglodytæ, and by the people situated next to them, extending to the confines of the Cinnamon country.[791]

We are now to describe the remaining parts contiguous to these nations, and situated about the Nile. We shall then give an account of Africa, which remains to complete this treatise on Geography.

And here we must previously adduce the opinions of Eratosthenes.

2. He says, that the Nile is distant from the Arabian Gulf towards the west 1000 stadia, and that it resembles (in its course) the letter N reversed. For after flowing, he says, about 2700 stadia from Meroë towards the north, it turns again to the south, and to the winter sunset, continuing its course for about 3700 stadia, when it is almost in the latitude of the places about Meroë. Then entering far into Africa, and having made another bend, it flows towards the north, a distance of 5300 stadia, to the great cataract;[792] and inclining a little to the east, traverses a distance of 1200 stadia to the smaller cataract at Syene,[793] and 5300 stadia more to the sea.[794]

[CAS. 786]Two rivers empty themselves into it, which issue out of some lakes towards the east, and encircle Meroë, a considerable [Pg 219]island.[795] One of these rivers is called Astaboras,[796] flowing along the eastern side of the island. The other is the Astapus, or, as some call it, Astasobas. But the Astapus[797] is said to be another river, which issues out of some lakes on the south, and that this river forms nearly the body of the (stream of the) Nile which flows in a straight line, and that it is filled by the summer rains; that above the confluence of the Astaboras and the Nile, at the distance of 700 stadia, is Meroë, a city having the same name as the island; and that there is another island above Meroë, occupied by the fugitive Egyptians, who revolted in the time of Psammitichus,[798] and are called Sembritæ, or foreigners. Their sovereign is a queen, but they obey the king of Meroë.

The lower parts of the country on each side Meroë, along the Nile towards the Red Sea, are occupied by Megabari and Blemmyes, who are subject to the Ethiopians, and border upon the Egyptians; about the sea are Troglodytæ. The Troglodytæ, in the latitude of Meroë, are distant ten or twelve days' journey from the Nile. On the left of the course of the Nile live Nubæ in Libya, a populous nation. They begin [Pg   220] [CAS. 787]from Meroë, and extend as far as the bends (of the river). They are not subject to the Ethiopians, but live independently, being distributed into several sovereignties.

The extent of Egypt along the sea, from the Pelusiac to the Canobic mouth, is 1300 stadia.

Such is the account of Eratosthenes.

3. We must, however, enter into a further detail of particulars. And first, we must speak of the parts about Egypt, proceeding from those that are better known to those which follow next in order.

The Nile produces some common effects in this and the contiguous tract of country, namely, that of the Ethiopians above it, in watering them at the time of its rise, and leaving those parts only habitable which have been covered by the inundation; it intersects the higher lands, and all the tract elevated above its current on both sides, which however are uninhabited and a desert, from an absolute want of water. But the Nile does not traverse the whole of Ethiopia, nor alone, nor in a straight line, nor a country which is well inhabited. But Egypt it traverses both alone and entirely, and in a straight line, from the lesser cataract above Syene and Elephantina, (which are the boundaries of Egypt and Ethiopia,) to the mouths by which it discharges itself into the sea. The Ethiopians at present lead for the most part a wandering life, and are destitute of the means of subsistence, on account of the barrenness of the soil, the disadvantages of climate, and their great distance from us.

Now the contrary is the case with the Egyptians in all these respects. For they have lived from the first under a regular form of government, they were a people of civilized manners, and were settled in a well-known country; their institutions have been recorded and mentioned in terms of praise, for they seemed to have availed themselves of the fertility of their country in the best possible manner by the partition of it (and by the classification of persons) which they adopted, and by their general care.

When they had appointed a king, they divided the people into three classes, into soldiers, husbandmen, and priests. The latter had the care of everything relating to sacred things (of the gods), the others of what related to man; some had the[Pg 221]management of warlike affairs, others attended to the concerns of peace, the cultivation of the ground, and the practice of the arts, from which the king derived his revenue.

The priests devoted themselves to the study of philosophy and astronomy, and were companions of the kings.

The country was at first divided into nomes.[799] The Thebaïs contained ten, the Delta ten, and the intermediate tract sixteen. But according to some writers, all the nomes together amounted to the number of chambers in the Labyrinth. Now these were less than thirty [six]. The nomes were again divided into other sections. The greater number of the nomes were distributed into toparchies, and these again into other sections; the smallest portions were the arouræ.

An exact and minute division of the country was required by the frequent confusion of boundaries occasioned at the time of the rise of the Nile, which takes away, adds, and alters the various shapes of the bounds, and obliterates other marks by which the property of one   person   is   distinguished [Pg   222] [CAS. 787]from that of another. It was consequently necessary to measure the land repeatedly. Hence it is said geometry originated here, as the art of keeping accounts and arithmetic originated with the Phœnicians, in consequence of their commerce.[800]

As the whole population of the country, so the separate population in each nome, was divided into three classes; the territory also was divided into three equal portions.

The attention and care bestowed upon the Nile is so great to cause industry to triumph over nature. The ground by nature, and still more by being supplied with water, produces a great abundance of fruits. By nature also a greater rise of the river irrigates a larger tract of land; but industry has completely succeeded in rectifying the deficiency of nature, so that in seasons when the rise of the river has been less than usual, as large a portion of the country is irrigated by means of canals and embankments, as in seasons when the rise of the river has been greater.

Before the times of Petronius there was the greatest plenty, and the rise of the river was the greatest when it rose to the height of fourteen cubits; but when it rose to eight only, a famine ensued. During the government of Petronius, however, when the Nile rose twelve cubits only, there was a most abundant crop; and once when it mounted to eight only, no famine followed. Such then is the nature of this provision for the physical state of the country. We shall now proceed to the next particulars.

4. The Nile, when it leaves the boundaries of Ethiopia, flows in a straight line towards the north, to the tract called the Delta, then "cloven at the head," (according to the expression of Plato,) makes this point the vertex, as it were, of a triangle, the sides of which are formed by the streams, which separate on each side, and extend to the sea, one on the right hand to Pelusium, the other on the left to Canobus and the neighbouring Heracleium, as it is called; the base is the coast lying between Pelusium and the Heracleium.

An island was therefore formed by the sea and by both streams of the river, which is called Delta from the resemblance of its shape to the letter (Δ) of that name. The spot at the vertex of the triangle has the same appellation, because it is [Pg 223]the beginning of the above-mentioned triangular figure. The village, also, situated upon it is called Delta.

These then are two mouths of the Nile, one of which is called the Pelusiac, the other the Canobic and Heracleiotic mouth. Between these are five other outlets, some of which are considerable, but the greater part are of inferior importance. For many others branch off from the principal streams, and are distributed over the whole of the island of the Delta, and form many streams and islands; so that the whole Delta is accessible to boats, one canal succeeding another, and navigated with so much ease, that some persons make use of rafts[801] floated on earthen pots, to transport them from place to place.

The whole island is about 3000 stadia in circumference, and is called, as also the lower country, with the land on the opposite sides of the streams, the Delta.

But at the time of the rising of the Nile, the whole country is covered, and resembles a sea, except the inhabited spots, which are situated upon natural hills or mounds; and considerable cities and villages appear like islands in the distant prospect.

The water, after having continued on the ground more than forty days in summer, then subsides by degrees, in the same manner as it rose. In sixty days the plain is entirely exposed to view, and dries up. The sooner the land is dry, so much the sooner the ploughing and sowing are accomplished, and it dries earlier in those parts where the heat is greater.

The country above the Delta is irrigated in the same manner, except that the river flows in a straight line to the distance of about 4000 stadia in one channel, unless where some island intervenes, the most considerable of which comprises the Heracleiotic Nome; or, where it is diverted by a canal into a large lake, or a tract of country which it is capable of irrigating, as the lake Mœris and the Arsinoïte Nome, or where the canals discharge themselves into the Mareotis.

[Pg   224]
[CAS. 789]

In short, Egypt, from the mountains of Ethiopia to the vertex of the Delta, is merely a river tract on each side of the Nile, and rarely if anywhere comprehends in one continued line a habitable territory of 300 stadia in breadth. It resembles, except the frequent diversions of its course, a bandage rolled out.[802]

The mountains on each side (of the Nile), which descend from the parts about Syene to the Egyptian Sea,[803] give this shape to the river tract of which I am speaking, and to the country. For in proportion as these mountains extend along that tract, or recede from each other, in the same degree is the river contracted or expanded, and they impart to the habitable country its variety of shape. But the country beyond the mountains is in a great measure uninhabited.

5. The ancients understood more by conjecture than otherwise, but persons in later times learnt by experience as eye-witnesses, that the Nile owes its rise to summer rains, which fall in great abundance in Upper Ethiopia, particularly in the most distant mountains. On the rains ceasing, the fulness of the river gradually subsides. This was particularly observed by those who navigated the Arabian Gulf on their way to the Cinnamon country, and by those who were sent out to hunt elephants, or for such other purposes as induced the Ptolemies, kings of Egypt, to despatch persons in that direction. These sovereigns had directed their attention to objects of this kind, particularly Ptolemy surnamed Philadelphus, who was a lover of science, and on account of bodily infirmities always in search of some new diversion and amusement. But the ancient kings paid little attention to such inquiries, although both they and the priests, with whom they passed the greater part of their lives, professed to be devoted to the study of philosophy. Their ignorance therefore is more surprising, both on this account and because Sesostris had traversed the whole of Ethiopia as far as the Cinnamon country, of which expedition monuments exist even to the present day, such as pillars and inscriptions. Cambyses also, when he was in possession of Egypt, had advanced with the Egyptians as far even as[Pg 225]Meroë; and it is said that he gave this name both to the island and to the city, because his sister, or according to some writers his wife, Meroë died there. For this reason therefore he conferred the appellation on the island, and in honour of a woman. It is surprising how, with such opportunities of obtaining information, the history of these rains should not have been clearly known to persons living in those times, especially as the priests registered with the greatest diligence in the sacred books all extraordinary facts, and preserved records of everything which seemed to

contribute to an increase of knowledge. And, if this had been the case, would it be necessary to inquire what is even still a question, what can possibly be the reason why rain falls in summer, and not in winter, in the most southerly parts of the country, but not in the Thebaïs, nor in the country about Syene? nor should we have to examine whether the rise of the water of the Nile is occasioned by rains, nor require such evidence for these facts as Poseidonius adduces. For he says, that Callisthenes asserts that the cause of the rise of the river is the rain of summer. This he borrows from Aristotle, who borrowed it from Thrasyalces the Thasian (one of the ancient writers on physics), Thrasyalces from some other person, and he from Homer, who calls the Nile "heaven-descended:"

"back to Egypt's heaven-descended stream."[804]

But I quit this subject, since it has been discussed by many writers, among whom it will be sufficient to specify two, who have (each) composed in our times a treatise on the Nile, Eudorus and Aristo the Peripatetic philosopher. [They differ little from each other] except in the order and disposition of the works, for the phraseology and execution is the same in both writers. (I can speak with some confidence in this matter), for when at a loss (for manuscripts) for the purpose of comparison and copy, I collated both authors.[805] But which of them surreptitiously substituted the other's account as his own, we may [Pg 226] [CAS. 790]go to the temple of Ammon to be informed. Eudorus accused Aristo, but the style is more like that of Aristo.

The ancients gave the name of Egypt to that country only which was inhabited and watered by the Nile, and the extent they assigned to it was from the neighbourhood of Syene to the sea. But later writers, to the present time, have included on the eastern side almost all the tract between the Arabian Gulf and the Nile (the Æthiopians however do not make much use of the Red Sea); on the western side, the tract extending to the Auases and the parts of the sea-coast from the Canobic mouth of the Nile to Catabathmus, and the kingdom of Cyrenæa. For the kings who succeeded the race of the Ptolemies had acquired so much power, that they became masters of Cyrenæa, and even joined Cyprus to Egypt. The Romans, who succeeded to their dominions, separated Egypt, and confined it within the old limits.

The Egyptians give the name of Auases (Oases) to certain inhabited tracts, which are surrounded by extensive deserts, and appear like islands in the sea. They are frequently met with in Libya, and there are three contiguous to Egypt, and dependent upon it.

This is the account which we have to give of Egypt in general and summarily. I shall now describe the separate parts of the country and their advantages.

6. As Alexandreia and its neighbourhood occupy the greatest and principal portion of the description, I shall begin with it.

In sailing towards the west, the sea-coast from Pelusium to the Canobic mouth of the Nile is about 1300 stadia in extent, and constitutes, as we have said, the base of the Delta. Thence to the island Pharos are 150 stadia more.

Pharos is a small oblong island, and lies quite close to the continent, forming towards it a harbour with a double entrance. For the coast abounds with bays, and has two promontories projecting into the sea. The island is situated between these, and shuts in the bay, lying lengthways in front of it.

Of the extremities of the Pharos, the eastern is nearest to the continent and to the promontory in that direction, called Lochias, which is the cause of the entrance to the port being narrow. Besides the narrowness of the passage, there are rocks, some under water, others rising above it, which at all times increase the violence of the waves rolling in upon them[Pg 227] from the open sea. This extremity itself of the island is a rock, washed by the sea on all sides, with a tower upon it of the same name as the island, admirably constructed of white marble, with several stories. Sostratus of Cnidus, a friend of the kings, erected it for the safety of mariners, as the inscription imports.[806] For as the coast on each side is low and without harbours, with reefs and shallows, an elevated and conspicuous mark was required to enable navigators coming in from the open sea to direct their course exactly to the entrance of the harbour.

The western mouth does not afford an easy entrance, but it does not require the same degree of caution as the other. It forms also another port, which has the name of Eunostus, or Happy Return: it lies in front of the artificial and close harbour. That which has its entrance at the above-mentioned tower of Pharos is the great harbour. These (two) lie contiguous in the recess called Heptastadium, and are separated from it by a mound. This mound forms a bridge from the continent to the island, and extends along its western side, leaving two passages only through it to the harbour of Eunostus, which are bridged over. But this work served not only as a bridge, but as an aqueduct also, when the island was inhabited. Divus Cæsar devastated the island, in his war against the people of Alexandreia, when they espoused the party of the kings. A few sailors live near the tower.

The great harbour, in addition to its being well enclosed by the mound and by nature, is of sufficient depth near the shore to allow the largest vessel to anchor near the stairs. It is also divided into several ports.

The former kings of Egypt, satisfied with what they possessed, and not desirous of foreign commerce, entertained a dislike to all mariners, especially the Greeks (who, on account of the poverty of their own country, ravaged and coveted the property of other nations), and stationed a guard here, who had orders to keep off all persons who approached. To the guard was assigned as a place of residence the spot called Rhacotis, which is now a part of the city of Alexandreia, situated above the arsenal. At that time, however, it was a village. The country about the village was given up to herdsmen,[Pg 228] [CAS. 792] who were also able (from their numbers) to prevent strangers from entering the country.

When Alexander arrived, and perceived the advantages of the situation, he determined to build the city on the (natural) harbour. The prosperity of the place, which ensued, was intimated, it is said, by a presage which occurred while the plan of the city was tracing. The architects were engaged in marking out the line of the wall with chalk, and had consumed it all, when the king arrived; upon which the dispensers of flour supplied the workmen with a part of the flour, which was provided for their own use; and this substance was used in tracing the greater part of the divisions of the streets. This, they said, was a good omen for the city.

7. The advantages of the city are of various kinds. The site is washed by two seas; on the north, by what is called the Egyptian Sea, and on the south, by the sea of the lake Mareia, which is also called Mareotis. This lake is filled by many canals from the Nile, both by those above and those at the sides, through which a greater quantity of merchandise is imported than by those communicating with the sea. Hence the harbour on the lake is richer than the maritime harbour. The exports by sea from Alexandreia exceed the imports. This any person may ascertain, either at Alexandreia or Dicæarchia, by watching the arrival and departure of the merchant vessels, and observing how much heavier or lighter their cargoes are when they depart or when they return.

In addition to the wealth derived from merchandise landed at the harbours on each side, on the sea and on the lake, its fine air is worthy of remark: this results from the city being on two sides surrounded by water, and from the favourable effects of the rise of the Nile. For other cities, situated near lakes, have, during the heats of summer, a heavy and suffocating atmosphere, and lakes at their margins become swampy by the evaporation occasioned by the sun's heat. When a large quantity of moisture is exhaled from swamps, a noxious vapour rises, and is the cause of pestilential disorders. But at Alexandreia, at the beginning of summer, the Nile, being full, fills the lake also, and leaves no marshy matter which is likely to occasion malignant exhalations. At the same period, the Etesian winds blow from the north, over a large expanse of sea, and the Alexandrines in consequence pass their summer very pleasantly.

[Pg 229]

8. The shape of the site of the city is that of a chlamys or military cloak. The sides, which determine the length, are surrounded by water, and are about thirty stadia in extent; but the isthmuses, which determine the breadth of the sides, are each of seven or eight stadia, bounded on one side by the sea, and on the other by the lake. The whole city is intersected by roads for the passage of horsemen and chariots. Two of these are very broad,

exceeding a plethrum in breadth, and cut one another at right angles. It contains also very beautiful public grounds and royal palaces, which occupy a fourth or even a third part of its whole extent. For as each of the kings was desirous of adding some embellishment to the places dedicated to the public use, so, besides the buildings already existing, each of them erected a building at his own expense; hence the expression of the poet may be here applied,

"one after the other springs."[807]

All the buildings are connected with one another and with the harbour, and those also which are beyond it.

The Museum is a part of the palaces. It has a public walk and a place furnished with seats, and a large hall, in which the men of learning, who belong to the Museum, take their common meal. This community possesses also property in common; and a priest, formerly appointed by the kings, but at present by Cæsar, presides over the Museum.

A part belonging to the palaces consists of that called Sema, an enclosure, which contained the tombs of the kings and that of Alexander (the Great). For Ptolemy the son of Lagus took away the body of Alexander from Perdiccas, as he was conveying it down from Babylon; for Perdiccas had turned out of his road towards Egypt, incited by ambition and a desire of making himself master of the country. When Ptolemy had attacked [and made him prisoner], he intended to [spare his life and] confine him in a desert island, but he met with a miserable end at the hand of his own soldiers, who rushed upon and despatched him by transfixing him with the long Macedonian spears. The kings who were with him, Aridæus, and the children of Alexander, and Roxana his wife, departed to Macedonia. Ptolemy carried away the body [Pg 230] [CAS. 794]of Alexander, and deposited it at Alexandreia in the place where it now lies; not indeed in the same coffin, for the present one is of hyalus (alabaster?) whereas Ptolemy had deposited it in one of gold: it was plundered by Ptolemy surnamed Cocce's son and Pareisactus, who came from Syria and was quickly deposed, so that his plunder was of no service to him.

9. In the great harbour at the entrance, on the right hand, are the island and the Pharos tower; on the left are the reef of rocks and the promontory Lochias, with a palace upon it: at the entrance, on the left hand, are the inner palaces, which are continuous with those on the Lochias, and contain numerous painted apartments and groves. Below lies the artificial and close harbour, appropriated to the use of the kings; and Antirrhodus a small island, facing the artificial harbour, with a palace on it, and a small port. It was called Antirrhodus, a rival as it were of Rhodes.

Above this is the theatre, then the Poseidium, a kind of elbow projecting from the Emporium, as it is called, with a temple of Neptune upon it. To this Antony added a mound, projecting still further into the middle of the harbour, and built at the extremity a royal mansion, which he called Timonium. This was his last act, when, deserted by his partisans, he retired to Alexandreia after his defeat at Actium, and intended, being forsaken by so many friends, to lead the [solitary] life of Timon for the rest of his days.

Next are the Cæsarium, the Emporium, and the Apostaseis, or magazines: these are followed by docks, extending to the Heptastadium. This is the description of the great harbour.

10. Next after the Heptastadium is the harbour of Eunostus, and above this the artificial harbour, called Cibotus (or the Ark), which also has docks. At the bottom of this harbour is a navigable canal, extending to the lake Mareotis. Beyond the canal there still remains a small part of the city. Then follows the suburb Necropolis, in which are numerous gardens, burial-places, and buildings for carrying on the process of embalming the dead.

On this side the canal is the Sarapium and other ancient sacred places, which are now abandoned on account of the erection of the temples at Nicopolis; for [there are situated] an amphitheatre and a stadium, and there are celebrated[Pg 231] quinquennial games; but the ancient and customs are neglected.

In short, the city of Alexandreia abounds with public and sacred buildings. The most beautiful of the former is the Gymnasium, with porticos exceeding a stadium in extent. In the middle of it are the court of justice and groves. Here also is a Paneium, an artificial mound of the shape of a fircone, resembling a pile of rock, to the top of which there is an ascent by a spiral path. From the summit may be seen the whole city lying all around and beneath it.

The wide street extends in length along the Gymnasium from the Necropolis to the Canobic gate. Next is the Hippodromos (or race-course), as it is called, and other buildings[808] near it, and reaching to the Canobic canal. After passing through the Hippodromos is the Nicopolis, which contains buildings fronting the sea not less numerous than a city. It is 30 stadia distant from Alexandreia. Augustus Cæsar distinguished this place, because it was here that he defeated Antony and his party of adherents. He took the city at the first onset, and compelled Antony to put himself to death, but Cleopatra to surrender herself alive. A short time afterwards, however, she also put an end to her life secretly, in prison, by the bite of an asp, or (for there are two accounts) by the application of a poisonous ointment. Thus the empire of the Lagidæ, which had subsisted many years, was dissolved.

11. Alexander was succeeded by Ptolemy the son of Lagus, the son of Lagus by Philadelphus, Philadelphus by Euergetes; next succeeded Philopator the lover[809] of Agathocleia, then Epiphanes, afterwards Philometor, the son (thus far) always succeeding the father. But Philometor was succeeded by his brother, the second Euergetes, who was also called Physcon. He was succeeded by Ptolemy surnamed Lathurus, Lathurus by Auletes of our time, who was the father of Cleopatra. All these kings, after the third Ptolemy, were corrupted by luxury and effeminacy, and the affairs of government were very badly administered by them; but worst of all by the fourth, the seventh, and the last (Ptolemy), Auletes (or the Piper), [Pg 232] [CAS. 796]who, besides other deeds of shamelessness, acted the piper; indeed he gloried so much in the practice, that he scrupled not to appoint trials of skill in his palace; on which occasions he presented himself as a competitor with other rivals. He was deposed by the Alexandrines; and of his three daughters, one, the eldest, who was legitimate, they proclaimed queen; but his two sons, who were infants, were absolutely excluded from the succession.

As a husband for the daughter established on the throne, the Alexandrines invited one Cybiosactes from Syria, who pretended to be descended from the Syrian kings. The queen after a few days, unable to endure his coarseness and vulgarity, rid herself of him by causing him to be strangled. She afterwards married Archelaus, who also pretended to be the son of Mithridates Eupator, but he was really the son of that Archelaus[810] who carried on war against Sylla, and was afterwards honourably treated by the Romans. He was grandfather of the last king of Cappadocia in our time, and priest of Comana in Pontus.[811] He was then (at the time we are speaking of) the guest of Gabinius, and intended to accompany him in an expedition against the Parthians,[812] but unknown to Gabinius, he was conducted away by some (friends) to the queen, and declared king.

At this time Pompey the Great entertained Auletes as his guest on his arrival at Rome, and recommended him to the senate, negotiated his return, and contrived the execution of most of the deputies, in number a hundred, who had undertaken to appear against him: at their head was Dion the academic philosopher.

Ptolemy (Auletes) on being restored by Gabinius, put to death both Archelaus and his daughter;[813] but not long after[814] he was reinstated in his kingdom, he died a natural death, leaving two sons and two daughters, the eldest of whom was Cleopatra.

The Alexandrines declared as sovereigns the eldest son and Cleopatra. But the adherents of the son excited a sedition,[Pg 233] and banished Cleopatra, who retired with her sister into Syria.[815]

It was about this time that Pompey the Great, in his flight from Palæpharsalus,[816]came to Pelusium and Mount Casium. He was treacherously slain by the king's party. When Cæsar arrived, he put the young prince to death, and sending for Cleopatra from her place of exile, appointed her queen of Egypt, declaring also her surviving brother, who was very young, and herself joint sovereigns.

After the death of Cæsar and the battle at Pharsalia, Antony passed over into Asia; he raised Cleopatra to the highest dignity, made her his wife, and had children by her. He was present with her at the battle of Actium, and accompanied her in her flight. Augustus Cæsar pursued them, put an end to their power, and rescued Egypt from misgovernment and revelry.

12. At present Egypt is a (Roman) province, pays considerable tribute, and is well governed by prudent persons, who are sent there in succession. The governor thus sent out has the rank of king. Subordinate to him is the administrator of justice, who is the supreme judge in many causes. There is another officer, who is called Idiologus, whose business it is to inquire into property for which there is no claimant, and which of right falls to Cæsar. These are accompanied by Cæsar's freedmen and stewards, who are intrusted with affairs of more or less importance.

Three legions are stationed in Egypt, one in the city, the rest in the country. Besides these there are also nine Roman cohorts, three quartered in the city, three on the borders of Ethiopia in Syene, as a guard to that tract, and three in other parts of the country. There are also three bodies of cavalry distributed in convenient posts.

Of the native magistrates in the cities, the first is the expounder of the law, who is dressed in scarlet; he receives the customary honours of the country, and has the care of providing what is necessary for the city. The second is the writer of records, the third is the chief judge. The fourth is the commander of the night guard. These magistrates existed in the time of the kings, but in consequence of the bad administration of affairs by the latter, the prosperity of the city was ruined by[Pg 234] [CAS. 797]licentiousness. Polybius expresses his indignation at the state of things when he was there: he describes the inhabitants of the city to be composed of three classes; the (first) Egyptians and natives, acute but indifferent citizens, and meddling with civil affairs. The second, the mercenaries, a numerous and undisciplined body; for it was an ancient custom to maintain foreign soldiers, who, from the worthlessness of their sovereigns, knew better how to govern than to obey. The third were the Alexandrines, who, for the same reason, were not orderly citizens;[817] but still they were better than the mercenaries, for although they were a mixed race, yet being of Greek origin, they retained the customs common to the Greeks. But this class was extinct nearly about the time of Euergetes Physcon, in whose reign Polybius came to Alexandreia. For Physcon, being distressed by factions, frequently exposed the multitude to the attacks of the soldiery, and thus destroyed them. By such a state of things in the city the words of the poet (says Polybius) were verified:

"The way to Egypt is long and vexatious."[818]

13. Such then, if not worse, was the condition of the city under the last kings. The Romans, as far as they were able, corrected, as I have said, many abuses, and established an orderly government, by appointing vice-governors, nomarchs, and ethnarchs, whose business it was to superintend affairs of minor importance.

The greatest advantage which the city possesses arises from its being the only place in all Egypt well situated by nature for communication with the sea by its excellent harbour, and with the land by the river, by means of which everything is easily transported and collected together into this city, which is the greatest mart in the habitable world.

These may be said to be the superior excellencies of the city. Cicero, in one of his orations,[819] in speaking of the revenues of Egypt, states that an annual tribute of 12,500 talents was paid to (Ptolemy) Auletes, the father of Cleopatra. If then a king, who administered his government in the worst possible manner, and with the greatest negligence, obtained so large a revenue, what must we suppose it to be at present, [Pg 235]when affairs are administered with great care, and when the commerce with India and with Troglodytica has been so greatly increased? For formerly not even twenty vessels ventured to navigate the Arabian Gulf, or advance to the smallest distance beyond the straits at its mouth; but now large fleets are despatched as far as India and the extremities of Ethiopia, from which places the most valuable freights are brought to Egypt, and are thence exported to other parts, so that a double amount of custom is collected, arising from imports on the one hand, and from exports on the other. The most expensive description of goods is charged with the heaviest impost; for in fact Alexandreia has a monopoly of trade, and is almost the only receptacle for this kind of merchandise and place of supply for foreigners. The natural convenience of the situation is still more apparent to persons travelling through the country, and particularly along the coast which commences at the Catabathmus; for to this place Egypt extends.

Next to it is Cyrenæa, and the neighbouring barbarians, the Marmaridæ.

14. From the Catabathmus[820] to Parætonium is a run of 900 stadia for a vessel in a direct course. There is a city and a large harbour of about 40 stadia in extent, by some called the city Parætonium,[821] by others, Ammonia. Between these is the village of the Egyptians, and the promontory Ænesisphyra, and the Tyndareian rocks, four small islands, with a harbour; then Drepanum a promontory, and Ænesippeia an island with a harbour, and Apis a village, from which to Parætonium are 100 stadia; [from thence] to the temple of Ammon is a journey of five days. From Parætonium to Alexandreia are about 1300 stadia. Between these are, first, a promontory of white earth, called Leuce-Acte, then Phœnicus a harbour, and Pnigeus a village; after these the island Sidonia (Pedonia?) with a harbour; then a little further off from the sea, Antiphræ. The whole of this country produces no wine of a good quality, and the earthen jars contain more sea-water than wine, which is called Libyan;[822] this and beer are the [Pg 236] [CAS. 799]principal beverage of the common people of Alexandria. Antiphræ in particular was a subject of ridicule (on account of its bad wine).

Next is the harbour Derrhis,[823] which has its name from an adjacent black rock, resembling δέῤῥις, a hide. The neighbouring place is called Zephyrium. Then follows another harbour, Leucaspis (the white shield), and many others; then the Cynossema (or dog's monument); then Taposeiris, not that situated upon the sea; here is held a great public festival. There is another Taposeiris,[824] situated at a considerable distance beyond the city (Alexandreia). Near this, and close to the sea, is a rocky spot, which is the resort of great numbers of people at all seasons of the year, for the purpose of feasting and amusement. Next is Plinthine,[825] and the village of Nicium, and Cherronesus a fortress, distant from Alexandreia and the Necropolis about 70 stadia.

The lake Mareia, which extends as far as this place, is more than 150 stadia in breadth, and in length less than 300 stadia. It contains eight islands. The whole country about it is well inhabited. Good wine also is produced here, and in such quantity that the Mareotic wine is racked in order that it may be kept to be old.[826]

15. The byblus[827] and the Egyptian bean grow in the marshes and lakes; from the latter the ciborium is made.[828] [Pg 237]The stalks of the bean are nearly of equal height, and grow to the length of ten feet. The byblus is a bare stem, with a tuft on the top. But the bean puts out leaves and flowers in many parts, and bears a fruit similar to our bean, differing only in size and taste. The bean-grounds present an agreeable sight, and afford amusement to those who are disposed to recreate themselves with convivial feasts. These entertainments take place in boats with cabins; they enter the thickest part of the plantation, where they are overshadowed with the leaves, which are very large, and serve for drinking-cups and dishes, having a hollow which fits them for the purpose. They are found in great abundance in the shops in Alexandreia, where they are used as vessels. One of the sources of land revenue is the sale of these leaves. Such then is the nature of this bean.

The byblus does not grow here in great abundance, for it is not cultivated. But it abounds in the lower parts of the Delta. There is one sort inferior to the other.[829]The best is the hieratica. Some persons intending to augment the revenue, employed in this case a method which the Jews practised with the palm, especially the caryotic, and with the balsamum.[830] In many places it is not allowed to be cultivated, and the price is enhanced by its rarity: the revenue is indeed thus increased, but the general consumption [of the article] is injured.

16. On passing through the Canobic gate of the city, on the right hand is the canal leading to Canobus, close to the lake. They sail by this canal to Schedia, to the great river, and to Canobus, but the first place at which they arrive is Eleusis. This is a settlement near Alexandreia and Nicopolis, and situated on the Canobic canal. It has houses of entertainment which command beautiful views, and hither [Pg 238]

resort men and women who are inclined to indulge in noisy revelry, a prelude to Canobic life, and the dissolute manners of the people of Canobus.

At a little distance from Eleusis, on the right hand, is the canal leading towards Schedia. Schedia is distant four schœni from Alexandreia. It is a suburb of the city, and has a station for the vessels with cabins, which convey the governors when they visit the upper parts of the country. Here is collected the duty on merchandise, as it is transported up or down the river. For this purpose a bridge of boats is laid across the river, and from this kind of bridge the place has the name of Schedia.

Next after the canal leading to Schedia, the navigation thence to Canobus is parallel to the sea-coast, extending from Pharos to the Canobic mouth. For between the sea and the canal, is a narrow band of ground, on which is situated the smaller Taposeiris, which lies next after Nicopolis, and Zephyrium a promontory, on which is a small temple dedicated to Venus Arsinoë.

Anciently, it is said, a city called Thonis stood there, which bears the name of the king, who entertained as his guests Menelaus and Helen. The poet thus speaks of the drugs which were given to Helen,

"the potent drugs, which Polydamna, the wife of Thon, gave to Helen."[831]

17. Canobus is a city, distant by land from Alexandreia 120 stadia. It has its name from Canobus, the pilot of Menelaus, who died there. It contains the temple of Sarapis, held in great veneration, and celebrated for the cure of diseases; persons even of the highest rank confide in them, and sleep there themselves on their own account, or others for them. Some persons record the cures, and others the veracity of the oracles which are delivered there. But remarkable above everything else is the multitude of persons who resort to the public festivals, and come from Alexandreia by the canal. For day and night there are crowds of men and women in boats, singing and dancing, without restraint, and with the utmost licentiousness. Others, at Canobus itself, keep hostelries situated on the banks of the canal, which are well adapted for such kind of diversion and revelry.

18. Next to Canobus is Heracleium, in which is a temple [Pg 239]of Hercules; then follows the Canobic mouth,[832] and the commencement of the Delta.

On the right of the Canobic canal is the Menelaïte Nome, so called from the brother of the first Ptolemy, but certainly not from the hero (Menelaus), as some writers assert, among whom is Artemidorus.

Next to the Canobic mouth is the Bolbitine, then the Sebennytic, and the Phatnitic, which is the third in magnitude compared with the first two, which form the boundaries of the Delta. For it branches off into the interior, not far from the vertex of the Delta. The Mendesian is very near the Phatnitic mouth; next is the Tanitic, and lastly the Pelusiac mouth. There are others, which are of little consequence, between these, since they are as it were false mouths.

The mouths have entrances which are not capable of admitting large vessels, but lighters only, on account of the shallows and marshes. The Canobic mart is principally used as a mart for merchandise, the harbours at Alexandreia being closed, as I have said before.

After the Bolbitine mouth there runs out to a great distance a low and sandy promontory. It is called Agnu-ceras (or Willow Point). Then follows the watch-tower of Perseus,[833] and the fortress of the Milesians. For in the time of Psammitichus, and when Cyaxares was king of the Medes, some Milesians with 30 vessels steered into the Bolbitine mouth, disembarked there, and built the above-mentioned fortress. Some time afterwards they sailed up to the Saïtic Nome, and having conquered Inarus in an engagement at sea, founded the city Naucratis, not far above Schedia.

Next after the fortress of the Milesians, in proceeding towards the Sebennytic mouth, are lakes, one of which is called Butice, from the city Butus; then the city Sebennytice and Sais, the capital of the lower country; here Minerva is worshipped. In the temple there of this goddess, is the tomb of Psammitichus. Near Butus is Hermopolis, situated in an island, and at Butus is an oracle of Latona.

[Pg   240]
19. In the interior above the Sebennytic and Phatnitic mouths is Xoïs, both an

island and a city in the Sebennytic Nome. There are also Hermopolis, Lycopolis, and Mendes, where Pan[834] is worshipped, and of animals a goat. Here, according to Pindar, goats have intercourse with women.

Near Mendes are Diospolis, and the lakes about it, and Leontopolis; then further on, the city Busiris,[835] in the Busirite Nome, and Cynospolis.

Eratosthenes says, "That to repel strangers is a practice common to all barbarians, but that this charge against the Egyptians is derived from fabulous stories related of (one) Busiris and his people in the Busirite Nome, as some persons in later times were disposed to charge the inhabitants of this place with inhospitality, although in truth there was neither king nor tyrant of the name of Busiris: that besides there was a common saying,

'The way to Egypt is long and vexatious,'[836]

which originated in the want of harbours, and in the state of the harbour at Pharos, which was not of free access, but watched and guarded by herdsmen, who were robbers, and attacked those who attempted to sail into it. The Carthaginians drown [he says] any strangers who sail past, on their voyage to Sardinia or to the Pillars. Hence much of what is related of the parts towards the west is discredited. The Persians also were treacherous guides, and conducted the ambassadors along circuitous and difficult ways."

20. Contiguous to the Busirite Nome are the Athribite Nome and the city Athribis; next the Prosopite Nome, in which latter is Aphroditopolis (the city of Venus). Above the Mendesian and the Tanitic mouths are a large lake, and the Mendesian and Leontopolite Nomes, and a city of Aphrodite (or Venus) and the Pharbetite Nome. Then follows the Tanitic, which some call the Saïtic mouth, and the Tanite Nome,[837] and in it Tanis a large city.

21. Between the Tanitic and the Pelusiac mouths are lakes [Pg 241]and large and continuous marshes, among which are numerous villages. Pelusium itself has many marshes lying around it, which some call Barathra (or water holes), and swamps. It is situated at a distance of more than 20 stadia from the sea. The circumference of the wall is 20 stadia. It has its name from the mud (πηλοῦ) of the swamps.[838] On this quarter Egypt is difficult of access, i. e. from the eastern side towards Phœnicia and Judæa, and on the side of Arabia Nabatæa, which is contiguous; through which countries the road to Egypt lies.

The country between the Nile and the Arabian Gulf is Arabia, and at its extremity is situated Pelusium. But the whole is desert, and not passable by an army. The isthmus between Pelusium and the recess of the Arabian Gulf near Heroopolis is 1000 stadia; but, according to Poseidonius, less than 1500 stadia in extent. Besides its being sandy and without water, it abounds with reptiles, which burrow in the sand.

22. In sailing up the river from Schedia to Memphis,[839] on the right hand, are a great many villages extending as far as the lake Mareia, among which is that called the village of Chabrias. Upon the river is Hermopolis, then Gynæcopolis, and the Gynæcopolite Nome; next Momemphis and the Momemphite Nome. Between these places are many canals, which empty themselves into the lake Mareotis. The Momemphitæ worship Venus, and a sacred cow is kept there, as Apis is maintained at Memphis, and Mneyis[840] at Heliopolis. [Pg 242] [CAS. 803]These animals are regarded as gods, but there are other places, and these are numerous, both in the Delta and beyond it, in which a bull or a cow is maintained, which are not regarded as gods, but only as sacred.

23. Above Momemphis are two nitre mines, which furnish nitre in large quantities, and the Nitriote Nome. Here Sarapis is worshipped, and they are the only people in Egypt who sacrifice a sheep. In this nome and near this place is a city called Menelaus. On the left hand in the Delta, upon the river, is Naucratis. At the distance of two schœni from the river is Saïs,[841] and a little above it the asylum of Osiris, [Pg 243]in which it is said Osiris is buried. This, however, is questioned by many persons, and particularly by the inhabitants of Philæ, which is situated above Syene and Elephantina. These people tell this tale, that Isis placed coffins of Osiris in various places, but that one only contained the body of Osiris, so that no one knew which of them it was; and that she did this with the intention of concealing it from Typhon,[842] who might come and cast the body out of its place of deposit.

24. This is the description of the country from Alexandreia to the vertex of the Delta.

Artemidorus says, that the navigation up the river is 28 schœni, which amount to 840 stadia, reckoning the schœnus at 30 stadia. When we ourselves sailed up the river, schœni of different measures were used at different places in giving the distances, so that sometimes the received schœnus was a measure of 40 stadia and even more. That the measure of the schœnus was unsettled among the Egyptians, Artemidorus himself shows in a subsequent place. In reckoning the distance from Memphis to Thebaïs, he says that each schœnus consists of 120 stadia, and from the Thebaïs to Syene of 60 stadia. In sailing up from Pelusium to the same vertex of the Delta, is a distance, he says, of 25 schœni, or 750 stadia, and he employs the same measure.

On setting out from Pelusium, the first canal met with is that which fills the lakes, "near the marshes," as they are called. There are two of these lakes, situated upon the left hand of the great stream above Pelusium in Arabia. He mentions other lakes also, and canals in the same parts beyond the Delta.

The Sethroïte Nome extends along one of the two lakes. He reckons this as one of the ten nomes in the Delta. There are two other canals, which discharge themselves into the same lakes.

25. There is another canal also, which empties itself into the Red Sea, or Arabian Gulf, near the city Arsinoë, which some call Cleopatris.[843] It flows through the Bitter Lakes, as [Pg 244] [CAS. 804]they are called, which were bitter formerly, but when the above-mentioned canal was cut, the bitter quality was altered by their junction with the river, and at present they contain excellent fish, and abound with aquatic birds.

The canal was first cut by Sesostris before the Trojan times, but according to other writers, by the son of[844] Psammitichus, who only began the work, and afterwards died; lastly, Darius the First succeeded to the completion of the undertaking, but he desisted from continuing the work, when it was nearly finished, influenced by an erroneous opinion that the level of the Red Sea was higher than Egypt, and that if the whole of the intervening isthmus were cut through, the country would be overflowed by the sea. The Ptolemaïc kings however did cut through it, and placed locks upon the canal,[845] so that they sailed, when they pleased, without obstruction into the outer sea, and back again [into the canal].

We have spoken of the surfaces of bodies of water in the first part of this work.[846]

26. Near Arsinoë are situated in the recess of the Arabian Gulf towards Egypt, Heroopolis and Cleopatris; harbours, [Pg 245]suburbs, many canals, and lakes are also near. There also is the Phagroriopolite Nome, and the city Phagroriopolis. The canal, which empties itself into the Red Sea, begins at the village Phaccusa, to which the village of Philon is contiguous. The canal is 100 cubits broad, and its depth sufficient to float a vessel of large burden. These places are near the apex of the Delta.

27. There also are the city Bubastus[847] and the Bubastite Nome, and above it the Heliopolite Nome. There too is Heliopolis, situated upon a large mound. It contains a temple of the sun, and the ox Mneyis, which is kept in a sanctuary, and is regarded by the inhabitants as a god, as Apis is regarded by the people of Memphis. In front of the mound are lakes, into which the neighbouring canal discharges itself. At present the city is entirely deserted. It has an ancient temple constructed after the Egyptian manner, bearing many proofs of the madness and sacrilegious acts of Cambyses, who did very great injury to the temples, partly by fire, partly by violence, mutilating [in some] cases, and applying fire [in others]. In this manner he injured the obelisks, two of which, that were not entirely spoilt, were transported to Rome.[848] There are others both here and at Thebes, the present Diospolis, some of which are standing, much corroded by fire, and others lying on the ground.

28. The plan of the temples is as follows.

At the entrance into the temenus is a paved floor, in breadth about a plethrum, or even less; its length is three or four times as great, and in some instances even more. This part is called Dromos, and is mentioned by Callimachus,

"this is the Dromos, sacred to Anubis."

Throughout the whole length on each side are placed stone sphinxes, at the distance of 20 cubits or a little more from each other, so that there is one row of sphinxes on the right hand, and another on the left. Next after the sphinxes is a large propylon, then on proceeding further, another propylon, and then another. Neither the number of the propyla nor of the sphinxes is determined by any rule. They are different in different temples, as well as the length and breadth of the Dromi.

[Pg 246]

[CAS. 805]Next to the propyla is the naos, which has a large and considerable pronaos; the sanctuary in proportion; there is no statue, at least not in human shape, but a representation of some of the brute animals. On each side of the pronaos project what are called the wings. These are two walls of equal height with the naos. At first the distance between them is a little more than the breadth of the foundation of the naos.[849] As you proceed onwards, the [base] lines incline towards one another till they approach within 50 or 60 cubits. These walls have large sculptured figures, very much like the Tyrrhenian (Etruscan) and very ancient works among the Greeks.

There is also a building with a great number of pillars, as at Memphis, in the barbaric style; for, except the magnitude and number and rows of pillars, there is nothing pleasing nor easily described,[850] but rather a display of labour wasted.

29. At Heliopolis we saw large buildings in which the priests lived. For it is said that anciently this was the principal residence of the priests, who studied philosophy and astronomy. But there are no longer either such a body of persons or such pursuits. No one was pointed out to us on the spot, as presiding over these studies, but only persons who performed sacred rites, and who explained to strangers [the peculiarities of] the temples.

A person of the name of Chæremon accompanied the governor, Ælius Gallus, in his journey from Alexandreia into Egypt, and pretended to some knowledge of this kind, but he was generally ridiculed for his boasting and ignorance. The houses of the priests, and the residences of Plato and of Eudoxus, were shown to us. Eudoxus came here with Plato, and, according to some writers, lived thirteen years in the society of the priests. For the latter were distinguished for their knowledge of the heavenly bodies, but were mysterious and uncommunicative, yet after a time were prevailed upon by courtesy to acquaint them with some of the principles of their science, but the barbarians concealed the greater part of them. They had, however, communicated the knowledge of the additional [Pg 247]portions of the day and night, in the space of 365 days, necessary to complete the annual period; and, at that time, the length of the year was unknown to the Greeks, as were many other things, until later astronomers received them from the persons who translated the records of the priests into the Greek language, and even now derive knowledge from their writings and from those of the Chaldeans.[851]

30. After Heliopolis is the "Nile above the Delta." The country on the right hand, as you go up the Nile, is called Libya, as well as that near Alexandreia and the lake Mareotis; the country on the left hand is called Arabia. The territory belonging to Heliopolis is in Arabia, but the city Cercesura is in Libya, and situated opposite to the observatory of Eudoxus. For there is shown an observing station in front of Heliopolis, as there is in front of Cnidus, where Eudoxus marked certain motions of the heavenly bodies. This is the Letopolite Nome.

In sailing up the river we meet with Babylon, a strong fortress, built by some Babylonians who had taken refuge there, and had obtained permission from the kings to establish a settlement in that place. At present it is an encampment for one of the three legions which garrison Egypt. There is a mountainous ridge, which extends from the encampment as far as the Nile. At this ridge are wheels and screws, by which water is raised from the river, and one hundred and fifty prisoners are [thus] employed.

The pyramids on the other side [of the river] at Memphis may be clearly discerned from this place, for they are not far off.

31. Memphis itself also, the residence of the kings of Egypt, is near, being only three schœni distant from the Delta. It contains temples, among which is that of Apis, who is the same as Osiris. Here the ox Apis is kept in a sort of sanctuary, and is held, as I have said, to be a god. The forehead and some other small parts of its body are white; the other parts are

119

black. By these marks the fitness of the successor[Pg 248] [CAS. 807] is always determined, when the animal to which they pay these honours dies. In front of the sanctuary is a court, in which there is another sanctuary for the dam of Apis. Into this court the Apis is let loose at times, particularly for the purpose of exhibiting him to strangers. He is seen through a door in the sanctuary, and he is permitted to be seen also out of it. After he has frisked about a little in the court, he is taken back to his own stall.

The temple of Apis is near the Hephæsteium (or temple of Vulcan); the Hephæsteium[852] itself is very sumptuously constructed, both as regards the size of the naos and in other respects. In front of the Dromos is a colossal figure consisting of a single stone. It is usual to celebrate bull-fights in this Dromos; the bulls are bred expressly for this purpose, like horses. They are let loose, and fight with one another, the conqueror receiving a prize.

At Memphis also there is a temple of Venus, who is accounted a Grecian deity. But some say that it is a temple dedicated to Selene, or the moon.[853]

32. There is also a temple of Sarapis, situated in a very sandy spot, where the sand is accumulated in masses by the wind. Some of the sphinxes which we saw were buried in this sand up to the head, and one half only of others was visible. Hence we may conceive the danger, should any one, in his way to the temple, be surprised by a [sand] storm.

The city is large and populous; it ranks next to Alexandreia, and, like that place, is inhabited by mixed races of people. There are lakes in front of the city and of the palaces, which at present are in ruins and deserted. They are situated upon an eminence, and extend as far as the lower part of the city.

Close to this place are a grove and a lake.

33. At the distance of 40 stadia from Memphis is a brow [Pg 249]of a hill, on which are many pyramids, the tombs of the kings.[854] Three of them are considerable. Two of these are reckoned among the seven wonders [of the world]. They are a stadium in height, and of a quadrangular shape. Their height somewhat exceeds the length of each of the sides.[855] One pyramid is a little larger than the other. At a moderate height in one of the sides[856] is a stone, which may be taken out; when that is removed, there is an oblique passage [leading] to the tomb. They are near each other, and upon the same level. Farther on, at a greater height of the mountain, is the third pyramid, which is much less than the two others, but constructed at much greater expense; for from the foundation[Pg 250] [CAS. 808] nearly as far as the middle, it is built of black stone. Mortars are made of this stone, which is brought from a great distance; for it comes from the mountains of Ethiopia, and being hard and difficult to be worked, the labour is attended with great expense. It is said to be the tomb of a courtesan, built by her lovers, and whose name, according to Sappho the poetess, was Doriche. She was the mistress of her brother Charaxus, who traded to the port of Naucratis with wine of Lesbos. Others call her Rhodopis.[857]

[Pg 251]A story is told of her, that, when she was bathing, an eagle snatched one of her sandals from the hands of her female attendant and carried it to Memphis; the eagle soaring over the head of the king, who was administering justice at the time, let the sandal fall into his lap. The king, struck with the shape of the sandal, and the singularity of the accident, [Pg   252] [CAS. 808]sent over the country to discover the woman to whom it belonged. She was found in the city of Naucratis, and brought to the king, who made her his wife. At her death she was honoured with the above-mentioned tomb.

34. One extraordinary thing which I saw at the pyramids must not be omitted. Heaps of stones from the quarries lie in front of the pyramids. Among these are found pieces which in shape and size resemble lentils.[858] Some contain substances like grains half peeled. These, it is said, are the remnants of the workmen's food converted into stone; which is not probable.[859] For at home in our country (Amasia), there is a long hill in a plain, which abounds with pebbles of a porus stone,[860] resembling lentils. The pebbles of the sea-shore and of rivers suggest somewhat of the same difficulty [respecting their origin]; some explanation may indeed be found in the motion [to which these are subject] in flowing waters, but the investigation of the above fact presents more difficulty. I have said elsewhere,[861] that in sight of the pyramids, on the other side in Arabia, and near the stone

quarries from which they are built, is a very rocky mountain, called the Trojan mountain; beneath it there are caves, and near the caves and the river a village called Troy, an ancient settlement of the captive Trojans who had accompanied Menelaus and settled there.[862]

[Pg 253]35. Next to Memphis is the city Acanthus, situated also in Libya, and the temple of Osiris, and the grove of the Thebaïc acantha, from which gum is procured. Next is the Aphroditopolite Nome, and the city in Arabia of the same name, where is kept a white cow, considered sacred. Then follows the Heracleote Nome, in a large island, near which is the canal on the right hand, which leads into Libya, in the direction of the Arsinoïte Nome; so that the canal has two entrances, a part of the island on one side being interposed between them.[863] This nome is the most considerable of all in appearance, natural properties, and embellishment. It is the only nome planted with large, full-grown olive trees, which bear fine fruit. If the produce were carefully collected, good oil might be obtained; but this care is neglected, and although a large quantity of oil is obtained, yet it has a disagreeable smell. (The rest of Egypt is without the olive tree, except the gardens near Alexandreia, which are planted with olive trees, but do not furnish any oil.) It produces wine in abundance, corn, pulse, and a great variety of other grains. It has also the remarkable lake Mœris, which in extent is a sea, and the colour of its waters resembles that of the sea. Its borders also are like the sea-shore, so that we may make the same suppositions respecting these as about the country near Ammon. For they are not very far distant from one another and from Parætonium; and we may conjecture from a multitude of proofs, that as the temple of Ammon was once situated upon the sea, so this tract of country also bordered on the sea at some former period. But Lower Egypt and the country as far as the Lake Sirbonis were sea, and confluent perhaps [Pg 254] [CAS. 809]with the Red Sea at Heroopolis, and the Ælanitic recess of the gulf.

36. We have treated these subjects at length in the First Book of the Geography. At present we shall make a few remarks on the operations of nature and of Providence conjointly.—On the operations of nature, that all things converge to a point, namely, the centre of the whole, and assume a spherical shape around it. The earth is the densest body, and nearer the centre than all others: the less dense and next to it is water; but both land and water are spheres, the first solid, the second hollow, containing the earth within it.—On the operations of Providence, that it has exercised a will, is disposed to variety, and is the artificer of innumerable works. In the first rank, as greatly surpassing all the rest, is the generation of animals, of which the most excellent are gods and men, for whose sake the rest were formed. To the gods Providence assigned heaven; and the earth to men, the extreme parts of the world; for the extreme parts of the sphere are the centre and the circumference. But since water encompasses the earth, and man is not an aquatic, but a land-animal, living in the air, and requiring much light, Providence formed many eminences and cavities in the earth, so that these cavities should receive the whole or a great part of the water which covers the land beneath it; and that the eminences should rise and conceal the water beneath them, except so much as was necessary for the use of the human race, the animals and plants about it.

But as all things are in constant motion, and undergo great changes, (for it is not possible that things of such a nature, so numerous and vast, could be otherwise regulated in the world,) we must not suppose the earth or the water always to continue in this state, so as to retain perpetually the same bulk, without increase or diminution, or that each preserves the same fixed place, particularly as the reciprocal change of one into the other is most consonant to nature from their proximity; but that much of the land is changed into water, and a great portion of water becomes land, just as we observe great differences in the earth itself. For one kind of earth crumbles easily, another is solid and rocky, and contains iron; and so of others. There is also a variety in the quality of water; for some waters are saline, others sweet and potable, others[Pg 255] medicinal, and either salutary or noxious, others cold or hot. Is it therefore surprising that some parts of the earth which are now inhabited should formerly have been occupied by sea, and that what are now seas should formerly have been inhabited land? so also fountains once existing have failed, and others have burst forth; and similarly in the case of rivers and lakes: again, mountains and plains have been converted reciprocally one into the other. On this subject I have spoken before at length,[864] and now let this be said:

37. The lake Mœris, by its magnitude and depth, is able to sustain the superabundance of water which flows into it at the time of the rise of the river, without overflowing the inhabited and cultivated parts of the country. On the decrease of the water of the river, it distributes the excess by the same canal at each of the mouths; and both the lake and the canal preserve a remainder, which is used for irrigation. These are the natural and independent properties of the lake, but in addition, on both mouths of the canal are placed locks, by which the engineers store up and distribute the water which enters or issues from the canal.

We have here also the Labyrinth, a work equal to the Pyramids, and adjoining to it the tomb of the king who constructed the Labyrinth.[865] After proceeding beyond the first entrance of the canal about 30 or 40 stadia, there is a table-shaped plain, with a village and a large palace composed of as many palaces as there were formerly nomes. There are an equal number of aulæ, surrounded by pillars, and contiguous to one another, all in one line and forming one building, like a long wall having the aulæ in front of it. The entrances into the aulæ are opposite to the wall. In front of the entrances there are long and numerous covered ways, with winding passages communicating with each other, so that no stranger could find his way into the aulæ or out of them without a guide. The (most) surprising circumstance is that the roofs of these dwellings consist of a single stone each, and that the covered ways through their whole range were roofed in the same manner with single slabs of stone of extraordinary size, without the intermixture of timber or of any other material. On ascending the roof,—which is not of great height, for it [Pg 256] [CAS. 811]consists only of a single story,—there may be seen a stone-field, thus composed of stones. Descending again and looking[866] into the aulæ, these may be seen in a line supported by twenty-seven pillars, each consisting of a single stone. The walls also are constructed of stones not inferior in size to these.

At the end of this building, which occupies more than a stadium, is the tomb, which is a quadrangular pyramid, each side of which is about four plethra in length, and of equal height. The name of the person buried there is Imandes.[867]They built, it is said, this number of aulæ, because it was the custom for all the nomes to assemble there together according to their rank, with their own priests and priestesses, for the purpose of performing sacrifices and making offerings to the gods, and of administering justice in matters of great importance. Each of the nomes was conducted to the aula appointed for it.

38. Sailing along to the distance of 100 stadia, we come to the city Arsinoë, formerly called Crocodilopolis; for the inhabitants of this nome worship the crocodile. The animal is accounted sacred, and kept apart by himself in a lake; it is tame, and gentle to the priests, and is called Suchus. It is fed with bread, flesh, and wine, which strangers who come to see it always present. Our host, a distinguished person, who was our guide in examining what was curious, accompanied us to the lake, and brought from the supper table a small cake, dressed meat, and a small vessel containing a mixture of honey and milk. We found the animal lying on the edge of the lake. The priests went up to it; some of them opened its mouth, another put the cake into it, then the meat, and afterwards poured down the honey and milk. The animal then leaped into the lake, and crossed to the other side. When another stranger arrived with his offering, the priests took it, and running round the lake, caught the crocodile, and gave him what was brought, in the same manner as before.

39. Next after the Arsinoïte and Heracleotic Nomes, is the city of Hercules, in which the ichneumon is worshipped, in opposition to the Arsinoïtes, who worship crocodiles; [Pg 257]hence the canal and the lake Mœris is full of these animals; for they venerate them, and are careful to do them no harm: but the Heracleotæ worship the ichneumon, which is most destructive both to crocodiles and asps. The ichneumons destroy not only the eggs of the latter, but the animals themselves. The ichneumons are protected by a covering of mud, in which they roll, and then dry themselves in the sun. They then seize the asps by the head or tail, and dragging them into the river, so kill them.

They lie in wait for the crocodiles, when the latter are basking in the sun with their mouths open; they then drop into their jaws, and eating through their intestines and belly, issue out of the dead body.

40. Next follows the Cynopolite Nome and Cynopolis, where they worship the dog Anubis, and pay certain honours to dogs; a subsistence is there provided for them, as sacred animals.

On the other side of the river is the city Oxyrynchus,[868] and a nome of the same name. They worship the oxyrynchus, and have a temple dedicated to this animal; but all the other Egyptians worship the oxyrynchus.[869] For all the Egyptians worship in common certain animals; three among the land animals, the ox, the dog, and the cat; two among the winged tribe, the hawk and the ibis; and two of the aquatic animals, the fish lepidotus and the oxyrynchus. There are also other animals which each people, independently of others, worship; as the Saïtæ and Thebaïtæ, a sheep; the Latopolitæ, the latus, a fish inhabiting the Nile; the people of Lycopolis, a wolf; those of Hermopolis,[870] the cynocephalus; those of Babylon,[871] near Memphis, a cephus, which has the countenance of a satyr, and in other respects is between a dog and a bear; it is bred in Ethiopia. The inhabitants of Thebes worship an eagle; the Leontopolitæ, a lion; the Mendesians, a male and female goat; the Athribitæ, a shrew-mouse; different people worshipping different animals. They do not, however, assign the same reasons for this difference of worship.

[Pg 258]

[CAS. 813]41. Then follows the Hermopolite Castle, a place where is collected the toll on merchandise brought down from the Thebaïs. At this place begins the reckoning by schœni of sixty stadia each, which is continued to Syene and Elephantina. Next is the Thebaïc Keep, and a canal leading to Tanis. Then follow Lycopolis, Aphroditopolis, and Panopolis, an old settlement belonging to masons and weavers of linen.

42. Then follows Ptolemaïs,[872] the largest city in the Thebaïs, not inferior to Memphis, with a form of government after the Grecian mode. Above this city is Abydos, where is the palace of Memnon, constructed in a singular manner, entirely of stone,[873] and after the plan of the Labyrinth, which we have described, but not composed of many parts. It has a fountain situated at a great depth. There is a descent to it through an arched passage built with single stones, of remarkable size and workmanship.

There is a canal which leads to this place from the great river. About the canal is a grove of Egyptian acanthus, dedicated to Apollo. Abydos seems once to have been a large city, second to Thebes. At present it is a small town. But if, as they say, Memnon is called Ismandes by the Egyptians, the Labyrinth might be a Memnonium, and the work of the same person who constructed those at Abydos and at Thebes; for in those places, it is said, are some Memnonia. In the latitude of Abydos is the first Auasis (Oasis) of the three which are said to be in Africa. It is distant from Abydos a journey of seven days through a desert. It is an inhabited place, well supplied with good water and wine, and sufficiently provided with other articles. The second is that near the lake Mœris. The third is that at the oracle of Ammon: these are considerable settlements.

43. Having before spoken at length of the temple of Ammon, we wish to add this only, that in ancient times divination in general and oracles were held in greater esteem than at present. Now they are greatly neglected; for the Romans are satisfied with the oracles of the Sibyl, and with Tyrrhenian divination by the entrails of animals, the flight of birds, and portentous appearances. Hence the oracle of Ammon, which was formerly held in great esteem, is now nearly deserted. This [Pg 259]appears chiefly from the historians who have recorded the actions of Alexander, adding, indeed, much that has the appearance of flattery, but yet relating what is worthy of credit. Callisthenes, for instance, says that Alexander was ambitious of the glory of visiting the oracle, because he knew that Perseus and Hercules had before performed the journey thither. He set out from Parætonium, although the south winds were blowing, and succeeded in his undertaking by vigour and perseverance. When out of his way on the road, he escaped being overwhelmed in a sand-storm by a fall of rain, and by the guidance of two crows, which directed his course. These things are stated by way of flattery, as also what follows: that the priest permitted the king alone to pass into the temple in his usual dress, whereas the others changed theirs; that all heard the oracles on the

outside of the temple, except Alexander, who was in the interior of the building; that the answers were not given, as at Delphi and at Branchidæ, in words, but chiefly by nods and signs, as in Homer;

"the son of Saturn nodded with his sable brows,"[874]

the prophet imitating Jupiter. This, however, the man told the king, in express terms, that he was the son of Jupiter. Callisthenes adds, (after the exaggerating style of tragedy,) that when Apollo had deserted the oracle among the Branchidæ, on the temple being plundered by the Branchidæ (who espoused the party of the Persians in the time of Xerxes,) and the spring had failed, it then reappeared (on the arrival of Alexander); that the ambassadors also of the Milesians carried back to Memphis numerous answers of the oracle respecting the descent of Alexander from Jupiter, and the future victory which he should obtain at Arbela, the death of Darius, and the political changes at Lacedæmon. He says also that the Erythræan Athenais, who resembled the ancient Erythræan Sibyl, had declared the high descent of Alexander. Such are the accounts of historians.

44. At Abydos Osiris is worshipped; but in the temple of Osiris no singer, nor player on the pipe, nor on the cithara, is permitted to perform at the commencement of the ceremonies celebrated in honour of the god, as is usual in rites celebrated in honour of the other gods. Next to Abydos is [Pg 260] [CAS. 814]the lesser Diospolis,[875] then the city Tentyra,[876] where the crocodile is held in peculiar abhorrence, and is regarded as the most odious of all animals. For the other Egyptians, although acquainted with its mischievous disposition, and hostility towards the human race, yet worship it, and abstain from doing it harm. But the people of Tentyra track and destroy it in every way. Some however, as they say of the Psyllians of Cyrenæa, possess a certain natural antipathy to snakes, and the people of Tentyra have the same dislike to crocodiles, yet they suffer no injury from them, but dive and cross the river when no other person ventures to do so. When crocodiles were brought to Rome to be exhibited, they were attended by some of the Tentyritæ. A reservoir was made for them with a sort of stage on one of the sides, to form a basking-place for them on coming out of the water, and these persons went into the water, drew them in a net to the place, where they might sun themselves and be exhibited, and then dragged them back again to the reservoir. The people of Tentyra worship Venus. At the back of the fane of Venus is a temple of Isis; then follow what are called the Typhoneia, and the canal leading to Coptos,[877] a city common both to the Egyptians and Arabians.

45. Then follows the isthmus, extending to the Red Sea near Berenice,[878] which has no harbour, but good landing-places, because the isthmus is conveniently situated. Philadelphus is said to be the first person that opened, by means of his army, this road, which had no supply of water, and to have provided stations.[879]This he did because the navigation of the Red Sea was difficult, particularly to those who set out from the recess of the bay. Experience showed the great utility of this plan, and at present all the Indian, Arabian, and such Ethiopian merchandise as is imported by the Arabian Gulf is carried to Coptos, which is the mart for such commodities. Not far from Berenice is Myos Hormus,[880] a city with a naval station[Pg 261]for vessels which navigate this sea; at no great distance from Coptos is the city of Apollo, so that two cities are the boundaries of the isthmus, one on each side. But at present Coptos and Myos Hormus are in repute, and they are frequented.

Formerly, the camel-merchants travelled in the night, directing their course by observing the stars, and, like mariners, carried with them a supply of water. But now watering-places are provided: water is also obtained by digging to a great depth, and rain-water is found, although rain rarely falls, which is also collected in reservoirs. It is a journey of six or seven days.

On this isthmus are mines, in which the emeralds and other precious stones are found by the Arabians, who dig deep subterraneous passages.

46. Next to the city of Apollo is Thebes, now called Diospolis,

"with her hundred gates, through each of which issue two hundred men, with horses and chariots,"[881] according to Homer, who mentions also its wealth;

"not all the wealth the palaces of Egyptian Thebes contain."[882]

Other writers use the same language, and consider Thebes as the metropolis of Egypt. Vestiges of its magnitude still exist, which extend 80 stadia in length. There are a great number of temples, many of which Cambyses mutilated. The spot is at present occupied by villages. One part of it, in which is the city, lies in Arabia; another is in the country on the other side of the river, where is the Memnonium. Here are two colossal figures near one another, each consisting of a single stone. One is entire; the upper parts of the other, from the chair, are fallen down, the effect, it is said, of an earthquake. It is believed, that once a day a noise as of a slight blow issues from the part of the statue which remains [Pg 262] [CAS. 816]in the seat and on its base. When I was at those places with Ælius Gallus, and numerous friends and soldiers about him, I heard a noise at the first hour (of the day), but whether proceeding from the base or from the colossus, or produced on purpose by some of those standing around the base, I cannot confidently assert. For from the uncertainty of the cause, I am disposed to believe anything rather than that stones disposed in that manner could send forth sound.

Above the Memnonium are tombs of kings in caves, and hewn out of the stone, about forty in number; they are executed with singular skill, and are worthy of notice. Among the tombs[883] are obelisks with inscriptions, denoting the wealth of the kings of that time, and the extent of their empire, as reaching to the Scythians, Bactrians, Indians, and the present Ionia; the amount of tribute also, and the number of soldiers, which composed an army of about a million of men.

The priests there are said to be, for the most part, astronomers and philosophers. The former compute the days, not by the moon, but by the sun, introducing into the twelve months of thirty days each five days every year. But in order to complete the whole year, because there is (annually) an excess of a part of a day, they form a period from out of whole days and whole years, the supernumerary portions of which in that period, when collected together, amount to a day.[884] They [Pg 263]ascribe to Mercury all knowledge of this kind. To Jupiter, whom they worship above all other deities, a virgin of the greatest beauty and of the most illustrious family (such persons the Greeks call pallades) is dedicated. She prostitutes herself with whom she pleases, until the time occurs for the natural purification of the body; she is afterwards married; but before her marriage, and after the period of prostitution, they mourn for her as for one dead.

47. Next after Thebes is the city Hermonthis, in which both Apollo and Jupiter are worshipped. They also keep an ox there (for worship).

Next is the city of Crocodiles, the inhabitants of which worship this animal; then Aphroditopolis (the city of Venus),[885] and next to it, Latopolis, where Minerva is worshipped, and the (fish) Latus; next, the city of Eileithyia, and a temple. In the country on the other side of the river is Hieraconpolis (the city of hawks), where a hawk is worshipped; then Apollonopolis, the inhabitants of which are at war with crocodiles.

48. Syene is a city situated on the borders of Ethiopia and Egypt. Elephantina is an island in the Nile, at the distance of half a stadium in front of Syene; in this island is a city with a temple of Cnuphis, and a nilometer like that at Memphis. The nilometer is a well upon the banks of the Nile, constructed of close-fitting stones, on which are marked the greatest, least, and mean risings of the Nile; for the water in the well and in the river rises and subsides simultaneously. Upon the wall of the well are lines, which indicate the complete rise of the river, and other degrees of its rising. Those [Pg 264] [CAS. 817]who examine these marks communicate the result to the public for their information. For it is known long before, by these marks, and by the time[886]elapsed from the commencement, what the future rise of the river will be, and notice is given of it. This information is of service to the husbandmen with reference to the distribution of the water; for the purpose also of attending to the embankments, canals, and other things of this kind. It is of use also to the governors, who fix the revenue; for the greater the rise of the river, the greater it is expected will be the revenue.

At Syene there is a well which indicates the summer solstice, because these places lie under the tropical circle,[887] [and occasions the gnomons to cast no shadows at mid-day].[888] For on proceeding from the places in our country, in Greece I mean, towards the south, the sun is there first over [Pg 265]our head, and occasions the gnomons to be without shadows at noon. When the sun is vertical to us, it must necessarily cast its rays down wells, however deep they may be, to the water. For we ourselves stand in a perpendicular position, and wells are dug perpendicular to the surface.

Here are stationed three Roman cohorts as a guard.

49. A little above Elephantine is the lesser cataract, where the boatmen exhibit a sort of spectacle to the governors.

The cataract is in the middle of the river, and is formed by a ridge of rock, the upper part [or commencement] of which is level, and thus capable of receiving the river, but terminating in a precipice, where the water dashes down. On each side towards the land there is a stream, up which is the chief ascent for vessels. The boatmen sail up by this stream, and, dropping down to the cataract, are impelled with the boat to the precipice, the crew and the boats escaping unhurt.

A little above the cataract is Philæ, a common settlement, like Elephantina, of Ethiopians and Egyptians, and equal in size, containing Egyptian temples, where a bird, which they call hierax, (the hawk,) is worshipped; but it did not appear to me to resemble in the least the hawks of our country nor of Egypt, for it was larger, and very different in the marks of its plumage. They said that the bird was Ethiopian, and is brought from Ethiopia when its predecessor dies, or before its death. The one shown to us when we were there was sick and nearly dead.

50. We came from Syene to Philæ in a waggon, through a very flat country, a distance of about 100 stadia.[889] Along the whole road on each side we could see, in many places, very high rocks, round, very smooth, and nearly spherical, of black hard stone, of which mortars are made: each rested upon a greater stone, and upon this another: they were like hermæa.[890] Sometimes these stones consisted of one mass. The largest was not less than twelve feet in diameter, and all of them exceeded this size by one half. We crossed over to the island in a pacton, which is a small boat made of rods, [Pg 266] [CAS. 818]whence it resembles woven-work. Standing then in the water, (at the bottom of the boat,) or sitting upon some little planks, we easily crossed over, with some alarm indeed, but without good cause for it, as there is no danger if the boat is not overloaded.

51. Throughout the whole of Egypt, the palm tree is of a bad species, and produces no good edible fruit in the places about the Delta and Alexandreia; yet the best kind is found in the Thebaïs. It is a subject of surprise how countries in the same latitude as Judæa, and bordering upon the Delta and Alexandreia, should be so different; for Judæa, in addition to other kinds of date-palms, produces the caryotic, which is not inferior to the Babylonian. There are, however, two kinds of dates in the Thebaïs and in Judæa, the caryotic and another. The Thebaïc is firmer, but the flavour is more agreeable. There is an island remarkable for producing the best dates, and it also furnishes the largest revenue to the governors. It was appropriated to the kings, and no private person had any share in the produce; at present it belongs to the governors.

52. Herodotus[891] and other writers trifle very much when they introduce into their histories the marvellous, like (an interlude of) music and song, or some melody; for example, in asserting that the sources of the Nile are near the numerous islands, at Syene and Elephantina, and that at this spot the river has an unfathomable depth. In the Nile there are many islands scattered about, some of which are entirely covered, others in part only, at the time of the rise of the waters. The very elevated parts are irrigated by means of screw-pumps.

53. Egypt was from the first disposed to peace, from having resources within itself, and because it was difficult of access to strangers. It was also protected on the north by a harbourless coast and the Egyptian Sea; on the east and west by the desert mountains of Libya and Arabia, as I have said before.[892] The remaining parts towards the south are occupied by Troglodytæ, Blemmyes, Nubæ, and Megabari, Ethiopians above Syene. These are nomades, and not numerous nor warlike, but accounted so by the ancients, because[Pg

267] frequently, like robbers, they attacked defenceless persons. Neither are the Ethiopians, who extend towards the south and Meroë, numerous nor collected in a body; for they inhabit a long, narrow, and winding tract of land on the riverside, such as we have before described; nor are they well prepared either for war or the pursuit of any other mode of life.

At present the whole country is in the same pacific state, a proof of which is, that the upper country is sufficiently guarded by three cohorts, and these not complete. Whenever the Ethiopians have ventured to attack them, it has been at the risk of danger to their own country. The rest of the forces in Egypt are neither very numerous, nor did the Romans ever once employ them collected into one army. For neither are the Egyptians themselves of a warlike disposition, nor the surrounding nations, although their numbers are very large.

Cornelius Gallus, the first governor of the country appointed by (Augustus) Cæsar, attacked the city Heroopolis, which had revolted,[893] and took it with a small body of men. He suppressed also in a short time an insurrection in the Thebaïs, which originated as to the payment of tribute. At a later period Petronius resisted, with the soldiers about his person, a mob of myriads of Alexandrines, who attacked him by throwing stones. He killed some, and compelled the rest to desist.

We have before[894] related how Ælius Gallus, when he invaded Arabia with a part of the army stationed in Egypt, exhibited a proof of the unwarlike disposition of the people; and if Syllæus had not betrayed him, he would have conquered the whole of Arabia Felix.

54. The Ethiopians, emboldened in consequence of a part of the forces in Egypt being drawn off by Ælius Gallus, who was engaged in war with the Arabs, invaded the Thebaïs, and attacked the garrison, consisting of three cohorts, near Syene; surprised and took Syene, Elephantina, and Philæ, by a sudden inroad; enslaved the inhabitants, and threw down the statues of Cæsar. But Petronius, marching with less than 10,000 infantry and 800 horse against an army of 30,000 men, first compelled them to retreat to Pselchis, an Ethiopian city. He then sent deputies to demand restitution of what they had taken, and the reasons which had induced them to begin the [Pg 268] [CAS. 820]war. On their alleging that they had been ill treated by the nomarchs, he answered, that these were not the sovereigns of the country, but Cæsar. When they desired three days for consideration, and did nothing which they were bound to do, Petronius attacked and compelled them to fight. They soon fled, being badly commanded, and badly armed; for they carried large shields made of raw hides, and hatchets for offensive weapons; some, however, had pikes, and others swords. Part of the insurgents were driven into the city, others fled into the uninhabited country; and such as ventured upon the passage of the river escaped to a neighbouring island, where there were not many crocodiles on account of the current. Among the fugitives, were the generals of Candace, queen of the Ethiopians in our time, a masculine woman, and who had lost an eye. Petronius, pursuing them in rafts and ships, took them all and despatched them immediately to Alexandreia. He then attacked Pselchis[895] and took it. If we add the number of those who fell in battle to the number of prisoners, few only could have escaped.

From Pselchis Petronius went to Premnis,[896] a strong city, travelling over the hills of sand, beneath which the army of Cambyses was overwhelmed by the setting in of a whirlwind. He took the fortress at the first onset, and afterwards advanced to Napata.[897] This was the royal seat of Candace; and her son was there, but she herself was in a neighbouring stronghold. When she sent ambassadors to treat of peace, and [Pg 269]to offer the restitution of the prisoners brought from Syene, and the statues, Petronius attacked and took Napata, from which her son had fled, and then razed it. He made prisoners of the inhabitants, and returned back again with the booty, as he judged any farther advance into the country impracticable on account of the roads. He strengthened, however, the fortifications of Premnis, and having placed a garrison there, with two years' provisions for four hundred men, returned to Alexandreia. Some of the prisoners were publicly sold as booty, and a thousand were sent to Cæsar, who had lately returned from the Cantabrians,[898] others died of various diseases.

In the mean time Candace[899] attacked the garrison with an army of many thousand men. Petronius came to its assistance, and entering the fortress before the approach of the enemy, secured the place by many expedients. The enemy sent ambassadors, but he ordered them to repair to Cæsar: on their replying, that they did not know who Cæsar was, nor where they were to find him, Petronius appointed persons to conduct them to his presence. They arrived at Samos, where Cæsar was at that time, and from whence he was on the point of proceeding into Syria, having already despatched Tiberius into Armenia. The ambassadors obtained all that they desired, and Cæsar even remitted the tribute which he had imposed.

CHAPTER II.

1. IN the preceding part[200] of this work we have spoken at length of Ethiopia, so that its description may be said to be included in that of Egypt.

In general, then, the extreme parts of the habitable world adjacent to the intemperate region, which is not habitable by reason either of heat or cold, must necessarily be defective and inferior, in respect to physical advantages, to the temperate[Pg 270] [CAS. 821] region. This is evident from the mode of life of the inhabitants, and their want of what is requisite for the use and subsistence of man. For the mode of life [of the Ethiopians] is wretched; they are for the most part naked, and wander from place to place with their flocks. Their flocks and herds are small in size, whether sheep, goats, or oxen; the dogs also, though fierce and quarrelsome, are small.[201] It was perhaps from the diminutive size of these people, that the story of the Pygmies originated, whom no person, worthy of credit has asserted that he himself has seen.

2. They live on millet and barley, from which also a drink is prepared. They have no oil, but use butter and fat instead.[202] There are no fruits, except the produce of trees in the royal gardens. Some feed even upon grass, the tender twigs of trees, the lotus, or the roots of reeds. They live also upon the flesh and blood of animals, milk, and cheese. They reverence their kings as gods, who are for the most part shut up in their palaces.

Their largest royal seat is the city of Meroë, of the same name as the island. The shape of the island is said to be that of a shield. Its size is perhaps exaggerated. Its length is about 3000, and its breadth 1000 stadia. It is very mountainous, and contains great forests. The inhabitants are nomades, who are partly hunters and partly husbandmen. There are also mines of copper, iron, gold, and various kinds of precious stones. It is surrounded on the side of Libya by great hills of sand, and on that of Arabia by continuous precipices. In the higher parts on the south, it is bounded by the confluent[203] streams of the rivers Astaboras,[204] Astapus,[205] and Astasobas. On the north is the continuous course of the Nile to Egypt, with its windings, of which we have spoken before.

[Pg 271]The houses in the cities are formed by interweaving split pieces of palm wood or of bricks.[206] They have fossil salt, as in Arabia. Palm, the persea[207] (peach), ebony, and carob trees are found in abundance. They hunt elephants, lions, and panthers. There are also serpents, which encounter elephants, and there are many other kinds of wild animals, which take refuge, from the hotter and parched districts, in watery and marshy districts.

3. Above Meroë is Psebo,[208] a large lake, containing a well-inhabited island. As the Libyans occupy the western bank of the Nile, and the Ethiopians the country on the other side of the river, they thus dispute by turns the possession of the islands and the banks of the river, one party repulsing the other, or yielding to the superiority of its opponent.

The Ethiopians use bows of wood four cubits long, and hardened in the fire. The women also are armed, most of whom wear in the upper lip a copper ring. They wear sheep-skins, without wool; for the sheep have hair like goats. Some go naked, or wear small skins or girdles of well-woven hair round the loins.

They regard as God one being who is immortal, the cause of all things; another who is mortal, a being without a name, whose nature is not clearly understood.

In general they consider as gods benefactors and royal persons, some of whom are their kings, the common saviours and guardians of all; others are private persons, esteemed as gods by those who have individually received benefits from them.

Of those who inhabit the torrid region, some are even supposed not to acknowledge any god, and are said to abhor even the sun, and to apply opprobrious names to him, when they behold him rising, because he scorches and tortures them with his heat; these people take refuge in the marshes.

The inhabitants of Meroë worship Hercules, Pan, and Isis, besides some other barbaric deity.[202]

Some tribes throw the dead into the river; others keep them in the house, enclosed in hyalus (oriental alabaster?). [Pg 272] [CAS. 822]Some bury them around the temples in coffins of baked clay. They swear an oath by them, which is reverenced as more sacred than all others.

Kings are appointed from among persons distinguished for their personal beauty, or by their breeding of cattle, or for their courage, or their riches.

In Meroë the priests anciently held the highest rank, and sometimes sent orders even to the king, by a messenger, to put an end to himself, when they appointed another king in his place. At last one of their kings abolished this custom, by going with an armed body to the temple where the golden shrine is, and slaughtering all the priests.

The following custom exists among the Ethiopians. If a king is mutilated in any part of the body, those who are most attached to his person, as attendants, mutilate themselves in the same manner, and even die with him. Hence the king is guarded with the utmost care. This will suffice on the subject of Ethiopia.

4. To what has been said concerning Egypt, we must add these peculiar products; for instance, the Egyptian bean, as it is called, from which is obtained the ciborium,[210] and the papyrus, for it is found here and in India only; the persea (peach) grows here only, and in Ethiopia; it is a lofty tree, and its fruit is large and sweet; the sycamine, which produces the fruit called the sycomorus, or fig-mulberry, for it resembles a fig, but its flavour is not esteemed. The corsium also (the root of the Egyptian lotus) grows there, a condiment like pepper, but a little larger.

There are in the Nile fish in great quantity and of different kinds, having a peculiar and indigenous character. The best known are the oxyrynchus,[211] and the lepidotus,[212] the latus,[213] the alabes,[214] the coracinus,[215] the chœrus, the phagrorius, called also the phagrus. Besides these are the silurus, the citharus,[216]the thrissa,[217] the cestreus,[218] the lychnus, the physa, the bous (or ox), and large shell-fish which emit a sound like that of wailing.

[Pg 273]The animals peculiar to the country are the ichneumon and the Egyptian asp, having some properties which those in other places do not possess. There are two kinds, one a span in length, whose bite is more suddenly mortal than that of the other; the second is nearly an orguia[219] in size, according to Nicander, the author of the Theriaca.

Among the birds, are the ibis and the Egyptian hawk, which, like the cat, is more tame than those elsewhere. The nycticorax is here peculiar in its character; for with us it is as large as an eagle, and its cry is harsh; but in Egypt it is the size of a jay, and has a different note. The tamest animal, however, is the ibis; it resembles a stork in shape and size. There are two kinds, which differ in colour; one is like a stork, the other is entirely black. Every street in Alexandreia is full of them. In some respects they are useful; in others troublesome. They are useful, because they pick up all sorts of small animals and the offal thrown out of the butchers' and cooks' shops. They are troublesome, because they devour everything, are dirty, and with difficulty prevented from polluting in every way what is clean and what is not given to them.

5. Herodotus[220] truly relates of the Egyptians, that it is a practice peculiar to them to knead clay with their hands, and the dough for making bread with their feet. Caces is a peculiar kind of bread which restrains fluxes. Kiki (the castor-oil bean) is a kind of fruit sowed in furrows. An oil is expressed from it which is used for lamps almost generally throughout the country, but for anointing the body only by the poorer sort of people and labourers, both men and women.

The coccina are Egyptian textures made of some plant,[221] woven like those made of rushes, or the palm-tree.

[Pg 274] [CAS. 824]Barley beer is a preparation peculiar to the Egyptians. It is common among many tribes, but the mode of preparing it differs in each.

This, however, of all their usages is most to be admired, that they bring up all children that are born. They circumcise the males, and spay the females, as is the custom also among the Jews, who are of Egyptian origin, as I said when I was treating of them.[222]

According to Aristobulus, no fishes ascend the Nile from the sea, except the cestreus, the thrissa, and dolphins, on account of the crocodiles; the dolphin, because it can get the better of the crocodile; the cestreus, because it is accompanied by the chœri along the bank, in consequence of some physical affinity subsisting between them. The crocodiles abstain from doing any hurt to the chœri, because they are of a round shape, and have spines on their heads, which are dangerous to them. The cestreus runs up the river in spring, when in spawn; and descends a little before the setting of the pleiad, in great numbers, when about to cast it, at which time they are taken in shoals, by falling into inclosures (made for catching them). Such also, we may conjecture, is the reason why the thrissa is found there.

So much then on the subject of Egypt.

CHAPTER III.

1. WE shall next describe Africa, which is the remaining portion of the whole description of the earth.

We have before said much respecting it; but at present I shall further describe what suits my purpose, and add what has not been previously mentioned.[923]

[Pg 275]The writers who have divided the habitable world according to continents, divide it unequally. But a threefold division denotes a division into three equal parts. Africa, however, wants so much of being a third part of the habitable world, that, even if it were united to Europe, it would not be equal to Asia; perhaps it is even less than Europe; in resources it is very much inferior, for a great part of the inland and maritime country is desert. It is spotted over with small habitable parts, which are scattered about, and mostly belonging to nomade tribes. Besides the desert state of the country, its being a nursery of wild beasts is a hindrance to settlement in parts which could be inhabited. It comprises also a large part of the torrid zone.

All the sea-coast in our quarter, situated between the Nile and the Pillars, particularly that which belonged to the Carthaginians, is fertile and inhabited. And even in this tract, some spots destitute of water intervene, as those about the Syrtes, the Marmaridæ, and the Catabathmus.

The shape of Africa is that of a right-angled triangle, if we imagine its figure to be drawn on a plane surface. Its base is the coast opposite to us, extending from Egypt and the Nile to Mauretania and the Pillars; at right angles to this is a side formed by the Nile to Ethiopia, which side we continue to the ocean; the hypothenuse of the right angle is the whole tract of sea-coast lying between Ethiopia and Mauretania.

As the part situated at the vertex of the above-mentioned figure, and lying almost entirely under the torrid zone, is inaccessible, we speak of it from conjecture, and therefore cannot say what is the greatest breadth of the country. In a former[924] part of this work we have said, that the distance proceeding from Alexandreia southwards to Meroë, the royal seat of the Ethiopians, is about 10,000 stadia; thence in a straight line to the borders of the torrid zone and the habitable country, 3000 stadia. The sum, therefore, may be assumed as the greatest breadth of Africa, which is 13,000 or 14,000 stadia: its length may be a little less than double this sum. So much then on the subject of Africa in general. I am now to describe its several parts, beginning from the most celebrated on the west.

[Pg 276]

[CAS. 825]

2. Here dwell a people called by the Greeks Maurusii, and by the Romans and the natives Mauri, a populous and flourishing African nation, situated opposite to Spain, on the other side of the strait, at the Pillars of Hercules, which we have frequently mentioned before. On proceeding beyond the strait at the Pillars, with Africa on the left hand, we come to a mountain which the Greeks call Atlas, and the barbarians Dyris. Thence projects into the sea a point formed by the foot of the mountain towards the west of Mauretania, and called the Coteis.[925] Near it is a small town, a little above the sea, which the barbarians call Trinx; Artemidorus, Lynx; and Eratosthenes, Lixus.[926] It lies on the side of the strait opposite to Gadeira,[927] from which it is separated by a passage of 800 stadia, the width of the strait at the Pillars between both places. To the south, near Lixus and the Coteis, is a bay

called Emporicus,[228] having upon it Phœnician mercantile settlements. The whole coast continuous with this bay abounds with them. Subtracting these bays, and the projections of land in the triangular figure which I have described, the continent may rather be considered as increasing in magnitude in the direction of south and east. The mountain which extends through the middle of Mauretania, from the Coteis to the Syrtes, is itself inhabited, as well as others running parallel to it, first by the Maurusii, but deep in the interior of the country by the largest of the African tribes, called Gætuli.

3. Historians, beginning with the voyage of Ophelas (Apellas?),[229] have invented a great number of fables respecting the sea-coast of Africa beyond the Pillars. We have mentioned them before, and mention them now, requesting our readers [Pg 277]to pardon the introduction of marvellous stories, whenever we may be compelled to relate anything of the kind, being unwilling to pass them over entirely in silence, and so in a manner to mutilate our account of the country.

It is said, that the Sinus Emporicus (or merchants' bay) has a cave which admits the sea at high tide to the distance even of seven stadia, and in front of this bay a low and level tract with an altar of Hercules upon it, which, they say, is not covered by the tide. This I, of course, consider to be one of the fictitious stories. Like this is the tale, that on other bays in the succeeding coast there were ancient settlements of Tyrians, now abandoned, which consisted of not less than three hundred cities, and were destroyed by the Pharusii[230] and the Nigritæ. These people, they say, are distant thirty days' journey from Lynx.

4. Writers in general are agreed that Mauretania is a fertile country, except a small part which is desert, and is supplied with water by rivers and lakes. It has forests of trees of vast size, and the soil produces everything. It is this country which furnishes the Romans with tables, formed of one piece of wood, of the largest dimensions, and most beautifully variegated. The rivers are said to contain crocodiles and other kinds of animals similar to those in the Nile. Some suppose that even the sources of the Nile are near the extremities of Mauretania. In a certain river leeches are bred seven cubits in length, with gills, pierced through with holes, through which they respire. This country is also said to [Pg 278] [CAS. 826]produce a vine, the girth of which two men can scarcely compass, and bearing bunches of grapes of about a cubit in size. All plants and pot-herbs are tall, as the arum and dracontium;[231] the stalks of the staphylinus,[232] the hippomarathum,[233] and the scolymus[234] are twelve cubits in height, and four palms in thickness. The country is the fruitful nurse of large serpents, elephants, antelopes, buffaloes, and similar animals; of lions also, and panthers. It produces weasels (jerboas?) equal in size and similar to cats, except that their noses are more prominent; and multitudes of apes, of which Poseidonius relates, that when he was sailing from Gades to Italy, and approached the coast of Africa, he saw a forest low upon the sea-shore full of these animals, some on the trees, others on the ground, and some giving suck to their young. He was amused also with seeing some with large dugs, some bald, others with ruptures, and exhibiting to view various effects of disease.

5. Above Mauretania, on the exterior sea (the Atlantic), is the country of the western Ethiopians, as they are called, which, for the most part, is badly inhabited. Iphicrates[235] says, that camel-leopards are bred here, and elephants, and the animals called rhizeis,[236] which in shape are like bulls, but in manner of living, in size, and strength in fighting, resemble elephants. He speaks also of large serpents, and says that even grass grows upon their backs; that lions attack the young of the elephants, and that when they have wounded them, they fly on the approach of the dams; that the latter, when they see their young besmeared with blood, kill them; and that the lions return to the dead bodies, and devour them; that Bogus king of the Mauretanians, during his expedition against the western Ethiopians, sent, as a present to his wife, canes similar to the Indian canes, each joint of which contained eight chœnices,[237] and asparagus of similar magnitude.

6. On sailing into the interior sea, from Lynx, there are Zelis[238] a city and Tingis,[239] then the monuments of the Seven Brothers,[240] and the mountain lying below, of the name of Abyle,[241] [Pg 279]abounding with wild animals and trees of a great size. They say, that the length of the strait at the pillars is 120 stadia, and the least breadth at Elephas[242] 60 stadia. On sailing further along the coast, we find cities and many rivers, as far as the river Molochath,[243] which is the boundary between the territories of the Mauretanians

and of the Masæsyli. Near the river is a large promontory, and Metagonium,<u>244</u> a place without water and barren. The mountain extends along the coast, from the Coteis nearly to this place. Its length from the Coteis to the borders of the Masæsylii<u>245</u> is 5000 stadia. Metagonium is nearly opposite to New Carthage.<u>246</u> Timosthenes is mistaken in saying that it is opposite to Massalia.<u>247</u> The passage across from New Carthage to Metagonium is 3000 stadia, but the voyage along the coast to Massalia is above 6000 stadia.

7. Although the Mauretanians inhabit a country, the greatest part of which is very fertile, yet the people in general continue even to this time to live like nomades. They bestow care to improve their looks by plaiting their hair, trimming their beards, by wearing golden ornaments, cleaning their teeth, and paring their nails; and you would rarely see them [Pg 280]
[CAS. 828]touch one another as they walk, lest they should disturb the arrangement of their hair.

They fight for the most part on horseback, with a javelin; and ride on the bare back of the horse, with bridles made of rushes. They have also swords. The foot-soldiers present against the enemy, as shields, the skins of elephants. They wear the skins of lions, panthers, and bears, and sleep in them. These tribes, and the Masæsylii next to them, and for the most part the Africans in general, wear the same dress and arms, and resemble one another in other respects; they ride horses which are small, but spirited and tractable, so as to be guided by a switch. They have collars<u>248</u>made of cotton or of hair, from which hangs a leading-rein. Some follow, like dogs, without being led.

They have a small shield of leather, and small lances with broad heads. Their tunics are loose, with wide borders; their cloak is a skin, as I have said before, which serves also as a breastplate.

The Pharusii and Nigretes, who live above these people, near the western Ethiopians, use bows and arrows, like the Ethiopians. They have chariots also, armed with scythes. The Pharusii rarely have any intercourse with the Mauretanians in passing through the desert country, as they carry skins filled with water, fastened under the bellies of their horses. Sometimes, indeed, they come to Cirta,<u>249</u> passing through places abounding with marshes and lakes. Some of them are said to live like the Troglodytæ, in caves dug in the ground. It is said that rain falls there frequently in summer, but that during the winter drought prevails. Some of the barbarians in that quarter wear the skins of serpents and fishes, and use them as coverings for their beds. Some say that the Mauretanians<u>250</u> are Indians, who accompanied Hercules hither. A little before my time, the kings Bogus and Bocchus, allies of the [Pg 281]Romans, possessed this country; after their death, Juba succeeded to the kingdom, having received it from Augustus Cæsar, in addition to his paternal dominions. He was the son of Juba who fought, in conjunction with Scipio, against divus Cæsar. Juba died<u>251</u> lately, and was succeeded by his son Ptolemy, whose mother was the daughter of Antony and Cleopatra.

8. Artemidorus censures Eratosthenes for saying that there is a city called Lixus, and not Lynx, near the extremities of Mauretania; that there are a very great number of Phœnician cities destroyed,<u>252</u> of which no traces are to be seen; and that among the western Ethiopians, in the evenings and the mornings, the air is misty and dense;—for how could this take place where there is drought and excessive heat? But he himself relates of these same parts what is much more liable to objection. For he speaks of some tribes of Lotophagi, who had left their own country, and might have occupied the tract destitute of water; whose food might be a lotus, a sort of herb, or root, which would supply the want of drink; that these people extend as far as the places above Cyrene, and that they live there on milk and flesh, although they are situated in the same latitude.

Gabinius, the Roman historian, indulges in relating marvellous stories of Mauretania. He speaks of a sepulchre of Antæus at Lynx, and a skeleton of sixty feet in length, which Sertorius exposed, and afterwards covered it with earth.<u>253</u>His stories also about elephants are fabulous. He says, that other animals avoid fire, but that elephants resist and fight against it, because it destroys the forests; that they engage with men in battle, and send out scouts before them; that when they perceive their enemies fly, they take to flight themselves; and that when they are wounded, they hold out as suppliants branches of a tree, or a plant, or throw up dust.

9. Next to Mauretania is the country of the Masæsylii, beginning from the river Molocath, and ending at the promontory which is called Tretum,[254] the boundary of the country of [Pg 282] [CAS. 829]the Masæsyli and of the Masylies. From Metagonium to Tretum are 6000 stadia; according to others, the distance is less.

Upon the sea-coast are many cities and rivers, and a country which is very fertile. It will be sufficient to mention the most renowned. The city of Siga,[255] the royal seat of Syphax, is at the distance of 1000 stadia from the above-mentioned boundaries. It is now razed. After Syphax, the country was in the possession of Masanasses, then of Micipsa, next of his successors, and in our time of Juba, the father of the Juba who died lately. Zama,[256] which was Juba's palace, was destroyed by the Romans. At the distance of 600 stadia from Siga is Theon-limen (port of the gods);[257] next are some other obscure places.

Deep in the interior of the country are mountainous and desert tracts scattered here and there, some of which are inhabited and occupied by Gætuli extending to the Syrtes. But the parts near the sea are fertile plains, in which are numerous cities, rivers, and lakes.

10. Poseidonius says, but I do not know whether truly, that Africa is traversed by few, and those small rivers; yet he speaks of the same rivers, namely those between Lynx and Carthage, which Artemidorus describes as numerous and large. This may be asserted with more truth of the interior of the country, and he himself assigns the reason of it, namely, that in the northern parts of Africa (and the same is said of Ethiopia) there is no rain; in consequence therefore of the drought, pestilence frequently ensues, the lakes are filled with mud only, and locusts appear in clouds.

Poseidonius besides asserts that the eastern parts are moist, because the sun quickly changes its place after rising; and that the western parts are dry, because the sun there turns in his course. Now, drought and moisture depend upon the abundance or scarcity of water, and on the presence or absence of the sun's rays. But Poseidonius means to speak of the effects produced by the sun, which all writers determine by the latitude, north or south; but east and west, as applied to the residence of men, differ in different places, according to [Pg 283]the position of each inhabited spot and the change of horizon; so that it cannot be asserted generally of places indefinite in number, that those lying to the east are moist, and those to the west dry: but as applied to the whole earth and such extremes of it as India and Spain, his expressions (east and west) may be just; yet what truth or probability is there in his (attempted) explanation (of the causes of drought and moisture)? for in the continuous and unceasing circuit of the sun, what turn can there be in his course? The rapidity too of his passage through every part is equal. Besides, it is contrary to evidence to say, that the extreme parts of Spain or Mauretania towards the west are drier than all other places, when at the same time they are situated in a temperate climate and have water in great abundance. But if we are to understand the turning of the sun in this way, that there at the extremities of the habitable world he is above the earth, how does that tend to produce drought? for there, and in other places situated in the same latitude, he leaves them for an equal portion of the night and returns again and warms the earth.

11. Somewhere there, also, are copper mines; and a spring of asphaltus; scorpions of enormous size,[258] both with and without wings, are said to be found there, as well as tarantulas, remarkable for their size and numbers. Lizards also are mentioned of two cubits in length. At the base of the mountains precious stones are said to be found, as those called the Lychnitis (the ruby) and the Carchedonius (the carbuncle?). In the plains are found great quantities of oyster and mussel shells, similar to those mentioned in our description of Ammon. There is also a tree called melilotus, from which a wine is made. Some obtain two crops from the ground and have two harvests, one in the spring, the other in the summer. The straw is five cubits in height, and of the thickness of the little finger; the produce is 250-fold. They do not sow in the spring, but bush-harrow the ground with bundles of the paliurus, and find the seed-grain sufficient which falls from[Pg 284] [CAS. 831] the sheaves during harvest to produce the summer crop. In consequence of the

number of reptiles, they work with coverings on the legs; other parts of the body also are protected by skins.

12. On this coast was a city called Iol,[259] which Juba, the father of Ptolemy, rebuilt and changed its name to Cæsarea. It has a harbour and a small island in front of it. Between Cæsarea and Tretum[260] is a large harbour called Salda,[261] which now forms the boundary between the territories subject to Juba and the Romans; for the country has been subject to many changes, having had numerous occupants; and the Romans, at various times, have treated some among them as friends, others as enemies, conceding or taking away territories without observing any established rule.

The country on the side of Mauretania produced a greater revenue and was more powerful, whilst that near Carthage and of the Masylies was more flourishing and better furnished with buildings, although it suffered first in the Carthaginian wars, and subsequently during the war with Jugurtha, who successfully besieged Adarbal in Ityca (Utica),[262] and put him to death as a friend of the Romans, and thus involved the whole country in war. Other wars succeeded one another, of which the last was that between divus Cæsar and Scipio, in which Juba lost his life. The death of the leaders was accompanied by the destruction of the cities Tisiæus,[263]Vaga,[264] Thala,[265] Capsa[266] (the treasure-hold of Jugurtha), Zama,[267] and Zincha. To these must be added those cities in the neighbourhood of which divus Cæsar obtained victories over Scipio, namely, first at Ruspinum,[268] then at Uzita, then at Thapsus and the neighbouring lake, and at many others. Near are the free [Pg 285]cities Zella and Acholla.[269] Cæsar also captured at the first onset the island Cercinna,[270]and Thena, a small city on the sea-coast. Some of these cities utterly disappeared, and others were abandoned, being partly destroyed. Phara was burnt by the cavalry of Scipio.

13. After Tretum follows the territory of the Masylies, and that of the Carthaginians which borders upon it. In the interior is Cirta, the royal residence of Masanasses and his successors. It is a very strong place and well provided with everything, which it principally owes to Micipsa, who established a colony of Greeks in it, and raised it to such importance, that it was capable of sending out 10,000 cavalry and twice as many infantry. Here, besides Cirta, are the two cities Hippo,[271] one of which is situated near Ityca, the other further off near Tretum, both royal residences. Ityca is next to Carthage in extent and importance. On the destruction of Carthage it became a metropolis to the Romans, and the head quarters of their operations in Africa. It is situated in the very bay itself of Carthage, on one of the promontories which form it, of which the one near Ityca is called Apollonium, the other Hermæa. Both cities are in sight of each other. Near Ityca flows the river Bagradas.[272] From Tretum to Carthage are 2,500 stadia, but authors are not agreed upon this distance, nor on the distance (of Carthage) from the Syrtes.

14. Carthage is situated upon a peninsula, comprising a circuit of 360 stadia, with a wall, of which sixty stadia in length are upon the neck of the peninsula, and reach from sea to sea. Here the Carthaginians kept their elephants, it being a wide open place. In the middle of the city was the acropolis, which they called Byrsa, a hill of tolerable height with dwellings round it. On the summit was the temple of Esculapius, which was destroyed when the wife of Asdrubas burnt herself to death there, on the capture of the city. Below the Acropolis were the harbours and the Cothon, a circular island, surrounded by a canal communicating with the sea (Euripus), and on every side of it (upon the canal) were situated sheds for vessels.

[Pg   286]

[CAS. 832]15. Carthage was founded by Dido, who brought her people from Tyre. Both this colony and the settlements in Spain and beyond the Pillars proved so successful to the Phœnicians, that even to the present day they occupy the best parts on the continent of Europe and the neighbouring islands. They obtained possession of the whole of Africa, with the exception of such parts as could only be held by nomade tribes. From the power they acquired they raised a city to rival Rome, and waged three great wars against her. Their power became most conspicuous in the last war, in which they were vanquished by Scipio Æmilianus, and their city was totally destroyed. For at the commencement of this war, they possessed 300 cities in Africa, and the population of Carthage amounted to 700,000 inhabitants. After being besieged and compelled to surrender, they delivered up 200,000 complete suits of armour and 3000[273] engines for throwing projectiles, apparently with the intention of abandoning all hostilities; but having resolved to recommence the war, they at once began to manufacture arms, and daily deposited in store 140 finished shields, 300 swords, 500 lances, and 1000 projectiles for the engines, for the use of which the women-servants contributed their hair. In addition to this, although at this moment they were in possession of only twelve ships, according to the terms of the treaty concluded in the second war, and had already taken refuge in a body at the Byrsa, yet in two months they equipped 120 decked vessels; and, as the mouth of the Cothon was closed against them, cut another outlet (to the sea) through which the fleet suddenly made its appearance. For wood had been collected for a long time, and a multitude of workmen were constantly employed, who were maintained at the public expense.

Carthage, though so great, was yet taken and levelled to the ground.

The Romans made a province of that part of the country which had been subject to Carthage, and appointed ruler of the rest Masanasses and his descendants, beginning with Micipsa. For the Romans paid particular attention to Masanasses on account of his great abilities and friendship for them. For [Pg 287]he it was who formed the nomades to civil life, and directed their attention to husbandry. Instead of robbers he taught them to be soldiers. A peculiarity existed among these people; they inhabited a country favoured in everything except that it abounded with wild beasts; these they neglected to destroy, and so to cultivate the soil in security; but turning their arms against each other, abandoned the country to the beasts of prey. Hence their life was that of wanderers and of continual change, quite as much as that of those who are compelled to it by want and barrenness of soil or severity of climate. An appropriate name was therefore given to the Masæsylii, for they were called Nomades.[274] Such persons must necessarily be sparing livers, eaters of roots more than of flesh, and supported by milk and cheese. Carthage remained a desolate place for a long time, for nearly the same period, indeed, as Corinth, until it was restored about the same time (as the latter city) by divus Cæsar, who sent thither such Romans to colonize it as elected to go there, and also some soldiers. At present it is the most populous city in Africa.

16. About the middle of the gulf of Carthage is the island Corsura.[275] On the other side of the strait opposite to these places is Sicily and Lilybæum,[276] at the distance of (about) 1500 stadia; for this is said to be the distance from Lilybæum to Carthage. Not far from Corsura and Sicily are other islands, among which is Ægimurus.[277] From Carthage there is a passage of 60 stadia to the nearest opposite coast, from whence there is an ascent of 120 stadia to Nepheris, a fortified city built upon a rock. On the same gulf as Carthage, is situated a city Tunis; hot springs and stone quarries are also found there; then the rugged promontory   Hermæa,[278] [Pg   288]
[CAS. 834]on which is a city of the same name; then Neapolis; then Cape Taphitis,[279] on which is a hillock named Aspis, from its resemblance (to a shield), at which place Agathocles, tyrant of Sicily, collected inhabitants when he made his expedition against Carthage. These cities were destroyed by the Romans, together with Carthage. At the distance of 400 stadia from Taphitis is an island Cossuros, with a city of the same name, lying opposite to the river Selinus in Sicily. Its circuit is 150 stadia, and its distance from Sicily about 600 stadia. Melite,[280] an island, is 500 stadia distant from Cossuros. Then follows the city Adrumes,[281] with a naval arsenal; then the Taracheiæ, numerous small islands; then the city Thapsus,[282] and near it Lopadussa,[283] an island situated far from the coast; then the promontory of Ammon Balithon, near which is a look-out for[284] the approach of thunny; then the city Thena, lying at the entrance of the Little Syrtis.[285] There are many small cities in the intervening parts, which are not worthy of notice. At the entrance of the Syrtis, a long island stretches parallel to the coast, called Cercinna; it is of considerable size, with a city of the same name; there is also another smaller island Cercinnitis.

17. Close, in the neighbourhood (of these islands), is the Little Syrtis, which is also called the Syrtis Lotophagitis (or the lotus-eating Syrtis). The circuit of this gulf is 1600, and the breadth of the entrance 600 stadia; at each of the promontories which form the entrance and close to the mainland is an island, one of which, just mentioned, is Cercinna, and the other Meninx;[286] they are nearly equal in size. Meninx is supposed to be the "land of the lotus-eaters"[287] mentioned by Homer. Certain tokens (of this) are shown, such as an altar of

135

Ulysses and the fruit itself. For the tree called the lotus-tree is found in abundance in the island, and the fruit is very sweet to the taste. There are many small cities in it, one of which bears the same name as the island. On the coast of the Syrtis itself are also some small cities. In the recess (of [Pg 289]the Syrtis) is a very considerable mart for commerce, where a river discharges itself into the gulf. The effects of the flux and reflux of the tides extend up to this point, and at the proper moment the neighbouring inhabitants eagerly rush (to the shore) to capture the fish (thrown up).

18. After the Syrtis, follows the lake Zuchis, 400 stadia (in circuit?), with a narrow entrance, where is situated a city of the same name, containing factories for purple dyeing and for salting of all kinds; then follows another lake much smaller; after this the city Abrotonon[288] and some others. Close by is Neapolis, which is also called Leptis.[289] From hence the passage across to the Locri Epizephyrii[290] is a distance of 3600 stadia. Next is the river [Cinyps].[291] Afterwards is a walled dam, constructed by the Carthaginians, who thus bridged over some deep swamps which extend far into the country. There are some places here without harbours, although the rest of the coast is provided with them. Next is a lofty wooded promontory, which is the commencement of the Great Syrtis, and called Cephalæ (The Heads),[292] from whence to Carthage is a distance of a little more than 5000 stadia.

19. Above the sea-coast from Carthage to Cephalæ (on the one hand) and to the territory of the Masæsyli (on the other) lies the territory of the Libo-Phœnicians, extending (into the interior) to the mountainous country of the Gætuli, which belongs to Africa Proper. Above the Gætuli is the country of the Garamantes, lying parallel to the former, and from whence are brought the Carthaginian pebbles (carbuncles). The Garamantes are said to be distant from the Ethiopians, who live on the borders of the ocean, nine or ten days' journey, and from the temple of Ammon fifteen days. Between the Gætuli and the coast of our sea (the Mediterranean) there are many plains and many mountains, great lakes and rivers, some of which sink into the earth and disappear. The inhabitants are simple in their mode of life and in their dress; they marry numerous wives, and have a numerous offspring; in other respects they resemble the nomade Arabians. The [Pg 290] [CAS. 835]necks both of horses and oxen are longer than in other countries.

The breeding of horses is most carefully attended to by the kings (of the country); so much so, that the number of colts is yearly calculated at 100,000. Sheep are fed with milk and flesh, particularly near Ethiopia. These are the customs of the interior.

20. The circuit of the Great Syrtis is about 3930 stadia,[293] its depth to the recess is 1500 stadia, and its breadth at the mouth is also nearly the same. The difficulty of navigating both these and the Lesser Syrtis [arises from the circumstances of] the soundings in many parts being soft mud. It sometimes happens, on the ebbing and flowing of the tide, that vessels are carried upon the shallows, settle down, and are seldom recovered. Sailors therefore, in coasting, keep at a distance (from the shore), and are on their guard, lest they should be caught by a wind unprepared, and driven into these gulfs. Yet the daring disposition of man induces him to attempt everything, and particularly the coasting along a shore. On entering the Great Syrtis on the right, after passing the promontory Cephalæ, is a lake of about 300 stadia in length, and 70 stadia in breadth, which communicates with the gulf, and has at its entrance small islands and an anchorage. After the lake follows a place called Aspis, and a harbour, the best of all in the Syrtis. Near this place is the tower Euphrantas, the boundary between the former territory of Carthage and Cyrenaïca under Ptolemy (Soter). Then another place, called Charax,[294] which the Carthaginians frequented as a place of commerce, with cargoes of wine, and loaded in return with silphium and its juice, which they received from merchants who brought it away clandestinely from Cyrene; then the Altars of the Philæni;[295]after these Automola, a fortress defended by a garrison, and situated in the recess of the whole gulf. The parallel passing through this recess is more to the south than that passing through [Pg 291]Alexandreia by 1000 stadia, and than that passing through Carthage by less than 2000 stadia; but it would coincide with the parallel passing, on one side, through Heroopolis, which is situated in the recess[296] of the Arabian Gulf, and passing, on the other, through the interior of the territory of the Masæsylii and the Mauretanians. The rest of the sea-coast, to the city Berenice,[297] is 1500 stadia in length. Above this length of coast, and extending to the Altars of the Philæni, are situated an

African nation called Nasamones. The intervening distance (between the recess of the Syrtis and Berenice) contains but few harbours, and watering-places are rare.

On a promontory called Pseudopenias is situated Berenice, near a lake Tritonis, in which is to be observed a small island with a temple of Venus upon it. There also is a lake of the Hesperides, into which flows a river (called) Lathon. On this side of Berenice is a small promontory called Boreion[998] (or North Cape), which with Cephalæ forms the entrance of the Syrtis. Berenice lies opposite to the promontories of Peloponnesus, namely, those called Ichthys[999] and [Chelonatas],[1000] and also to the island Zacynthus,[1001] at an interval of 3600 stadia. Marcus Cato marched from this city, round the Syrtis, in thirty days, at the head of an army composed of more than 10,000 men, separated into divisions on account of the watering-places; his course lay through deep sand, under burning heat. After Berenice is a city Taucheira,[1002] called also Arsinoë; then Barca,[1003] formerly so called, but now Ptolemaïs;[Pg 292] [CAS. 837] then the promontory Phycus,[1004] which is low, but extends further to the north than the rest of the African coast; it is opposite to Tænarum,[1005] in Laconia, at the distance[1006] of 2800 stadia; on it there is also a small town of the same name as the promontory. Not far from Phycus, at a distance of about 170 stadia, is Apollonias, the naval arsenal of Cyrene; from Berenice it is distant 1000 stadia, and 80 stadia from Cyrene, a considerable city situated on a table-land, as I observed it from the sea.

21. Cyrene was founded by the inhabitants of Thera,[1007] a Lacedæmonian island which was formerly called Calliste, as Callimachus says,

"Calliste once its name, but Thera in later times, the mother of my home, famed for its steeds."

The harbour of Cyrene is situated opposite to Criu-Metopon,[1008] the western cape of Crete, distant 2000 stadia. The passage is made with a south-south-west wind. Cyrene is said to have been founded by Battus,[1009] whom Callimachus claims to have been his ancestor. The city flourished from the excellence of the soil, which is peculiarly adapted for breeding horses, and the growth of fine crops. It has produced many men of distinction, who have shown themselves capable of worthily maintaining the freedom of the place, and firmly resisting the barbarians of the interior; hence the city was independent in ancient times, but subsequently[1010] it was attacked [successfully] by the Macedonians, (who had conquered Egypt, and thus increased their power,) under the command of Thibron the murderer of Harpalus: having continued for some time to be governed by kings, it finally came under the power of the Romans, and with Crete forms a single province. In the neighbourhood of Cyrene are Apollonia, Barca, Taucheira, Berenice, and other small towns close by.

22. Bordering upon Cyrenaïca is the district which produces silphium, and the juice called Cyrenaic, which the silphium discharges from incisions made in it. The plant was once [Pg 293]nearly lost, in consequence of a spiteful incursion of barbarians, who attempted to destroy all the roots. The inhabitants of this district are nomades.

Remarkable persons of Cyrene were Aristippus,[1011] the Socratic philosopher, who established the Cyrenaïc philosophy, and his daughter named Arete, who succeeded to his school; she again was succeeded by her son Aristippus, who was called Metrodidactos, (mother-taught,) and Anniceris, who is supposed to have reformed the Cyrenaïc sect, and to have introduced in its stead the Anniceric sect. Callimachus and Eratosthenes[1012] were also of Cyrene, both of whom were held in honour by the kings of Egypt; the former was both a poet and a zealous grammarian; the latter followed not only these pursuits, but also philosophy, and was distinguished above all others for his knowledge of mathematics. Carneades[1013] also came from [Pg 294] [CAS. 838]thence, who by common consent was the first of the Academic philosophers, and Apollonius Cronos, the master of Diodorus the Dialectician, who was also called Cronos, for the epithet of the master was by some transferred to the scholar.

The rest of the sea-coast of Cyrene from Apollonia to Catabathmus is 2200 stadia in length; it does not throughout afford facilities for coasting along it; for harbours, anchorage, habitations, and watering-places are few. The places most in repute along the coast are the Naustathmus,[1014] and Zephyrium with an anchorage, also another Zephyrium, and a promontory called Chersonesus,[1015] with a harbour situated opposite to and to the south of Corycus[1016] in Crete, at the distance of 2500 stadia; then a temple of Hercules, and above it a village Paliurus; then a harbour Menelaus, and a low promontory Ardanixis, (Ardanis,)[1017] with an anchorage; then a great harbour, which is situated opposite to Chersonesus in Crete, at a distance of about 3000 (2000?) stadia; for the whole of Crete, which is (a) long and narrow (island), lies opposite and nearly parallel to this coast. After the great harbour is another harbour, Plynos, and about it Tetra-pyrgia (the four towers). The place is called Catabathmus.[1018] Cyrenæa extends to this point; the remainder (of the coast) to Parætonium,[1019] and from thence to Alexandreia, we have spoken of in our account of Egypt.[1020]

23. The country deep in the interior, and above the Syrtis and Cyrenæa, a very sterile and dry tract, is in the possession of Libyans. First are the Nasamones, then Psylli, and some Gætuli, then Garamantes; somewhat more towards the east (than the Nasamones) are the Marmaridæ, who are situated for the most part on the boundaries of Cyrenæa, and extend to the temple of Ammon. It is asserted, that persons directing their course from the recess of the Great Syrtis, (namely,) from about the neighbourhood of Automala,[1021] in the direction of the winter [Pg 295]sunrise, arrive on the fourth day at Augila.[1022] This place resembles Ammon, and is productive of palm trees, and is well supplied with water. It is situated beyond Cyrenæa to the south: for 100 stadia the soil produces trees; for another 100 stadia the land is only sown, but from excessive heat does not grow rice.

Above these parts is the district which produces silphium, then follows the uninhabited tract, and the country of the Garamantes. The district which produces silphium is narrow, long, and dry, extending in an easterly direction about 1000 stadia, but in breadth 300 stadia, or rather more, at least as far as has been ascertained. For we may conjecture that all countries which lie on the same parallel (of latitude) have the same climate, and produce the same plants; but since many deserts intervene, we cannot know every place. In like manner, we have no information respecting the country beyond (the temple of) Ammon, nor of the oases, as far as Ethiopia, nor can we state distinctly what are the boundaries of Ethiopia, nor of Africa, nor even of the country close upon Egypt, still less of the parts bordering on the ocean.

24. Such, then, is the disposition of the parts of the world which we inhabit.[1023]But since the Romans have surpassed (in power) all former rulers of whom we have any record, and possess the choicest and best known parts of it, it will be suitable to our subject briefly to refer to their Empire.

It has been already stated[1024] how this people, beginning from [Pg 296] [CAS. 839]the single city of Rome, obtained possession of the whole of Italy, by warfare and prudent administration; and how, afterwards, following the same wise course, they added the countries all around it to their dominion.

Of the three continents, they possess nearly the whole of Europe, with the exception only of the parts beyond the Danube, (to the north,) and the tracts on the verge of the ocean, comprehended between the Rhine and the Tanaïs (Don).

Of Africa, the whole sea-coast on the Mediterranean is in their power; the rest of that country is uninhabited, or the inhabitants only lead a miserable and nomade life.

Of Asia likewise, the whole sea-coast in our direction (on the west) is subject to them, unless indeed any account is to be taken of the Achæi, Zygi, and Heniochi,[1025] who are robbers and nomades, living in confined and wretched districts. Of the interior, and of the parts far inland, the Romans possess one portion, and the Parthians, or the barbarians beyond them, the other; on the east and north are Indians, Bactrians, and Scythians; then (on the south) Arabians and Ethiopians; but territory is continually being abstracted from these people by the Romans.

Of all these countries some are governed by (native) kings, but the rest are under the immediate authority of Rome, under the title of provinces, to which are sent governors and collectors of tribute; there are also some free cities, which from the first sought the friendship of Rome, or obtained their freedom as a mark of honour. Subject to her also are some princes, chiefs of tribes, and priests, who (are permitted) to live in conformity with their national laws.

25. The division into provinces has varied at different periods, but at present it is that established by Augustus Cæsar; for after the sovereign power had been conferred upon him by his country for life, and he had become the arbiter of peace and war, he divided the whole empire into two parts, one of which he reserved to himself, the other he assigned to the (Roman) people. The former consisted of such parts as required military defence, and were barbarian, or bordered upon nations not as yet subdued, or were barren and uncultivated, which though ill provided with everything else, were yet well furnished with strongholds, and might thus dispose [Pg 297]the inhabitants to throw off the yoke and rebel. All the rest, which were peaceable countries, and easily governed without the assistance of arms, were given over to the (Roman) people. Each of these parts was subdivided into several provinces, which received respectively the titles of "provinces of Cæsar" and "provinces of the People."

To the former provinces Cæsar appoints governors and administrators, and divides the (various) countries sometimes in one way, sometimes in another, directing his political conduct according to circumstances.

But the people appoint commanders and consuls to their own provinces, which are also subject to divers divisions when expediency requires it.

(Augustus Cæsar) in his first organization of (the Empire) created two consular governments, namely, (1.) the whole of Africa in possession of the Romans, excepting that part which was under the authority, first of Juba, but now of his son Ptolemy; and (2.) Asia within the Halys and Taurus, except the Galatians and the nations under Amyntas, Bithynia, and the Propontis. He appointed also ten consular governments in Europe and in the adjacent islands. Iberia Ulterior (Further Spain) about the river Bætis[1026] and Celtica Narbonensis[1027] (composed the two first). The third was Sardinia, with Corsica; the fourth Sicily; the fifth and sixth Illyria, districts near Epirus, and Macedonia; the seventh Achaia, extending to Thessaly, the Ætolians, Acarnanians, and the Epirotic nations who border upon Macedonia; the eighth Crete, with Cyrenæa; the ninth Cyprus; the tenth Bithynia, with the Propontis and some parts of Pontus.

Cæsar possesses other provinces, to the government of which he appoints men of consular rank, commanders of armies, or knights;[1028] and in his (peculiar) portion (of the empire) there are and ever have been kings, princes, and (municipal) magistrates.

FOOTNOTES:

[1]Book xii. c. iii. 39. Vol. ii. page 311, 312.

[2]Book xiii. c. iv. § 8. Vol. ii. page 405.

[3]Book x. c. iv. § 10, and book xii. c. iii. § 33. Vol. ii. pp. 197, 307, of this Translation.

[4]Book xiv. c. i. § 48. Vol. iii. p. 26.

[5]Book xiv. c. v. § 4. Vol. iii. p. 53.

[6]Book xii. c. iii. § 16. Vol. ii. p. 296, 380.

[7]c. ii. § 24. Vol. iii. p. 173.

[8]Book ii. c. v. § 10. Vol. i. p. 176, of this Translation.

[9]Chap. i. § 20.

[10]Chap. i. § 13.

[11]Chap. i. § 20.

[12]Ibid.

[13]Book ii. c. 3, § 6. Vol. i. p. 154.

[14]Herodotus iv. 85, 86.

[15]Book i. c. iv. § 6. Vol. i. p. 102, of the Translation.

[16]Book ii. c. i. § 20. Vol. i. p. 119, of the Translation.

[17]Book xiii. c. i. § 54, vol. ii. p. 380.

[18]"A Reply to the Calumnies of the Edinburgh Review against Oxford," page 98, by Dr. Copleston, late Bishop of Landaff. Oxford, 1810.

[19]That is, the maritime parts of Asia Minor, from Cape Coloni opposite Mitilini to Bajas, the ancient Issus. The coast of Ionia comprehended between Cape Coloni and the Mæander (Bojuk Mender Tschai) forms part of the modern pachalics, Saruchan and Soghla;

Caria and Lycia are contained in the pachalic, Mentesche; Pamphylia and Lycia in those of Teke and Itsch-ili. Mount Taurus had its beginning at the promontory Trogilium, now Cape Samsoun, or Santa Maria opposite Samos.

[20]Jenikoi.

[21]Cape Arbora.

[22]Karadscha-Fokia.

[23]Gedis-Tschai.

[24]Derekoi.

[25]Lebedigli, Lebeditzhissar.

[26]A portion of this poem by Mimnermus is quoted in Athenæus, b. xi. 39, p. 748 of the translation, Bohn's Class. Library.

[27]Pliny, v. 29, says the distance is 20 stadia.

[28]The Branchidæ were descendants of Branchus, who himself was descended from Macæreus, who killed Neoptolemus, son of Achilles. According to Herodotus, the temple was burnt by order of Darius, Herod. v. 36; vi. 19.

[29]Pliny, v. 29, says that the distance is 180 stadia.

[30]According to Pausanias, vii. 2, a friend of Sarpedon, named Miletus, conducted the colony from Crete, founded Miletus, and gave his name to it. Before his arrival the place bore the name of Anactoria, and more anciently Lelegis.

[31]More than 80, according to Pliny, v. 29.

[32]To be well.

[33]Hence the English weal, the mark of a stripe.

[34]Od. xxiv. 402.

[35]Coraÿ, who is followed by Groskurd, supposes the words "and Cadmus" to be here omitted. Kramer considers this correction to be very doubtful; see b. i. c. ii. § 6.

[36]Chandler says that the Tragææ were sand-banks or shallows.

[37]Bafi.

[38]Il. ii. 868.

[39]ἐν ὕψει, according to Groskurd's emendation, in place of ἐν ὄψει.

[40]Derekoi.

[41]Two other towns, Percote and Palæscepsis, were also given to Themistocles, the first to supply him with dress, the second with bed-room furniture.—*Plutarch, Life of Themistocles.*

[42]Aineh-Basar.

[43]Samsun.

[44]Samsun Dagh.

[45]Cape Santa Maria.

[46]The Furni islands.

[47]Stapodia.

[48]According to Pliny, it is 716 stadia.

[49]In b. x. ch. ii. § 17, Strabo informs us that Samos was first called Melamphylus, then Anthemis, and afterwards Parthenia. These names appear in this passage in a reversed but, as appears from Pliny, b. v. 31, in their true chronological order.

[50]Either an error of our author, or he speaks of its wine in comparison with that of other islands.

[51]After the death of Pericles.

[52]Among distinguished natives of Samos, Strabo has omitted to mention Melissus the philosopher, who commanded the fleet of the island, and was contemporary with Pericles.—*Plutarch, Life of Pericles.*

[53]Before called Drepanum.

[54]Ischanli.

[55]Scala Nova.

[56]Pliny and Mela give a different origin and name to this town: by them it is called Phygela from Φυγὴ, flight or desertion of the sailors, who, wearied with the voyage, abandoned Agamemnon.

[57]Chersiphron was of Gnossus in Crete. The ground being marshy on which the temple was to be built, he prepared a foundation for it of pounded charcoal, at the suggestion of Theodorus, a celebrated statuary of Samos.

[58]The temple is said to have been burnt the night Alexander the Great was born.—*Cicero, de Nat. Deo.* ii. 27.

[59]Plutarch says that the artist offered Alexander to make a statue of Mount Athos, which should hold in the left hand a city, capable of containing 10,000 inhabitants, and pouring from the right hand a river falling into the sea.

[60]For the word κρήνη, a fountain, which occurs in the text before Penelope, and is here unintelligible, Kramer proposes to read κηρίνη. The translation of the passage, thus corrected, would be, "a figure in wax of Penelope." Kramer does not adopt the reading, on the ground that no figures in wax are mentioned by ancient authors.

[61]ὀνήιστος.

[62]Coraÿ is of opinion that the name of Artemidorus of Ephesus has been omitted by the copyist in this passage, before the name of Alexander. Kramer thinks that if the name had existed in the original manuscript, it would have been accompanied, according to the practice of Strabo, with some notice of the writings of Artemidorus. The omission of the name is remarkable, as Artemidorus is one of the geographers most frequently quoted by Strabo. He flourished about 100 B. C. His geography in eleven books is lost. An abridgement of this work was made by Marcianus, of which some portions still exist, relating to the Black Sea and its southern shore.

[63]It must have been in existence in the time of Strabo.—*Tacit. Ann.* ii. 54.

[64]Another explanation is given to the proverb, from the circumstance of Colophon having a casting vote in the deliberations of the twelve cities forming the Panionium.

[65]Lebedigli Lebeditz hissar.

[66]During the season when these actors, dancers, and singers were not on circuit at festivals.

[67]Budrun.

[68]Ouvriokasli.

[69]Ypsilo Nisi.

[70]Called by Livy, xxvii. 27, Portus Geræsticus.

[71]Which forms the Gulf of Smyrna.

[72]The district called Chalcitis by Pausanias, xii. 5, 12.

[73]Ritri.

[74]Sighadschik.

[75]Koraka, or Kurko.

[76]Called in Thucyd. viii. 34, Arginum.

[77]Karaburun-Dagh.

[78]Karaburun, which has the same meaning.

[79]Groskurd is of opinion that "of the same name" is omitted after "city."

[80]Cape Mastico.

[81]Porto Mastico.

[82]This name is doubtful. Coraÿ suggests Elæus; Groskurd, Lainus, which Kramer does not approve of, although this part of the coast is now called Lithi. It seems to be near a place called Port Aluntha.

[83]Cape Nicolo.

[84]Psyra.

[85]Ilias.

[86]Ion was a contemporary of Sophocles. Theopompus was the disciple of Socrates, and the author of an epitome of the history of Herodotus, of a history of Greece, of a history of Philip, father of Alexander the Great, and of other works. He was of the aristocratic or Macedonian party. Theocritus, his contemporary, was a poet, orator, and historian; he was of the democratic party. To these, among illustrious natives of Chios, may be added Œnopides the astronomer and mathematician, who was the discoverer of the obliquity of the ecliptic and the cycle of 59 years, for bringing the lunar and solar years into

accordance; Nessus the philosopher; his disciple Metrodorus (about B. C. 330) the sceptic, and master of Hippocrates; Scymnus the geographer, and author of a description of the earth.

[87]The Homeridæ may have been at first descendants of Homer; but in later times those persons went by the name Homeridæ, or Homeristæ, who travelled from town to town for the purpose of reciting the poems of Homer. They did not confine themselves to that poet alone, but recited the poetry of Hesiod, Archilochus, Mimnermus, and others; and finally passages from prose writers.—*Athenæus*, b. xiv. c. 13.

[88]Of the 283 vessels sent by the eight cities of Ionia in the war with Darius, one hundred came from Chios.

[89]Kelisman.

[90]Still to be found in collections of coins.

[91]Leokaes?

[92]B. xiii. c. iv. § 2.

[93]Ak-Hissar.

[94]Karadscha-Fokia.

[95]Marseilles, b. iv. ch. i. § 4.

[96]B. xiii. ch. i. § 2.

[97]Jenidscheh.

[98]Western Africa.

[99]Gumusch-dagh.

[100]According to Suidas, Daphnidas ridiculed oracles, and inquired of the oracle of Apollo, "Shall I find my horse?" when he had none. The oracle answered that he would find it. He was afterwards, by the command of Attalus, king of Pergamum, taken and thrown from a precipice called the Horse.

[101]The incursions of the Treres, with Cimmerians, into Asia and Europe followed after the Trojan war. The text is here corrupt. The translation follows the amendments proposed partly by Coraÿ, and partly by Kramer, τὸ δ' ἐξῆς Ἐφεσίους.

[102]These innovations or corruptions were not confined to the composition of pieces intended for the theatre, but extended also to the manner of their representation, to music, dancing, and the costume of the actors. It was an absolute plague, which corrupted taste, and finally destroyed the Greek theatre. We are not informed of the detail of these innovations, but from what we are able to judge by comparing Strabo with what is found in Athenæus, (b. xiv. § 14, p. 990, of Bohn's Classical Library,) Simodia was designated by the name of Hilarodia, (joyous song,) and obtained the name Simodia from one Simus, or Simon, who excelled in the art. The Lysiodi and Magodi, or Lysodia and Magodia, were the same thing, according to some writers. Under these systems decency appears to have been laid aside.

[103]Od. ix. 3.

[104]Aidin-Gusel-Hissar.

[105]The chain of mountains between the Caÿster and the Mæander, the different eminences of which bear the names of Samsun-dagh, Gumusch-dagh, Dsehuma-dagh, &c.

[106]Sultan-Hissar.

[107]The Tralli Thracians appear to have acted as mercenary soldiers, according to Hesychius.

[108]Groskurd supplies the word πρόσκεινται.

[109]Groskurd reads τοιούτων, for τοσούτων in the text. Coraÿ proposes νοσούντων.

[110]Meineke's conjecture is followed, λίπα ἀληλιμμένοι, for ἀπαληλιμμένοι.

[111]Groskurd's emendation of this corrupt passage is adopted, ὑπερβᾶσι τὴν Μεσωγίδα ἐπὶ τὰ πρὸς τὸν νότον μέρη Τμώλου τοῦ ὄρους.

[112]Il. ii. 461.

[113]Arpas-Kalessi.

[114]Mastauro.

[115]Adopting Kramer's correction of Καρίας for παραλίας.

[116]Cape Arbora.

[117]Schelidan Adassi islands, opposite Cape Chelidonia.

[118]Near Gudschek, at the bottom of the Gulf of Glaucus, now Makri.

[119]The Phœnix (Phinti?) rises above the Gulf of Saradeh.

[120]Alessa, or, according to others, Barbanicolo.

[121]Dalian.

[122]Doloman-Ischai.

[123]Kramer suggests the words ὑπομέλανας καὶ, for the corrupt reading, ἐπιμελῶς.

[124]Il. vi. 146.

[125]The Caunians were aborigines of Caria, although they affected to come from Crete.—*Herod.* i. 72.

[126]Castro Marmora. The gulf on which it stands is still called Porto Fisko.

[127]Chares flourished at the beginning of the third century B. C. The accounts of the height of the Colossus of Rhodes differ slightly, but all agree in making it 105 English feet. It was twelve years in erecting, (B. C. 292-280,) and it cost 300 talents. There is no authority for the statement that its legs extended over the mouth of the harbour. It was overthrown 56 years after its erection. The fragments of the Colossus remained on the ground 923 years, until they were sold by Moawiyeh, the general of the Caliph Othman IV., to a Jew of Emessa, who carried them away on 900 camels, A. D. 672. Hence Scaliger calculated the weight of the bronze at 700,000 pounds.—*Smith's Dict. of Biog. and Mythology.*

[128]Protogenes occupied seven years in painting the Jalysus, which was afterwards transferred to the Temple of Peace at Rome. The Satyr was represented playing on a flute, and was entitled, The Satyr Reposing.—*Plutarch, Demetr.*; *Pliny,* xxxv. 10.

[129]ὀψωνιασμοῦ, Kramer's proposed correction, is adopted for ὀψωνιαζόμενοι.

[130]Marseilles and Artaki.

[131]Bodrum.

[132]Il. ii. 662.

[133]Il. ii. 656.

[134]Il. ii. 678.

[135]Formerly, says Pliny, it was called Ophiussa, Asteria, Æthræa, Trinacria, Corymbia, Pœeessa, Atabyria, from a king of that name; then Macaria and Oloëssa. B. v. 31. To these names may be added Lindus and Pelagia. Meineke, however, suspects the name Stadia in this passage to be a corruption for Asteria.

[136]That is, Children of the Sun. They were seven in number, Cercaphus, Actis, Macareus, Tenages, Triopes, Phaethon, and Ochimus, born of the Sun and of a nymph, or, according to others, of a heroine named Rhodus.

[137]Il. ii. 656.

[138]Hippodamus of Miletus.

[139]Naples.

[140]Majorca.

[141]Negropont.

[142]Called light-armed probably from the use of the sling, common among the Rhodians, as it was also among the Cretans. The use of the sling tends to prove the Rhodian origin of the inhabitants of the Balearic islands. The Athenian expedition to Sicily (Thucyd. vi. 43) was accompanied by 700 slingers from Rhodes.

[143]Strabo here omits to mention the Rhodian origin of Agrigentum and Gela in Sicily.

[144]Il. ii. 668.

[145]Od. vii. 61.

[146]Lindo.

[147]According to Strabo, Alexandria and Rhodes were upon the same meridian.

[148]Camiro.

[149]Lanathi?

[150]Abatro.

[151]B. x. c. v. § 14.

[152]The original, which is a play upon words, cannot be rendered in English.
[153]Called before, Eleussa, c. ii. § 2.
[154]The Sea of Marmora.
[155]Capo Volpe, or Alepo Kavo, meaning the same thing.
[156]Isle of Symi.
[157]Crio.
[158]Indschirli, or Nisari.
[159]Keramo.
[160]The word ἔργον, "a work," suggests that there is some omission in the text. Coraÿ supposes that the name of the architect or architects is wanting. Groskurd would supply the words Σκόπα καὶ ἄλλων τεχνιτῶν, "the work of Scopas and other artificers." See Pliny, N. H. xxxvi., and Vitruvius Præf. b. vii.
[161]Coronata.
[162]Mela says, of Argives. B. i. c. xvi. § 19.
[163]Petera, or Petra Termera.
[164]Cape Kephala.
[165]Pascha-Liman.
[166]Assem-Kalessi.
[167]Cape Arbore.
[168]Mylassa, or Marmora.
[169]Eski-hissar.
[170]Arab-hissar.
[171]This is a parody on a passage in Aristophanes. Lysis. v. 1038.
[172]Of the golden rays (around the head).
[173]Cicero. *Brut.*, c. 91.
[174]Il. ii. 867, in which the reading is Νάστης, but Μέσθλης in Il. ii. 864.
[175]Od. i. 344.
[176]Il. xv. 80.
[177]Il. v. 222.
[178]βατταρίζειν, τραυλίζειν, ψελλίζειν.
[179]κελαρύζειν, κλαγγὴ, ψόφος, βοὴ, κρότος.
[180]Chelidoniæ, in this passage, is probably an error. Groskurd adopts the name Philomelium.
[181]Ilgun.
[182]At the base of Sultan-dagh.
[183]Ak-Schehr.
[184]Sultan Chan.
[185]Ak-Sera.
[186]Kaiserieh.
[187]Called Herpa, b. xii. ch. ii. § 6, pages 281, 283.
[188]Μετὰ τὴν Ῥοδίων Περαίαν, or, "After the Peræa of Rhodes." Peræa was the name of the coast of Caria opposite to Rhodes, which for several centuries formed a dependency of that opulent republic. In the time of Scylax, the Rhodians possessed only the peninsula immediately in face of their island. As a reward for their assistance in the Antiochian war, the Romans gave them a part of Lycia, and all Caria as far as the Mæander. By having adopted a less prudent policy in the second Macedonic war, they lost it all, including Caunus, the chief town of Peræa. It was not long, however, before it was restored to them, together with the small islands near Rhodes; and from this time Peræa retained the limits which Strabo has described, namely, Dædala on the east and Mount Loryma on the west, both included. Vespasian finally reduced Rhodes itself into the provincial form, and joined it to Caria.—*Leake.*
[189]Samsun.
[190]Eski Adalia, Old Attaleia; but the Greeks gave the name παλαιὰ Ἀττάλεια, Old Attaleia, to Perge.—*Leake.*
[191]Gunik.

[192]Patera.

[193]Minara.

[194]Duvar.

[195]Gillies, in his translation of Aristotle, makes use of this example of the Lycians to prove that representative government was not unknown to the ancients. The deputies sent from the twenty-three cities formed a parliament. The taxes and public charges imposed on the several towns were in proportion to the number of representatives sent from each city.—*Gillies*, vol. ii. p. 64, &c.

[196]Makri.

[197]Site unknown.

[198]Efta Kavi, the Seven Capes.

[199]Od. xix. 518.

[200]Kodscha.

[201]The passage in the original, in which all manuscripts agree, and which is the subject of much doubt, is— ὧν καὶ μεγίστη νῆσος καὶ πόλις ὁμώνυμος, ἡ Κισθήνη. Groskurd would read καὶ before ἡ, and translates,—"Among others is Megiste an island, and a city of the same name, and Cisthene." Later writers, says Leake, make no mention of Cisthene; and Ptolemy, Pliny, Stephanus, agree in showing that Megiste and Dolichiste were the two principal islands on the coast of Lycia: the former word Megiste, *greatest*, well describing the island Kasteloryzo or Castel Rosso, as the latter word (longest) does that of Kakava. Nor is Scylax less precise in pointing out Kasteloryzo as Megiste, which name is found in an inscription copied by M. Cockerell from a rock at Castel Rosso. It would seem, therefore, that this island was anciently known by both names, (Megiste and Cisthene,) but in later times perhaps chiefly by that of Megiste.

[202]Cape Chelidonia.

[203]Aboukir, nearly under the same meridian.

[204]Tschariklar.

[205]Garabasa.

[206]Tschiraly. Deliktasch.—*Leake*.

[207]Ianartasch.

[208]Tirikowa.

[209]Solyma-dagh.

[210]Gulik-Chan?

[211]Il. vi. 184.

[212]Duden-su.

[213]Adalia.

[214]Ernatia.

[215]Ak-su.

[216]Murtana.

[217]Tekeh.

[218]Kopru-su.

[219]Balkesu.

[220]Kislidscha-koi.

[221]Menavgat-su.

[222]Alara.

[223]Alaja, or Castel Ubaldo.

[224]Herod. vii. 91. According to this passage, therefore, the name Pamphylians is derived from πᾶν, "all," and φῦλον, "nation."

[225]Alaja.

[226]Syedra probably shared with Coraresium (Alaja), a fertile plain which here borders on the coast. But Syedra is Tzschucke's emendation of Arsinoë in the text.

[227]Not mentioned by any other author.

[228]Selindi.

[229]Charadran.

[230]Kara-Gedik.

[231]Inamur.

[232]Cape Kormakiti.

[233]Mesetlii.

[234]Softa-Kalessi.

[235]Mandane?

[236]Kilandria, or Gulnar.

[237]According to Pliny, Cilicia anciently commenced at the river Melas, which Strabo has just said belongs to Pamphylia. Ptolemy fixes upon Coracesium as the first place in Cilicia, which, according to Mela, was separated from Pamphylia by Cape Anemurium, which was near Nagidus.

[238]Nahr-el-Asy.

[239]B. xvi. c. ii. § 33.

[240]Selefke.

[241]Cape Lissan.

[242]Gok-su.

[243]Cape Cavaliere.

[244]Eurip. Hec. 1.

[245]Its distance (40 stadia) from the Calycadnus, if correct, will place it about Pershendi, at the north-eastern angle of the sandy plain of the Calycadnus.

[246]Anamur.

[247]Ianartasch; but, according to Leake, it still preserves its name.

[248]A sandy plain now connects Elæussa with the coast.—*Leake.*

[249]Lamas-su, of which Lamuzo-soui is an Italian corruption.

[250]Lamas.

[251]Tschirlay, or Porto Venetico.

[252]Mesetlii.

[253]Cape Zafra.

[254]What better inscription, said Aristotle, could you have for the tomb, not a king, but of an ox? Cicero, Tusc. Quæs. iii. 35.

[255]Mesarlyk-tschai.

[256]Strabo means to say, that the coast, from the part opposite Rhodes, runs E. in a straight line to Tarsus, and then inclines to the S.E.; that afterwards it inclines to the S., to Gaza, and continues in a westerly direction to the Straits of Gibraltar.

[257]The translation follows the reading proposed by Groskurd, παχυνευροῦσι καὶ ῥοϊζομένοις καὶ ποδαγριζομένοις, who quotes Vitruv. viii. 3, and Pliny xxxi. 8.

[258]Kramer does not approve of the corrections proposed in this passage by Groskurd. The translation follows the proposed emendation of Falconer, which Kramer considers the least objectionable.

[259]Augustus.

[260]Groskurd, with some probability, supposes the name of Achilles to be here omitted.

[261]Il. iii. 235.

[262]Dschehan-tschai.

[263]Chun.

[264]Ajas.

[265]Demir-Kapu.

[266]The ridge extending N. E., the parts of which bear various names, Missis, Durdan-dagh, &c.

[267]Deli-tschai.

[268]Arsus.

[269]Iskenderun.

[270]Its name under the Byzantine empire was corrupted to Mampsysta, or Mamista; of which names the modern Mensis appears to be a further corruption.—*Leake.*

[271]The passage is defended by the fortress of Merkes.

[272]Suveidijeh.

[273]Nahr-el-Asy.

[274]Groskurd is desirous of reading Tarsus for Issus. See above, c. v. § 11. But Strabo is here considering the two opinions held respecting the isthmus.

[275]Scymnus of Chios counts fifteen nations who occupied this peninsula, namely, three Greek and twelve barbarian. The latter were Cilicians, Lycians, Carians, Maryandini, Paphlagonians, Pamphylians, Chalybes, Cappadocians, Pisidians, Lydians, Mysians, and Phrygians. In this list the Bithynians, Trojans, and Milyæ are not mentioned; but in it are found the Cappadocians and Lydians—two nations whom, according to Strabo, Ephorus has not mentioned. This discrepancy is the more remarkable as Scymnus must have taken the list from Ephorus himself.

[276]Od. xi. 122.

[277]Apollodorus, like Scymnus, had probably found the Lydians mentioned in the list of Ephorus, as also the Cappadocians.

[278]Kramer says that he is unable to decide how this corrupt passage should be restored. The translation follows the conjectures of Coray.

[279]Il. ii. 862.

[280]Il. iii. 187.

[281]Isnik.

[282]Euphorion acquired celebrity as a voluminous writer. Vossius, i. 16, gives a catalogue of his works. According to Suidas, he was born in Chalcis, in Negropont, at the time Pyrrhus, king of Epirus, was defeated by the Romans. He acquired a considerable fortune by his writings and by his connexion with persons of eminent rank. He was invited to the court of Antiochus the Great, king of Syria, who intrusted him with the care of his library. According to Sallust, (Life of Tiberius,) he was one of the poets whom Tiberius took as his model in writing Greek verse. Fecit et Græca poemata, imitatus Euphorionem, et Rhianum et Parthenium.

[283]The Clides, off Cape Andrea.

[284]Cape Arnauti.

[285]Dschehan-Tschai.

[286]Kormakiti.

[287]Lapito.

[288]Near Artemisi.

[289]To the north of Tamagousta.

[290]Carpas.

[291]Lissan el Cape, in Cilicia.

[292]Near the present Larnaka.

[293]Limasol.

[294]Cape Gata.

[295]Cape Grego.

[296]Piscopia.

[297]Capo Bianco.

[298]Bisur.

[299]Point Zephyro.

[300]Jeroskipo.

[301]Solea.

[302]The Indian Ocean.

[303]Behul or Jelum.

[304]Beas.

[305]The island Cos, or Stanco, one of the earlier names of which was Meropis.

[306]ἢ κατ᾽ ἄλλους for καὶ ἄλλου.—*Groskurd.*

[307]See ch. i. § 73.

[308]Mekran.

[309]It is evident that the name Pillars misled Megasthenes or the writers from whom he borrowed the facts; for it is impossible to suppose that Tearcho, who reigned in Arabia,

or that Nabuchodonosor, who reigned at Babylon, ever conducted an army across the desert and through the whole breadth of Africa to the Straits of Gibraltar, to which place nothing invited them, and the existence of which, as well as that of the neighbouring countries, must have been unknown. The Egyptians, Arabians, and Babylonians directed their invasions towards the north, to Palestine, Syria, Mesopotamia, Armenia, Iberia, and Colchis. This was the line of march followed by Sesostris.

Ptolemy indicates the existence of "Pillars," which he calls "the Pillars of Alexander," above Albania and Iberia, at the commencement of the Asiatic Sarmatia. But as it is known that Alexander never penetrated into these regions, it is clear that the title "of Alexander" was added by the Greeks to the names of mountains, which separated a country partly civilized from that entirely occupied by hordes of savages. Everything therefore seems to show, that these Pillars near Iberia in Asia, and not the Pillars of Hercules in Europe, formed the boundary of the expeditions of Sesostris, Tearcho, and Nabuchodonosor.—*Gossellin.*

[310]As the Oxydraci are here meant, Groskurd adopts this name in the text. They were settled in Sagur and Outch, of the province of Lahore.

[311]Eurip. Bacchæ, v. 13.—*Wodehull.*

[312]Many cities and mountains bore the name of Nysa; but it is impossible to confound the mountain Nysa, spoken of by Sophocles, with the Nysa of India, which became known to the Greeks by the expedition only of Alexander, more than a century after the death of the poet.

[313]Probably interpolated.

[314]Il. vi. 132. Nysa in India was unknown to Homer, who here refers to Mount Nysa in Thrace.

[315]Strabo takes for the source of the Indus the place where it passes through the mountains to enter the Punjab. The site of Aornos seems to correspond with Renas.—*Gossellin.*

[316]The Sibæ, according to Quintus Curtius, who gives them the name of Sobii, occupied the confluent of the Hydaspes and the Acesines. This people appear to have been driven towards the east by one of those revolutions so frequent in all Asia. At least, to the north of Delhi, and in the neighbourhood of Hardouar, a district is found bearing the name of Siba.

[317]That is, the Macedonians transferred the name of the Caucasus, situated between the Black Sea and the Caspian, to the mountains of India. The origin of their mistake arose from the Indians giving, as at present, the name of Kho, which signifies "white," to the great chain of mountains covered with snow, from whence the Indus, and the greater part of the rivers which feed it, descend.

[318]This people occupied the Paropamisus, where the mountains now separate Candahar from Gaour.

[319]Book ii. c. i. 2.

[320]Under the name of Ariana, the ancients comprehended almost all the countries situated between the Indus and the meridian of the Caspian Gates. This large space was afterwards divided by them according to the position of the different nations which occupied it.—*Gossellin.* There can be no doubt the modern Iran represents the ancient Ariana. See *Smith*, art. Ariana, and b. ii. c. v. § 32, vol. i. p. 196, note[903].

[321]Eratosthenes and Strabo believed that the eastern parts of Asia terminated at the mouth of the Ganges, and that, consequently, this river discharged itself into the Eastern Ocean at the place where terminated the long chain of Taurus.

[322]According to Major Rennell, Emodus and Imaus are only variations of the same name, derived from the Sanscrit word Himmaleh, which signifies "covered with snow."

[323]In some MSS. the following diagram is to be found.

The

River Indus.

[324]The extremity of India, of which Eratosthenes speaks, is Cape Comorin, which he placed farther to the east than the mouth of the Ganges.

[325]Patelputer or Pataliputra near Patna, see b. ii. ch. i. §9.

[326]The reading is σχοινίοις, which Coraÿ changes to σχοίνοις, Schœni: see Herod. i. 66. The Schœnus was 40 stadia. B. xii. ch. ii. §12.

[327]Athenæus (b. xi. ch. 103, page 800, Bohn's Classical Library) speaks of Amyntas as the author of a work on the Stations of Asia. The Stathmus, or distance from station to station, was not strictly a measure of distance, and depended on the nature of the country and the capability of the beasts of burthen.

[328]The reading Coliaci in place of Coniaci has been proposed by various critics, and Kramer, without altering the text, considers it the true form of the name. The Coliaci occupied the extreme southern part of India. Cape Comorin is not precisely the promontory Colis, or Coliacum, which seems to answer to Panban, opposite the island Ramanan Kor.

[329]The Indian Caucasus.

[330]Book ii. ch. i. § 3.

[331]λίνον, probably the λίνον τὸ ἀπὸ δενδρέων, or cotton, of Arrian.

[332]βόσμορον. § 18.

[333]Ceylon.

[334]The voyage from the Ganges to Ceylon, in the time of Eratosthenes, occupied seven days, whence he concluded that Ceylon was seven days' sail from the continent.

[335]Groskurd reads 5000 stadia. B. ii. c. i. § 14.

[336]εἰδοποιήσουσι. Coraÿ.

[337]The text is, as Coraÿ observes, obscure, if not corrupt. The proposed emendations of Coraÿ and Kramer are followed.

[338]Herod. ii. 5.

[339]At the beginning of autumn.

[340]At the beginning of winter.

[341]Taxila seems to have been situated at some distance to the east of Attock.

[342]At the delta formed by the Indus.

[343]Towards the end of summer.

[344]The Chenab.

[345]The district between Moultan and the mountains.

[346]Herod. ii. 86. Velleraque ut foliis depectant tenuia Seres? Virg. Geor. ii. 121.

[347]Cloth of silk.

[348]The sugar-cane.

[349]C. i. § 33.

[350]The Banyan tree.

[351]Probably the Caroubba (Lotus Zizyphus), but it does not produce the effect here mentioned.

[352]The Ravee.

[353]Arist. Hist. An. vii. 4, who speaks however of five only.

[354]πεπλησμένως. Coraÿ.

[355]Od. ii. 157.

[356]That is to say, he crossed the Paropamisus, or Mount Ghergistan, from the western frontier of Cabul, by the pass of Bamian, to enter the district of Balk.

[357]The Attock.

[358]The river of Cabul.

[359]The Gandaræ were a widely extended people of Indian or Arianian origin, who occupied a district extending more or less from the upper part of the Punjab to the neighbourhood of Candahar, and variously called Gandaris and Gandaritis. See Prof. Wilson's *Ariana Antiqua.*

[360]Aspasii. Coraÿ.

[361]Peucela, in Arrian iv. 22. Rennell supposes it to be Puckholi, or Pehkely.

[362]Abisarus was king of the mountainous part of India, and, according to the conjecture of Vincent, which is not without some probability, his territory extended to Cashmir.

[363]India is bordered to the north, from Ariana to the Eastern Sea, by the extremities of Taurus, to which the aboriginal inhabitants give the different names of

Paropamisus, Emodon, Imaon, and others, while the Macedonians call them Caucasus. The Emodi mountains were the Western Himalaya. See *Smith*, art. Emodi Montes.

[364]The name of the modern city Lahore, anciently Lo-pore, recalls that of Porus. It is situated on the Hyarotis or Hydraotes (Ravee), which does not contradict our author; for, as Vincent observes, the modern Lahore represents the capital of the second Porus, whom Strabo will mention immediately; and the Lahore situate between the Hydaspes (the Behut or Jelum) and the Acesines (the Chenab), the exact position of which is unknown, was that of the first Porus. Probably these two districts, in which the two cities were situated, formed a single district only, one part of which was occupied and governed by Porus the uncle, and the other by Porus the nephew. It is probable, also, that these two princes took their name from the country itself, Lahore, as the prince of Taxila was called Taxiles, and the prince of Palibothra, Palibothrus.

[365]Strabo's Bucephalia was on the Hydaspes, between Beherat and Turkpoor, not far from Rotas. *Groskurd.* The exact site is not ascertained, but the probabilities seem to be in favour of Jelum, at which place is the ordinary passage of the river, or of Jellapoor, about 16 miles lower down.*Smith.*

[366]Ox-headed.

[367]Cercopitheces.

[368]Hence the Cathay of the Chinese and Modern Europe.

[369]So also Arrian, who takes the number from Megasthenes. Pliny says that nineteen rivers unite with the Indus.

[370]Probably an interpolation.

[371]The island Cos.

[372]B. xv. c. i. § 7.

[373]The Malli occupied a part of Moultan.

[374]The Sambus of Arrian. Porticanus is the Oxycanus of Arrian. Both Porticanus and Musicanus were chiefs of the cicar of Sehwan. *Vincent's Voyage of Nearchus*, p. 133.

[375]This number is too large. There is probably an error in the text. Groskurd reads 20; but Kramer refers to Arrian's expedition of Alexander, v. 20, and suggests that we may here read 100 (ρ) instead of 200 (σ).

[376]The Seres are here meant, whose country and capital still preserve the name of Serhend. It was the Serica India of the middle ages, and to this country Justinian sent to procure silkworms' eggs, for the purpose of introducing them into Europe. Strabo was not acquainted with the Seres of Scythia, whose territory is now called Serinagar, from whence the ancients procured the wool and fine fabrics which are now obtained from Cashmir; nor was he acquainted with the Seres who inhabited the peninsula of India, and whose territory and capital have retained the name of Sera. Pliny is the only ancient author who seems to have spoken of these latter Seres. *Gossellin.* The passage in brackets is supposed by Groskurd to be an interpolation. Meineke would retain it, by reading καί τοι for καὶ γὰρ.

[377]The passage is corrupt, and for κήτη, "whales or cetaceous animals," Groskurd proposes λέγει. The whole would therefore thus be translated, "and speaks of what he saw on it, of its magnitude," &c.

[378]The exaggeration of Megasthenes is nothing in comparison of Ælian, who gives to the Ganges a breadth of 400 stadia. Modern observations attribute to the Ganges a breadth of about three quarters of a geographical mile, or 30 stadia.

[379]About 120 feet.

[380]Hiranjavahu.

[381]B. ii. c. i. § 9.

[382]B. xvi. c. i. § 28.

[383]Herodotus iii. 102. The marmot?

[384]The passage is corrupt. Groskurd proposes to add the word ὥς before καὶ καμήλους, "as camels." Coraÿ changes the last word to ἀχαλίνους, which is adopted in the translation. See below, § 53.

[385]θρίσσα.

[386]κεστρεύς.

[387]καρίδες.

[388]In the text, μέχρι ὄρους, "to a mountain." Coraÿ changes the last word to the name of a people, Οὔρων, but Strabo does not appear to have been acquainted with them; Groskurd, to ὀρῶν. The translation adopts this correction, with the addition of the article, which, as Kramer observes, is wanting if we follow Groskurd.

[389]Groskurd proposes τειχῶν, "walls," in place of, τιμῶν, "prices."

[390]Κώδων, "a bell," or gong, or trumpet?

[391]The orguia was equal to four cubits, or six feet one inch.

[392]Men who slept on their ears. See b. i. c. ii. § 35.

[393]The Brahmins.

[394]Sarmanes, Clem. Alex. Strom. i. 305.

[395]Meineke's conjecture, ἐσθητοὺς φλοιῷ δενδρείῳ.

[396]According to Diodorus Siculus, xix. 33, an exception was made for women with child, or with a family; but otherwise, if she did not comply with this custom, she was compelled to remain a widow during the rest of her life, and to take no part in sacrifices or other rites, as being an impious person.

[397]By Arrian and Plutarch he is called Dandamis.

[398]By φιμοῖς, probably here is meant a circular segment, or band of iron, furnished with slightly raised points in the inside; it passes over the bone of the nose, and is fastened below by a cord which is continued as a bridle. Such a contrivance is still in use for mules and asses in the East.

[399]Coraÿ reads πόθος instead of κόρος in the text. The translation would then be, "who required nothing;" but ἐκείνου here refers to Alexander.

[400]On the day of his birth, Herod. ix. 109.

[401]Of Armenia.

[402]About 6 feet.

[403]The text is corrupt. Tzschucke's emendation is adopted, viz. βόνασοι. Groskurd translates the word by "hump-backed oxen," or zebus.

[404]Ælian de Nat. Animal. xvii. 21.

[405]Bird of paradise?

[406]Not far from the present Anopschir on the Ganges, south-east from Delhi. *Groskurd.*

[407]Pataliputer, b. ii. c. i. § 9.

[408]Probably the Iomanes.

[409]A subordinate town in the pachalic of Aleppo, and its modern name is still Antakieh. It was anciently distinguished as Antioch by the Orontes, because it was situated on the left bank of that river, where its course turns abruptly to the west, after running northwards between the ranges of Lebanon and Antilebanon, and also Antioch by Daphne, because of the celebrated grove of Daphne which was consecrated to Apollo, in the immediate neighbourhood.

[410]In Dion Cassius, liv. ix. he is called Zarmanus, a variation probably of Garmanus, see above, § 60. Chegas, or Sheik, seems to be the Tartar title Chan or Khan, which may be detected also in the names Musi-canus, Porti-canus, Oxy-canus, Assa-canus. *Vincent, Voyage of Nearchus,* p. 129. Groskurd writes Zarmanos Chanes.

[411]Bargosa is probably a corruption of Barygaza mentioned in Arrian's Periplus of the Red Sea. It was a large mart on the north of the river Nerbudda, now Baroatsch or Barutsch. *Groskurd.*

[412]"Beyond," as Strabo has just been speaking of India, with reference to which Ariana is to the west of the Indus.

[413]To the south of the great chain bearing that name, extending from west to east of Asia.

[414]The exact place corresponding with the Caspiæ Pylæ is probably a spot between Hark-a-Koh and Siah-Koh, about 6 parasangs from Rey, the name of the entrance of which is called Dereh. *Smith,* art. Caspiæ Pylæ.

[415]An extensive province of Asia along the northern side of the Persian Gulf, extending from Carpella (either C. Bombareek or C. Isack) on the E. to the river Bagradas (Nabend) on the W. According to Marcian the distance between these points was 4250 stadia. It appears to have comprehended the coast-line of the modern Laristan, Kirman, and Moghostan. It was bounded on the N. by Parthia and Ariana; on the E. by Drangiana and Gedrosia; on the S. by the Persian Gulf, and on the W. by Persis. *Smith*, art. Carmania.

[416]The Purali.

[417]Mekran.

[418]By the achronical rising of the Pleiades is meant the rising of this constellation, or its first becoming visible, after sunset. Vincent (Voyage of Nearchus) fixes on the 23rd October, 327 B. C., as the date of the departure of Alexander from Nicæa; August, 326 B. C., as the date of his arrival at Pattala; and the 2nd of October, 326 B. C., as the date of the departure of the fleet from the Indus.

[419]The pith in the young head-shoot of the palm-tree.

[420]Called Pura by Arrian.

[421]The Oritæ are no doubt here meant.

[422]By the line drawn from the Caspian Gates to Carmania.

[423]See above, c. i. § 12.

[424]Herat.

[425]Candahar.

[426]See b. xi. c. viii. § 9.

[427]The text is corrupt: ἐκ μέρους is probably taken from some other part of the text and here inserted.

[428]The same as Zarangæ; they probably dwelt on the lake Zarah, which undoubtedly retains its Zend name. *Wilson's* Ariana.

[429]Corresponding nearly with the present Hamadan.

[430]None is said to be found there at the present day.

[431]They were called Ariaspi; Cyrus, son of Cambyses, gave them the name Euergetæ, "benefactors," in consideration of the services which they had rendered in his expedition against the Scythians.

[432]At the beginning of winter.

[433]The text is corrupt; the words between brackets are supplied by Kramer's conjecture. See b. xi. c. xi. § 2.

[434]Theophrastus, iv. 5. The Pistatia-nut tree.

[435]Bamian, see b. xi. c. xi. § 2.

[436]In the text 19,000. Kramer's proposed reading is adopted of separating the amount.

[437]Ariana in the text. Groskurd proposes to read Carmania; Kramer, Bactriana.

[438]About 140 feet. Arrian says twenty-five orguiæ, or about 150 feet.

[439]Groskurd proposes to supply after "Sea" words which he thinks are here omitted; upon insufficient grounds, however, according to Kramer.

[440]The Arosis of Arrian, now the Tab.

[441]This passage is very corrupt, and many words, according to Kramer, appear to be omitted. See b. ii. c i. § 26. We read with Groskurd "Media" for "Caspian Gates" in the text: and insert "9000 stadia," here from b. ii. c. i. § 26, and, following the same authority, 3000 for 2000 stadia in the text below.

[442]Persæ, v. 17 and 118.

[443]Pasa or Fesa.

[444]Taug or Taüog, on the river Grâ.

[445]The Uxii occupied the district of Asciac.

[446]There seems little doubt that the Karun represents the ancient Eulæus (on which some authors state Susa to have been situated), and the Kerkhah the old Choaspes. See *Smith*, art. Choaspes.

[447]Groskurd adds 1000 stadia to this amount.

[448]Quin. Curtius, v. 10. Diod. Sic. xvii. 67.

[449]Ab-Zal.

[450]Hollow Persis.

[451]Bendamir.

[452]The capital of Parætacene is Ispahan.

[453]Probably the Ab-Kuren.

[454]Pasa or Fesa.

[455]Orxines, Quint. Cur. x. c. 1.

[456]For sacrifice to Cyrus. Arrian, vi. c. 29.

[457]Arrian adds, "Son of Cambyses."

[458]Groskurd reads, ἄλλεσθαι, hops or jumps up.

[459]Founded probably by the Macedonians.

[460]The Elymæi reached to the Persian Gulf. Ptolem. vi. 1. They appear to have left vestiges of their name in that of a gulf, and a port called Delem.

[461]The account of the Persians is taken from Herodotus, i. 131, &c.

[462]According to Herodotus, the priest who sacrificed was crowned.

[463]

Roused the sacred fire, as the law bids,

Touching the god with consecrated wand.

Athenæus xii. 40, p. 850. Bohn's Classical Library.

[464]i. e. "who kindle fire."

[465]i. e. places where fire is kindled.

[466]B. xi. c. viii. § 4.

[467]Not the same plant as mentioned above, c. i. § 10, but the pistacia terebinthus.

[468]An interpolation. The Cardaces were not Persians, but foreign soldiers. "Barbari milites quos Persæ Cardacas appellant," (Cornel. Nepos.) without doubt were Assyrian and Armenian Carduci. See b. xvi. c. i. § 24, and Xenoph. Anab. iv. 3. Later Gordyæi or Gordyeni, now the Kurds. *Groskurd.*

[469]Cardamum is probably the "lepidum perfoliatum" of Linnæus, or the "nasturtium orientale" of Tournefort. Xenophon also, Expedit. Cyr. iii. 5 and vii. 8, speaks of the great use made of this plant by the Persians.

[470]The length of the arms and the surname "Longhand" here given to Darius are assigned by others to Artaxerxes. It was in fact the latter to whom this surname was given, according to Plutarch, in consequence of the right arm being longer than the left. Therefore Falconer considers this passage an interpolation. *Coräj.*

[471]This, says Gossellin, may account for the rarity of the Persian Darius, badly struck, and coined long before the time of Alexander, and appearing to belong to a period anterior to the reign of Darius Hystaspes.

[472]Chalybon was the name of the modern Aleppo, but the wine of Damascus must have possessed the same qualities, and had the same name. "The Chalybonian wine, Posidonius says, is made in Damascus in Syria, from vines which were planted there by the Persians." *Athenæus*, b. i. page 46, Bohn's Classical Library.

[473]In the text "ten or eleven years," which reading is contrary to all other authorities, and is rejected by Kramer.

[474]This is only an approximation. From the conquest of the Medes by Cyrus to the death of Darius Codomanus, last king of Persia, is a period of 225 years.

[475]According to Dion Cassius, xviii. § 26, Aturia is synonymous with Assyria, and only differs from it by a barbarous pronunciation; which shows that the name Assyria belonged peculiarly to the territory of Nineveh.

[476]Aiaghi-dagh.

[477]It is to be remarked that the people bordering upon the Gordyæi are the only people of Mesopotamia here mentioned, for the whole of Mesopotamia, properly so called, is comprised under the name of Assyria.

[478]The bridge or passage at the foot of the modern fortress Roum-Kala.

[479]B. xii. c. iii. § 5; Herod. i. 6 and 72.

[480]Al. Lucan. b. xi. c. xii. § 4; b. xiv. c. v. § 18; b. xvi. c. ii. § 8.

[481]Probably walls built for the protection of certain districts. Such was the διατείχισμα Σεμιράμιδος, constructed between the Euphrates and the Tigris, and intended, together with canals brought from those rivers, to protect Babylon from the incursions of the Arabian Scenitæ or Medes. B. ii.

[482]κλίμακες, roads of steep ascent, with steps such as may be seen in the Alps of Europe; the word differs from ὁδοὶ, roads below, inasmuch as the former roads are only practicable for travellers on foot and beasts of burthen, the latter for carriages also.

[483]The union of these two names, says Kramer, is remarkable, and still more so is the insertion of the article τῆς before them: he, therefore, but with some hesitation, suggests that the word μάχης has been omitted in the text by the copyist.

[484]Assyrians.

[485]Erbil.

[486]Called also Zabus, Zabatus, and Zerbes, now the Great Zab.

[487]Adopting Kramer's reading, καὶ ᾱ.

[488]Probably a branch of the Karadgeh-dagh.

[489]The Little Zab, or Or.

[490]As the name Artacene occurs nowhere else, Groskurd, following Cellarius (v. Geogr. Ant. i. 771), suspects that here we ought to read Arbelene, and would understand by it the same district which is called Arbelitis by Ptolemy, vi. 1, and by Pliny, H. N. vi. 13, § 16, but as this form of the national name is nowhere to be found, it would appear improper to introduce it into the text. It is more probable, continues Kramer, that Strabo wrote Adiabene, of which Arbelitis was a part, according to Pliny, loco citato.

[491]The same, no doubt, as the goddess Anaïtis. B. xi. c. viii. § 4, and b. xv. c. iii. § 15.

[492]All manuscripts agree in giving this number, but critics agree also in its being an error for 365. The number of stadia in the wall, according to ancient authors, corresponded with the number of days in the year.

[493]That is, at a short distance from the Persian Gulf, a little more to the south than the modern town Basra.

[494]Some extensive ruins near the angle formed by the Adhem (the ancient Physcus) and the Tigris, and the remains of the Nahr-awan canal, are said to mark the site of Opis.

[495]The name Cœle-Syria, or Hollow Syria, which was properly applied to the district between Libanus and Antilibanus, was extended also to that part of Syria which borders upon Egypt and Arabia; and it is in this latter sense that Strabo here speaks of Cœle-Syria. So also Diodorus Siculus, i. § 30, speaks of "Joppa in Cœle-Syria;" and Polybius, v. 80, § 2, of "Rhinocolura, the first of the cities in Cœle-Syria;" and Josephus, Ant. Jud. xiii. 13, § 2, "of Scythopolis of Cœle-Syria."

[496]El-Arish.

[497]El-Kas near Sebakit-Bardoil, the ancient lake Serbonis.

[498]Barathra.

[499]Strabo has misunderstood the meaning of Eratosthenes, who had said that the excess of the waters of the Euphrates sunk into the ground and reappeared under the form of torrents, which became visible near "Rhinocolura in Cœle-Syria and Mt. Casius," the Casius near Egypt. Our author properly observes that the length and nature of the course contradicts this hypothesis: but, misled by the names Cœle-Syria and Casius, he forgets that the Casius of Egypt and the district bordering upon Egypt, improperly called Cœle-Syria, are here in question; he transfers the first name to Cœle-Syria of Libanus, and the second to Mt. Casius near Seleucia and Antioch, and adds that, according to the notion of Eratosthenes, the waters of the Euphrates would have to traverse Libanus, Antilibanus, and the Casius (of Syria), whilst Eratosthenes has not, and could not, say any such thing. The hypothesis of Eratosthenes could not, indeed, be maintained, but Strabo renders it absurd. The error of our author is the more remarkable, as the name of the city Rhinocolura ought necessarily to have suggested to him the sense in which the words Casius and Cœle-Syria should be understood.

[500]καὶ οὕτως πλημμυρεῖν. These words are, as Kramer proposes, transferred from below. There can be no meaning given to them as they stand in the text, which is here corrupt.

[501]Herod. i. 193.

[502]Herod. i. 194.

[503]Al-Madain.

[504]Strabo probably here refers to Hecatompylos, which, in b. xi. c. ix. § 1, he calls "the royal seat of the Parthians," and which shared with Ecbatana the honour of being a residence of the Parthian kings. The name Hyrcania has here a wide meaning; the proper name would have been Parthia.

[505]Cicero de Nat. Deor. i. § 5.

[506]Descura. *D'Anville.*

[507]Sus.

[508]Asciac part of Khosistan.

[509]Kerman.

[510]Groskurd here supposes an omission by the copyist of the words ἑσπέραν καὶ πρὸς before ἄρκτον.

[511]Parætacene, Cossæa, and Elymaïs occupied the mountainous parts of Irak Adjami.

[512]Aiaghi-dagh.

[513]Media extended partly into Irak Adjami, and partly into Kurdistan.

[514]ὕστερον in the text must be omitted, or altered to πρότερον, unless, as Kramer proposes, the words καὶ πρὸς τοὺς Πέρσας be introduced into the text. Strabo frequently mentions together the three successive governments of Persians, Macedonians, and Parthians. B. xi. c. xiii. § 4, and c. xiv. § 15.

[515]Mithridates I., son of Phraates, 163 B. C., and 124 years after the expedition of Antiochus.

[516]Probably the Djerrahi.

[517]On comparing this passage with others, (b. xi. c. xiv. § 12, and b. xvi. c. i. § 1, and c. i. § 8,) in which Strabo speaks of Adiabene, we perceive that he understood it to be a part of the country below the mountains of Armenia, and to the north of Nineveh, on both banks of the Tigris. Other authors have given a more extended meaning to the name, and applied it to the country on the north of the two rivers Zab, from whence (Amm. Marcel. xxiii. 5, 6) the name Adiabene appears to be derived. In this sense Adiabene may be considered the same as Assyria Proper.

[518]B. xi. c. xiv. § 15.

[519]Groskurd proposes reading Saulopodes, delicate walkers, in place of Saccopodes, sack-footed.

[520]Herod. i. 198. Almost all the details concerning the Babylonian customs are taken from Herodotus, who sets them forth with greater clearness; there are, however, some differences, as, for example, the disposal of young women in marriage, and the different tribunals, which prove that Strabo had other sources of information.

[521]Groskurd here suspects a corruption of the text, and for τούτου reads τοῦ πρώτου, "of the first," and for ἄλλου, "of another," δευτέρου, "of the second."

[522]Merkan.

[523]El-der.

[524]The Van. B. xi. c. xiv. § 8.

[525]In b. xi. c. xiv. § 8, Strabo says that this lake contains one kind of fish only.

[526]Now Roumkala, from the fortress which defends the passage of the river.

[527]Nisibin.

[528]Kara-dagh.

[529]Sered.

[530]Haran.

[531]Racca.

[532]B. C. 51.

[533]Gordyæa was the most northerly part of Assyria, or Kurdistan, near the lake Van. From Carduchi, the name of the inhabitants, is derived the modern name Kurds.

[534]Pliny, x. c. iii. and xxxvi. c. xix., calls it "Gagates lapis;" a name derived, according to Dioscorides, from a river Gagas in Lycia.

[535]Herod. vi. 199.

[536]These appear to be the rivers found in the neighbourhood of Roha or Orfa, the ancient Edessa. One of these rivers bears the name of Beles, and is perhaps the Basileios of Strabo. Chabur is the Aborrhas.

[537]Probably an interpolation.

[538]The passage of the Euphrates here in question was effected at the Zeugma of Commagene, called by Strabo the present passage. On passing the river you entered Anthemusia, a province which appears to have received, later on, the name of Osroene. It extended considerably towards the north, for in it the Aborrhas, according to Strabo, had its source; but it is doubtful whether it extended to the north of Mount Masius, where the latitudes, as given by Ptolemy, would place it. I do not exactly know whether Strabo intends to speak of a city or a province, for the position of the city is unknown; we only learn from a passage in Pliny, vi. c. xxvi., that it was not on the Euphrates. The word τόπος is not, I think, so applicable to a province as to a city, and in this last sense I have understood it, giving also to κατὰ the meaning of latitude, in which it is so often applied by Strabo; strictly speaking, the sense of "vis-à-vis," "opposite to," might be given to it.—*Letronne.*

[539]This is an error of the author or of the copyist. Edessa (now Orfa) is not to be confounded with Bambyce (Kara-Bambuche, or Buguk Munbedj) of Cyrrhestica in Syria, which obtained its Hellenic name from Seleucus Nicator.

[540]B. C. 54.

[541]The Parthians became masters of Syria under Pacorus, and of Asia Minor under Labienus. B. C. 38.

[542]Artavasdes, king of the Armenians. B. xi. c. xiii. § 4.

[543]The text would lead us to suppose that Phraates succeeded Pacorus, whereas below, § 8, Pacorus, the eldest son of the Parthian king, died before his father, Orodes. Letronne, therefore, and Groskurd suppose that the words, "the son of Orodes," are omitted after "Pacorus" above, and "his" in the translation would then refer to Orodes.

[544]See b. vi. c. iv. § 2, in which the motives for getting rid of these members of his family are not mentioned.

[545]Judging from Arrian (Anab. v. § 25; vii. § 9; iii. § 8), the historians of Alexander, as well as more ancient authors, gave the name of Syria to all the country comprehended between the Tigris and the Mediterranean. The part to the east of the Euphrates, afterwards named Mesopotamia, was called "Syria between the rivers;" that to the west was called by the general name Cœle-Syria, and although Phœnicia and Palestine were sometimes separated from it, yet it often comprehended the whole country as far as Egypt. Strabo below, c. ii. § 21, refers to this ancient division, when he says that the name Cœle-Syria extends to the whole country as far as Egypt and Arabia, although in its peculiar acceptation it applied only to the valley between Libanus and Antilibanus.

[546]B. C. 70.

[547]Antakieh.

[548]Modern conjecture has identified it with Shogh and Divertigi.

[549]Kulat-el-Mudik.

[550]Ladikiyeh.

[551]Mesopotamia in the text is no doubt an error of the copyist. We ought probably to read Commagene. Groskurd proposes to read "Commagene, like Mesopotamia, consisted of one satrapy." Groskurd's emendation of the text is followed, although not approved of, by Kramer.

[552]These four portions were no doubt formed by the four hills contained within the circuit of Antioch. The circuit wall existed in the time of Pococke. The detailed and exact description given of it by this learned traveller, as also his plan of Antioch, agree with Strabo's account.*Pococke, Descrip. of the East,* ii. p. 190.

[553]C. i. § 25.
[554]Mount Soldin.
[555]Orontes, or Nahr-el-Asy.
[556]Beit-el-ma.
[557]B. xii. c. viii. § 19; b. xiii. c. iv. § 6.
[558]Also Hierapolis, the modern Kara Bambuche.
[559]Berœa owes its name to Seleucus Nicator, and continued to be so called till the conquest of the Arabs under Abu Obeidah, A. D. 638, when it resumed its ancient name of Chaleb, or Chalybon.
[560]The territory subject to the town Cyrrhus, now Coro.
[561]Baghras.
[562]The modern names of the Arceuthus and Labotas are unknown.
[563]The Afreen.
[564]B. C. 145.
[565]A table.
[566]Called Phraates by Pseudo-Appian, in Parthicis, p. 72.
[567]Selefkeh.
[568]Posidi, on the southern side of the bay, which receives the Orontes.
[569]On Cape Ziaret.
[570]B. C. 40.
[571]The text is corrupt. The translation follows the proposed corrections of Letronne and Kramer.
[572]Shizar, on the Orontes.
[573]Cæcilius Bassus was besieged twice in Apameia, first by C. Antistius, afterwards by Marcus Crispus and Lucius Statius Marcius. Cassius succeeded in dispersing the troops of this rebel without much difficulty, according to Dion Cassius, xlvii. 27.
[574]Arethusa, now Restan, was founded by Seleucus Nicator. According to Appian, Pompey subdued Sampsiceramus, who was king of Arethusa. On this account Cicero, in his letters to Atticus (ii. 14, 16, 17, 23), calls Pompey in derision Sampsiceramus. Antony put Iamblicus, son of Sampsiceramus, to death; but Augustus restored the small state of Arethusa to another Iamblicus, son of the former.
[575]The people of Emesa, now Hems.
[576]Balbek and Kalkos.
[577]This Ptolemy, son of Mennæus, was master chiefly of Chalcis, at the foot of Libanus, from whence he made incursions on the territory of Damascus. Pompey was inclined to suppress his robberies, but Ptolemy softened his anger by a present of 1000 talents, which the Roman general applied to the payment of his troops. He remained in possession of his dominion until his death, and was succeeded by his son Lysanias, whom Cleopatra put to death, on the pretext that he had induced the Parthians to come into the country. *Josephus, Bell. Jud.*
[578]One of the branches of Antilibanus.
[579]This Alchædamnus is constantly called Alchaudonius by Dion Cassius, whom he calls the "Arabian dynast." Falconer therefore inferred that here we ought to read Ἀράβων instead of Ῥαμβαίων, but Letronne does not adopt this reading, and supposes the Rhambæi may have been a tribe of the Arabians.
[580]The text is here corrupt, and the passage, according to Kramer, probably introduced into the text from a marginal note.
[581]παραλία, but this is a correction for παλαιά, which Letronne proposes to correct for περαία, which is supported in § 13, below. The part of the continent opposite, and belonging to an island, was properly called Peræa, of which there are many examples. That part of Asia Minor which is opposite Rhodes was so called, b. xiv. c. v. § 11, as also the coast opposite Tenedos, b. xiii. c. i. § 46. Peræa was also adopted as a proper name. Livy, xxxiii. 18.
[582]Pococke places Paltus at Boldo; Shaw, at the ruins at the mouth of the Melleck, six miles from Jebilee, the ancient Gabala.

[583]Carnoon.

[584]Ain-el-Hiyeh.

[585]According to Pococke, the ruins of Aradus (Ruad) are half a mile to the north of Tortosa (Antaradus). It is remarkable that Strabo makes no mention of Antaradus, situated on the continent opposite Aradus; Pliny is the first author who speaks of it. Probably the place only became of note subsequent to the time of Strabo, and acquired power at the expense of some of the small towns here mentioned. Antaradus, reëstablished by Constantine, assumed the name of Constantia.

[586]Sumrah.

[587]Ortosa.

[588]Carnus.

[589]The resistance of the sea water to the ascent of the fresh water is cut off by this ingenious contrivance, and the fresh water rises above the level of the sea through the pipe, by natural causes, the head or source of the spring being in the upper ground of the mainland. This fountain is now known by the name of Ain Ibrahim, Abraham's fountain.

[590]B. xiv. c. v. § 2.

[591]Greego.

[592]If the words of the text, φέρει δὲ καὶ, "it produces also," refer to the lake, our author would contradict himself; for below, § 41, he says that Jericho alone produces it. They must therefore be referred to "a hollow plain" above; and the fact that they do so arises from the remarkable error of Strabo, in placing Judæa in the valley formed by Libanus and Antilibanus. From the manner in which he expresses himself, it is evident that he supposed the Jordan to flow, and the Lake Gennesaret to be situated, between these two mountains. As to the Lycus (the Nahr el Kelb), Strabo, if he had visited the country, would never have said that the Arabians transported upon it their merchandise. It is evident that he has confused the geography of all these districts, by transferring Judæa, with its lakes and rivers, to Cœle-Syria Proper; and here probably we may find the result of his first error in confounding Cœle-Syria Proper with Cœle-Syria understood in a wider meaning. See above, c. i. § 12.

[593]Nahr-el-Kelb.

[594]Iouschiah.

[595]Gebail.

[596]Beyrout.

[597]Nahr-Ibrahim.

[598]Josephus, i. 1.

[599]Above, c. ii. § 3.

[600]Ortosa.

[601]Tineh.

[602]Above, c. i. § 12, 15.

[603]Nahr-Damur.

[604]Sour.

[605]Tyre—daughter of Zidon. Isaiah xxiii. 12.

[606]In B. v. c. iii. § 7, Strabo tells us that Augustus prohibited houses being erected of more than 70 Roman feet in height.

[607]Josephus (Antiq. Jud. xv. 4, § 1) states, that Mark Antony gave Cleopatra all the coast of Phœnicia, from Eleutheria to Egypt, with the exception of Tyre and Sidon, which he left in the enjoyment of their ancient independence. But according to Dion Cassius (liv., 7), Augustus arrived in the East in the spring of the year 734, B. C., or eighteen years before the Christian era, and deprived the Tyrians and Sidonians of their liberty, in consequence of their seditious conduct. It follows therefore, that if Strabo had travelled in Phœnicia, he must have visited Tyre before the above date, because his account refers to a state of things anterior to the arrival of Augustus in Syria; and in this case the information he gives respecting the state of the neighbouring cities must belong to the same date; but he speaks above (§ 19) of the order reëstablished by Agrippa at Beyrout, which was effected four years after the coming of Augustus into Syria. We must conclude, therefore, that Strabo speaks

only by hearsay of the Phœnician cities, and that he had never seen the country itself.*Letronne.*

[608]Il. xxiii. 743.

[609]Probably under Zenarchus of Seleucia, the Peripatetic philosopher whose lectures he attended. B. xiv. c. v. § 4.

[610]Nahr-Quasmieh.

[611]Vestiges of the ancient city still remain. Here was the celebrated temple of the Phœnician Hercules, founded according to Herodotus, ii. 44, before 2700 B. C.

[612]Acre.

[613]Letronne estimates this at a penny.

[614]Athenæus, p. 742, Bohn's Class. Library.

[615]The Tower of Strato was an ancient city almost in ruins, which was repaired, enlarged, and embellished by Herod with magnificent buildings; for he found there excellent anchorage, the value of which was increased by the fact of its being almost the only one on that dangerous coast. He gave it the name of Cæsarea, in honour of Augustus, and raised it to the rank of a city of the first order. The repairs of the ancient city, the Tower of Strato, or rather the creation of the new city Cæsarea, took place about eight or nine years B. C.; so that this passage of Strabo refers to an earlier period.

[616]Josephus (Ant. Jud. xiv. 13, § 3) calls a district near Mount Carmel Drumos, employing the word Δρυμός, a forest, as a proper name.

[617]Jaffa.

[618]Van Egmont (Travels, vol. i. p. 297) considers it impossible, from the character of the intervening country, to see Jerusalem from Joppa. Pococke, on the contrary, says, that it would not be surprising to see from the heights of Joppa, in fine weather, the summit of one of the high towers of Jerusalem; and this is not so unlikely, for according to Josephus the sea was visible from the tower of Psephina at Jerusalem.

[619]Jebna.

[620]Ras-el-Kasaroun.

[621]Esdod.

[622]Asculan.

[623]Akaba or Akaba-Ila.

[624]Near Suez.

[625]Refah.

[626]B. C. 218.

[627]El Arish.

[628]Sebakı-Bardoıl.

[629]The passage through which the lake discharged itself into the sea.

[630]El-Cas.

[631]It appears that in the time of Strabo and Josephus the temple of Jupiter only remained; at a later period a town was built there, of which Steph. Byzant., Ammianus Marcellinus, and others speak, and which became the seat of a bishopric.

[632]B. xvi. c. iii. § 3.

[633]B. xiv. c. v. § 3.

[634]Arabia Petræa. Petra, now called Karac, was the capital.

[635]Josephus, Ant. Jud. xiii. 9. 1.

[636]§ 27, above.

[637]Jaffa.

[638]Rabbath-Ammon, or Amma.

[639]Herod rebuilt Samaria, and surrounded it with a vast enclosure. There also he erected a magnificent temple, and gave to the city the surname of Sebaste, in honour of Augustus.

[640]In b. xvii. c. ii. § 5, our author again says that the Jews were originally Egyptians. So also Josephus, xiv. 7. 2.

[641]"Judæi mente solâ, unumque numen intelligunt, summum illud et eternum, neque mutabile, neque interiturum." Tacitus, Hist. v. c. 5.

[642]Strabo here attributes to Moses the opinions of the Stoics.

[643]Strabo appears to have had little acquaintance with the Jewish history previous to the return from captivity, nor any exact knowledge until the arrival of the Romans in Judæa. Of the Bible he does not seem to have had any knowledge.

[644]Probably Strabo copies from accounts when the country was not well cultivated.

[645]αἱ γυναῖκες Ἰουδαϊκῶς ἐκτετμημέναι, below, c. iv. Section 9.

[646]Od. xix. 494.

[647]Diviners by the dead.

[648]Diviners by a dish into which water was poured and little waxen images made to float.

[649]Diviners by water.

[650]ὡροσκόποι is the reading of the text, which Groskurd supposes to be a corruption of the Latin word Haruspex. I adopt the reading οἰωνοσκόποι, approved by Kramer, although he has not introduced it into the text.

[651]According to Josephus, Johannes Hyrcanus dying, B. C. 107, was succeeded by Aristobulus, who took the title of king, this being the first instance of the assumption of that name among the Jews since the Babylonish captivity. Aristobulus, was succeeded by Alexander Jannæus, whose two sons were Hyrcanus II. and Aristobulus II., successively kings of Judæa, B. C. 67, 68.

[652]B. C. 63.

[653]Solomon's conduit was constructed on the hydraulic principle, that water rises to its own level. The Romans subsequently, being ignorant of this principle, constructed an aqueduct.

[654]Balsamodendron Giliadense. Pliny xii. 25.

[655]Medicago arborea.

[656]The pistachia, b. xv. c. ii. § 10.

[657]In. b. xvi. c. ii. § 16, our author says that it is found on the borders of the Lake Gennesareth.

[658]It yields, during the hot season, an immense quantity of toddy or palm wine.

[659]Obtained by boiling the branches of the balsamodendron in water, and skimming off the resin.

[660]Strabo here commits the singular error of confounding the Lake Asphaltites, or the Dead Sea, with the Lake Sirbonis. Letronne attempts to explain the origin of the error. According to Josephus, the Peræa, or that part of Judæa which is on the eastern side of the Jordan, between the lake of Tiberias and the Dead Sea, contained a district (the exact position of which is not well known, but which, according to Josephus, could not be far from the Lake Asphaltites) called Silbonitis. The resemblance of this name to Sirbonis probably misled our author.

[661]Specific gravity 1·211, a degree of density scarcely to be met with in any other natural water. Marcet's Analysis. Philos. Trans. part ii. page 298. 1807.

[662]By chrysocolla of the ancients is generally understood borax, which cannot however be meant in this passage. It may probably here mean uric acid, the colour of which is golden.

[663]A place near the Lake Asphaltites, called Masada by Josephus, de B. Jud. iv. 24, v. 3.

[664]Genesis xiv. and Wisdom x. 6: "the fire which fell down on the five cities."

[665]In this quotation from Eratosthenes we are probably to understand the Lake Sirbonis, and not the Dead Sea; a continuation, in fact, of Strabo's first error. The translator adopts Kramer's suggestion of Θετταλίαν for θάλατταν in the text.

[666]"The salting station," on the lake of Gennesareth.

[667]It has been a subject of dispute whether Herod was of Jewish or Idumæan origin.

[668]Herod went to Rome B. C. 38, and obtained from the senate the title of king. In the dispute between Octavius and Antony, he espoused the cause of the latter. Octavius not

only pardoned him and confirmed him in his title, but also added other cities to his dominions. B. C. 18.

[669]The chief promoters of the crimes of Herod were Salome his sister, who desired to gratify her hatred; and Antipater, who aimed at the throne. Herod, influenced by their misrepresentations, put to death Mariamne his wife, Aristobulus her brother, and Alexandra her mother; also his sons Aristobulus and Alexander, besides Antipater, a third son, who had conspired against his life.

[670]Augustus conferred on Archelaus the half of the kingdom of Herod with the title of ethnarch, promising to grant the title of king, should he prove worthy of it. The other half of the kingdom was separated into two tetrarchies, and divided between Philip and Antipas, two other sons of Herod.

[671]Augustus not only confirmed to Salome the legacy made to her by Herod, of the towns Jamneia, Azoth, and Phasaëlis, but granted to her also the royal palace and domains of Ascalon.

[672]This was Archelaus, whose tyranny was insupportable. He was accused by the chief Jews and Samaritans before Augustus, who exiled him to Vienne, to the south of Lyons, where he died the following year, A. D. 7.

[673]This refers to the journey of Philip and Antipas to Rome. At the death of Herod, Archelaus went to Rome, A. D. 2, to solicit the confirmation of his father's will, in which he had been named king. The two brothers, Antipas and Philip, also went there, and the kingdom of Herod was divided as above stated. After the exile of Archelaus, his dominions were administered by his two brothers. Strabo does not appear to have been acquainted with the history of the two brothers after their return to Judæa; for otherwise he would not have omitted to mention the exile of Antipas. This tetrarch, it is known, went to Rome A. D. 38, to intrigue against his brother, of whom he was jealous; but he was himself accused by Agrippa of having intelligence with the Parthians, and was exiled to Lyons, A. D.39.

[674]C. i. § 21.

[675]C. i. § 6.

[676]C. iii. § 4.

[677]The name Erythræan, or Red Sea, was extended to the whole of the Arabian Gulf, to the sea which surrounds Arabia to the south, and to a great part of the Persian Gulf.

[678]The cape Harmozi, or Harmozon, is the cape Kuhestek of Carmania, Kerman, situated opposite to the promontory Maceta, so called from the Macæ, an Arabian tribe living in the neighbourhood. This last promontory is now called Mocandon, and is the "Asaborum promontorium" of Ptolemy.

[679]For a long period the Euphrates has ceased to discharge itself directly into the Persian Gulf, and now unites with the Tigris above 100 miles from the sea.

[680]The reading followed, but not introduced into the text, by Kramer is that suggested by the corrections of Letronne and Groskurd, καὶ τὴν Ἀράβων παραλίαν παραπλεύσαντα καθ᾽ αὑτόν.

[681]Peludje, at the entrance of the Gulf of Gran.

[682]Heeren (Comment. Gotting. 1793. Vol. xi. pp. 66, 67) supposes that this city was founded by Chaldæans solely for the purpose of a depôt for the transit of goods to Babylon, the trade having for a long time been in the hands of the Phœnicians. He also conjectures that the most flourishing period of the town was when the Persians, for political reasons, destroyed the commerce of Babylon, and Gerrha then became the sole depôt for the maritime commerce of India.

[683]El-Der.

[684]The island Ormus, which before the year 1302 was called Turun or Gerun, from which the Greeks formed the names Tyros, Tyrine, Gyris, Gyrine, Ogyris, and Organa. *Gossellin.*

[685]Arek.

[686]Besides the islands Tyre and Aradus, there existed even in the time of Alexander, and near the present Cape Gherd, a city called Sidon or Sidodona, which was visited by Nearchus, as may be seen in his Periplus. The Phœnician inhabitants of these places appear

to have afterwards removed to the western side of the Persian Gulf, and to the islands Bahraïn, to which they gave the names Tylos, or Tyre, and Aradus. The latter name still exists; it was from this place that the Phœnicians moved, to establish themselves on the shores of the Mediterranean, and transferred the name of Sidon, their ancient capital, and those of Tyre and Aradus, to the new cities which they there founded. *Gossellin.*

[687]As Nearchus in his voyage kept along the coast, this distance must not be understood as so much to the south of Carmania in the open sea, but as the distance from Cape Jask, the commencement of Carmania.

[688]In Ptolemy, this island is called Vorochtha, now Vroct, or Kismis, or Dschisme.

[689]ἡ καυσία, a broad-brimmed Macedonian hat.

[690]Pliny, v. 21, mentions a place which he calls Massica, situated on the Euphrates, near the mouth of a canal which communicated with the Tigris near Seleucia. It is now called Masseib-khan, and is at a short distance above Babylon, on the borders of the desert. I do not know whether this is the Mæcene of Strabo. *Gossellin.*

[691]Strabo here refers to the marsh lakes now called Mesdjed Hosaïn, Rahémah, Hour, &c. The Chaldæans whom he mentions occupied the country along the banks of the Euphrates to the coast of the Persian Gulf.

[692]In Cashmir melons are now grown in the same manner. Humboldt remarks that the same contrivance is adopted in Mexico for the cultivation of vegetables.

[693]Letronne here proposes to read Erythræan or Ethiopian Sea.

[694]Mimosa Nilotica.

[695]This is remarkable.

[696]Carn Almanazil.

[697]Mariaba was not the name of a city, but the title of a city acquired by the residence of their sovereigns. "Mariana oppidum," says Pliny, vi. 32, "significat dominos omnium." The capital was called Saba, now Sabbea; and the country in which it is situated is called Sabieh.

[698]Yemen.

[699]The people of Hadramaüt.

[700]The extent was six times as large as the Delta.

[701]Ailah, or Hœle, or Acaba-Ila.

[702]C. ii. § 30.

[703]The ruins are still visible at Abu-Keyschid.

[704]Deire, or the "neck," so called from its position on a headland of the same name, was a town situated on the African shore of the straits of Bab-el-Mandeb, at their narrowest part.

[705]The Troglodytica extended along the western side of the Arabian Gulf, from about the 19th degree of latitude to beyond the strait. According to Pliny, vi. c. 34, Sesostris conducted his army as far as the promontory Mossylicus, which I think is Cape Mète of the modern kingdom of Adel.*Gossellin.*

[706]The 60 and 200 stadia assigned to the straits refer to the two passages there to be found. The 60 stadia agree with the distance of the eastern cape of Babelmandeb, the ancient Palindromos, to the island Mehun; and the 200 stadia to the distance of this island from the coast of Africa. In this last interval are the six islands of which Strabo speaks.

[707]This passage has sometimes been mistaken to mean, that the region producing myrrh and cinnamon refers to the southern coast of Arabia. Our author here speaks of the coast of Africa, which extends from the Strait of Babelmandeb to Cape Guardafui. This space in following the coast is 160 or 165 leagues, which are equivalent to 5000 olympic stadia.*Gossellin.*

[708]The long and interesting passage from § 5 to the end of § 20 is taken from Artemidorus, with the exception of a very few facts, which our author has taken from other sources, accompanied by observations of his own. On comparing this fragment of Artemidorus with the extracts of Agatharchides preserved by Photius, and the description of Arabia and Troglodytica which Diodorus Siculus (b. iii. 31) says he derived from Agatharchides, we find an identity, not only in almost all the details, but also in a great

number of the expressions. It is, therefore, evident that Artemidorus, for this part of his work, scarcely did anything more than copy Agatharchides. Agatharchides, in his youth, held the situation of secretary or reader to Heraclides Lembus, who (according to Suidas) lived in the reign of Ptolemy Philometor. This king died B. C. 146. He wrote a work on Asia in 10 books, and one on Europe in 49 books; a geographical work on the Erythræan Sea in 5 books; a treatise on the Troglodytæ in 5 books; and other works. He wrote in the Attic dialect. His style, according to Photius, was dignified and perspicuous, and abounded in sententious passages, which inspired a favourable opinion of his judgment. In the composition of his speeches he was an imitator of Thucydides, whom he equalled in dignity, and excelled in clearness. His rhetorical talents also are highly praised by Photius. He was acquainted with the language of the Ethiopians, and appears to have been the first who discovered the true cause of the inundations of the Nile. See Smith, art. Agatharchides.

[709]Ghela.

[710]Kosseir.

[711]Mouse Harbour, or Harbour of Venus.

[712]Meleagrides.

[713]Bender-el-Kebir.

[714]Zemorget or Zamargat. The "Agathonis Insula" of Ptolemy.

[715]Ptolemy Philadelphus.

[716]About 12 feet.

[717]The whole of this description is so vague that it would be difficult to recognise the position of the places mentioned by Strabo without the assistance of scattered notices by other authors. The result of many comparisons leads me to fix upon 16° 58′ as about the latitude of Ptolemaïs Epitheras. Mount Taurus was 22 leagues higher up, and the harbour of the goddess Soteira 12 leagues beyond. *Gossellin.*

[718]Letronne translates Πτολεμαῒς πρὸς τῇ θήρᾳ as Ptolemaïs Epitheras; see c. iv. § 4.

[719]Tacazze, which however does not appear to have such a branch.

[720]These islands are to the north of Arkiko.

[721]Gulf of Matzua.

[722]From the position here assigned to the fortress of Suchus, it is impossible to place it at Suachem, as is commonly done. *Gossellin.*

[723]An island Stratioton is mentioned in Pliny vi. 29, as though he had read in our author the word Στρατιωτῶν, "the island of soldiers." As the island of Strato is named only in this extract from Artemidorus, we might be tempted to correct the text of Strabo by the text of Pliny. But as it is not certain that the two authors speak of one and the same island, it is more prudent to make no change. *Du Theil.*

[724]I am not acquainted with this place. The ancients speak only of one town of the name of Saba (c. iv. § 19). Was there a town Saba which gave its name to the Sabaïtic Gulf? but the one in question does not appear to have been situated there. *Gossellin.*

[725]B. C. 658.

[726]The modern Senaar corresponds with the territory of the Sembritæ. See also b. xvii. c. i. § 2. Herodotus, b. ii. 30.

[727]Tacazze.

[728]The Blue Nile.

[729]ἀκροδρύων is expressed in the Periplus of Agatharchides by the words τὸν καρπὸν π ί π τ ο ν τ α ἀ π ὸ τῶν δένδρων, "the fruit falling from the trees." The Periplus adds another tribe, the Hylophagi, "wood-eaters," who subsisted on the tender branches of certain trees. Strabo refers to them, b. xvii. c. ii. § 2, but without giving their name. The pods of the Lotus Zizyphus are eatable, and may here be meant.

[730]Gymnetæ. Between the Spermophagi and the Creophagi, Agatharchides places another people called Cynegetæ. Strabo and Pliny do not mention them; but the sort of life the Gymnetæ, of which they both speak, lead resembles that of the Cynegetæ or Cynegi of Agatharchides and Diodorus Siculus (iii. 25). It seems therefore that these two authors, as well as Strabo and Pliny, meant here to speak of one and the same tribe of Ethiopian

Gymnetæ, which might have been distinguished by the particular name of Cynegetæ, or Cynegi. *Du Theil.*

[731]Above, c. ii. § 37.

[732]Milkers of bitches.

[733]This Berenice was also surnamed Epi Dire, because it was nearer the promontory Dire than the other cities of the same name. It is probably Bailul, about 12 leagues to the north-west of Assab.

[734]Assab or As-Sab.

[735]Below, Artemidorus calls it the harbour of Eumenes, § 13.

[736]Agatharchides, as quoted by Diodorus Sic. iii. 27, says expressly that this bird is the ostrich. May it be the cassowary?

[737]Groskurd supposes the name of this nation has been omitted in the text, and proposes Acridophagi, or Locust-eaters.

[738]According to Agatharchides and Diodorus Sic. iii. 28, the habit of living on locusts produced a kind of winged louse in the interior of the body; but this is denied by Niebuhr.

[739]Above, § 4.

[740]Pliny, xiii. 17; xv. 13.

[741]Perhaps Zeila. Strabo is here describing the coast of the modern kingdom of Adel.

[742]The Periplus of the Erythræan Sea indicates on this coast a place called Niloptolemæum, which appears to correspond with the mouth of the river Pedra. *Gossellin.*

[743]Phleus schæoris. *Linn.*

[744]Daphnus Parvus of the Periplus of the Erythræan Sea.

[745]Now Fellis or Fel, which signifies Elephant in Arabic.

[746]I think that there is something here omitted and wanting in the text of Strabo, as he seems to make Artemidorus say, that a little after Mount Elephas we find the Horn, or the Cape of the South; for this last appellation appears to have been applied to Cape Guardafui. But this cape, from the time of Philadelphus, and consequently before the period in which Artemidorus wrote, was known by the name of the Promontory of the Aromatics; this author therefore could not have confounded it with the Southern Horn. I have already come to the conclusion that the Southern Horn corresponds with the Southern Cape of Bandel-caus, where commences the desert coast of Ajan, the ancient Azania, respecting which Artemidorus confesses that he was unable to procure any information. It therefore appears to me, that the description which this author must have given of the coast of Africa, from Mount Elephas to the Southern Horn, and which Strabo should have copied, is now wanting in the text. This omission seems to have been noticed by some copyist, who thought to supply it by naming again, to the south of Mount Elephas, the altars of Pytholaus, Lichas, Pythangelus, and Leon, which Artemidorus had already spoken of, and which navigators meet with on the west, and before arriving at Mount Elephas. *Gossellin.*

[747]The text of this paragraph is corrupt; but the reading followed is that suggested in a note by Kramer.

[748]λέων μύρμηξ. Agatharchides calls them μυρμηκολέων, and Ælian simply μύρμηξ. What animal is intended by the name is uncertain. In b. xv. c. i. §44, the marmot seems to be described.

[749]What the words ἐπὶ σειρὰν mean is doubtful. Casaubon supposes that some words are wanting in the text; Groskurd proposes to read ἀπὸ κεφαλῆς ἐπὶ οὐρὰν, "from the head to the tail."

[750]The passage is corrupt, and some words are wanting to complete the sense. Groskurd proposes, "a span less."

[751]Pliny, viii. 29.

[752]Ancient authors, under the name of Sphinx, generally describe the ape, *Simia troglodyte* of Gmelin. *Du Theil.*

[753]Simia innuus.

[754]Simia cepus.

[755]The spotted hyæna.

[756]See b. xv. c. 1, § 45.

[757]The juice of the berries is a strong purge.

[758]Above, § 5.

[759]The bay of Heroopolis is the modern bay of Suez. In the text "Ælanitic bay," which is an error of the author or of the copyist.

[760]An altar to Poseidon (Neptune), which was erected by Aristo, whom one of the Ptolemies had sent to explore the Arabian Gulf.

[761]Φοινικών, a grove of palm trees, is taken as a proper name by Diodorus Siculus, b. iii. 41.

[762]Sheduan. The "Saspirene insula" of Ptolemy.

[763]Ras Mahomet, which terminates the south of the peninsula formed by the two bays, the Ælanitic running up to Petra, and that of Heroopolis running up to Suez. The meaning of Strabo seems to be, that this cape is in a direction due south of Petra and Palestine.

[764]There is a wide difference of opinion among geographers with regard to the position of this important tribe in the modern map of Arabia. See Smith, art. Minæi.

[765]The Maraneitæ appear to me to be the same people whom other geographers call Pharanitæ, and who received their name from their proximity to Cape Pharan, now Ras Mahomet. *Gossellin.*

[766]Diodorus Siculus, iii. § 41, following Agatharchides, narrates the fact with greater precision. The Garindæi took advantage of the absence of the greater part of the Maraneitæ, and put to death those that remained. They then laid in wait for and massacred all those who were returning from the festival.

[767]Gulf of Akaba.

[768]"Light vessels." Diodorus Sic.

[769]Thamud, formerly occupied by the ancient Thamudeni.

[770]Shaur and Iobab?

[771]Gibel Seik, Gibel el Hawene, and Gibel Hester.

[772]The harbour of Charmothas seems to be the ancient Iambo, the "Iambia" of Ptolemy, which now, from the accumulation of soil, is more than a day's journey into the interior of the country. It is in a fertile territory. The Arabs call it Iambo el Nakel, or Iambo of Palm Trees, to distinguish it from the new Iambo situated on an arid soil on the sea-coast. Al Charm, in Arabic, signifies a fissure or opening in the mountains. It seems as if the Greeks had formed the name Charmothas from this word, mistaking the epithet given to the narrow entrance of the harbour of Iambo for the name of the town itself. *Gossellin.*

[773]The Debæ occupied Sockia. The river which flows through the country is called Bætius by Ptolemy.

[774]τὰ πλείω is Kramer's correction for παλαιὰ .

[775]Some are called by Diodorus Siculus, iii. 44, and Agatharchides, Asilæi and Casandres or Gasandres.

[776]Instead of εὔομβρος, Groskurd reads πάμφορος, "produces everything," following the fragments the Agatharchides and Diodorus Sic. b. iii 44.

[777]Groskurd's correction, σιδήρου for ἀργύρου, in the text, is adopted. But the passage is probably corrupt, and after σιδήρου we may read καὶ δεκαπλάσιον τοῦ ἀργύρου, "for ten times the quantity of silver," according to Bochart, and approved by Kramer.

[778]The precise boundaries of Sabæa it is impossible to ascertain. The area we have presumed is comprised within the *Arabian Sea* W., the *Persian Gulf* E., the *Indian Ocean* S., and an irregular line skirting the desert, and running up in a narrow point to Idumæa N. See Smith, Dict. of Greek and Roman Geography, art. Saba. Milton appears to have been acquainted with the following passage from Diodorus Siculus, b. iii. 46, descriptive of Sabæa: "It is impossible to enumerate the peculiarities and nature of all these trees and plants, on account of the surpassing variety and body of perfume which fall upon and excite the senses, in a manner divine and beyond description. The mariner, as he sails even at a distance along the coast, has his share of enjoyment; for when the breezes of spring blow from off the land,

165

the fragrance of the trees and shrubs is carried down to the shore; nor is it of the kind with which we are acquainted, proceeding from old and stored aromatics, but fresh and in full perfection from new-blown flowers, striking the inmost sense."

[779]The same as Saba; see c. iv. § 2.

[780]The above details derived from Artemidorus, and by him from Agatharchides, would not be found in Eratosthenes, who lived before the time of Agatharchides.

[781]We must not confound this measure with the 5000 stadia mentioned in c. iv. § 4. The distance here in question is that taken along the southern coast of Arabia from the straits to Kesem, the ancient Cane, through which passes now, as in former times, the greater part of the perfumes collected in Hadramaut and Seger. But this harbour is about the middle, and not at the extremity of the cinnamon-bearing country. *Gossellin.*

[782]Cardinal Noris places these facts in the year of Rome 730, and quotes, besides Strabo, the historian Josephus. In following the last author, the Cardinal places the death of Obodas in the prefecture of C. Sentius Saturninus, about the year of Rome 740. After the death of Obodas, Æneas, afterwards called Aretas, took possession of the kingdom of the Nabatæans. Upon this Syllæus, the late king's minister, went to Rome, and declared before Augustus that Æneas, or Aretas, had no right to the kingdom. How this corrupt minister was punished by Augustus may be seen in Nicolas of Damascus and in Josephus. This Aretas must have reigned for a long time, to at least the last years of Tiberius. *Du Theil.*"The interest attaching to this expedition, which promises so much for the elucidation of the classical geography of Arabia, has hitherto served only still further to perplex it." The author of the article Marsyabæ in *Smith's Dict. of Greek and Roman Geography*, where the subject is discussed at some length.

[783]Called also Arsinoë, b. xvii. c. i. § 25. It was near Heroopolis, or Suez.

[784]Koft.

[785]This name is variously written in manuscripts. If Negra be adopted, as by Letronne, it is not the same town as the city of the Negrani above mentioned, which was in the interior; but, as Kramer observes, "Mire corrupta est hæc ultima libri pars."

[786]B. xvi. c. iv. § 2.

[787]See above, § 2.

[788]This reminds us of the prophet Elijah and John the Baptist.

[789]Od. iv. 84.

[790]This subject was discussed in b. i. c. ii. § 34.

[791]B. xvi. c. iv, § 2 and § 14.

[792]Genadil.

[793]Assouan.

[794]Thus Eratosthenes calculated, in following the windings of the Nile, 12,900 stadia, which is 7900 stadia more than he calculated in a straight line, as he made the distance between the same points (Meroë and Syene, b. ii. c. v. § 7) to be 5000 stadia. M. Falconer suspects that there is an error in the text; but the error lies further off. I believe that it is attributable to Eratosthenes himself, and that that geographer did nothing more than convert the days' marches into stadia. According to Pliny, Timosthenes, commander of the fleet of Ptolemy Philadelphus, and consequently anterior to Eratosthenes, said that from Syene to Meroë was a march of 60 days; and this statement agrees tolerably well with that of Herodotus, who calculated 56 days' march between Elephantina and Meroë, besides a small distance the extent of which he does not state.

Procopius, a learned writer, estimates a day's march at 210 stadia; and employment of this value, in the whole course of his history, proves that it was generally adopted. Now, if we multiply 60 by 210, we shall have 12,600 stadia, and dividing 12,900 by 60, we have 215 stadia, or nearly the amount of a day's march according to Procopius. I am therefore of opinion that Eratosthenes did nothing more than multiply 210 or 215 by the number of 60 days, furnished by Timosthenes; and as the excessive length of 12,900 stadia could not agree with the 5000 stadia, which he had calculated in a straight line for the same interval, he imagined this great difference arose from the excessive winding course of the Nile; consequently he supposed the Nile to change frequently the direction of its course.

This opinion had its influence in the construction of Ptolemy's map, which presents to us nearly all the inflexions which Eratosthenes imagined; in calculating the intervals of positions assigned by Ptolemy along the river, we find a total of 1260 minutes; and adding about 1/6 for the small windings, we have a total of 1470 minutes, which are equal to 12,400 stadia of the module (700 to the degree) adopted by that geographer.

According to this hypothesis, the distance in Strabo will be thus divided: Setting out from Meroë, the Nile runs,

	days.
1. 2700 stadia to the north	12·8
2. 3700 to the S. and S. W.	17·6
3. 5300 to the N. 1/4 E.	25·0
4. 1200 to the N.	5·7
	61·1

which nearly corresponds with the account of Timosthenes. The number of days corresponds tolerably well with the distance given by the explorers sent by Nero for the discovery of Meroë: they reported the distance to be 873 miles. If we divide this number by 60, we shall have for the day's mean march 14·55 Roman miles, or 11·64 geographical miles, which is in fact the day's mean march, according to Major Rennell.*Letronne.*

In carefully measuring, upon a large map of Egypt in 47 sheets, the course of the Nile through all its windings, and with the compass opened to 1000 metres, I find—

	metres.
From the middle of Syene to Luxor in the ancient territory of Thebes	218,900
From Luxor to Becous situated at the point of the Delta	727,500
From Becous following the Damietta branch to that city	234,000
	1,180,400

This measure reduced to mean degrees of the earth equals 637° 25′, and represents 5312 stadia of 500 (to the degree). I certainly did not expect to find such an agreement between the new and the ancient measures. The periodic rising of the Nile, I think, must have produced, since the time of Eratosthenes, some partial changes in the windings of the river; but we must acknowledge that these changes, for greater or for less, compensate one another on the whole.

We observe, moreover, as I have already often observed, that the use of the stadium of 500 to the degree is anterior to the Alexandrine school; for at the time of Eratosthenes the stadium of 700 was more particularly made use of in Egypt. *Gossellin.*

[795]Although generally described as an island, it was, like Mesopotamia, a district included between rivers: the city Meroë was situated in lat. 16° 44′.

[796]Tacazze.

[797]Bahr-el-Azrek, or Blue river.

[798]See b. xvi. c. iv. § 8, and Herod. ii. 30, who calls the Sembritæ, Automoloi, that is, persons who had voluntarily quitted their abode.

[799]The Nile valley was parcelled out into a number of cantons, varying in size and number. Each of these cantons was called a nome (νομὸς) by the Greeks, "præfectura oppidorum" by the Romans. Each had its civil governor, the Nomarch, who collected the crown revenues, and presided in the local capital and chief court of justice. Each nome too had its separate priesthood, its temple, chief and inferior towns, its magistrates, registration and peculiar creed, ceremonies and customs; and each was apparently independent of every other nome. At certain seasons, delegates from the various cantons met in the palace of the Labyrinth, for consultation on public affairs (b. xvii. c. i. § 37). According to Diodorus, the nomes date from Sesostris. But they did not originate from that monarch, but emanated probably from the distinctions of animal worship; and the extent of the local worship probably determined the boundary of the nome. Thus in the nome of Thebaïs, where the ram-headed deity was worshipped, the sheep was sacred, the goat was eaten and sacrificed: in that of Mendes, where the goat was worshipped, the sheep was a victim and an article of food. Again, in the nome of Ombos, divine honours were paid to the crocodile: in that of Tentyra, it was hunted and abominated: and between Ombos and Tentyra there existed an internecine   feud.

Ardet adhuc Ombos et Tentyra: summus utrinque Inde furor vulgo, quod numina vicinorum Odit uterque locus, cum solos credat habendos Esse deos, quos ipse colit. Juv. xv. 35. The extent and number of the nomes cannot be ascertained. They probably varied with the political state of Egypt. See *Smith*, art. Ægyptus.

[800]See b. xvi. c. ii. § 24.

[801]In the text ὀστράκινα πορθμεῖα "earthen-ware ferry boats." The translation is not literal, but a paraphrase.

Hac sævit rabie imbelle et inutile vulgus

Parvula fictilibus solitum dare vela phaselis,

Et brevibus pictæ remis incumbere testæ.

Juv. xv. 126.

[802]In the text κειρίᾳ ψυχομένη ἐπὶ μῆκος, which is evidently corrupt. Kramer proposes to read ἀναπτυσσομένη or ἀνεπτυγμένη, and Groskurd reads αὐξομένη for ψυχομένη "lengthened out." Alii alia proposuerunt, infelicia omnia.

[803]The Mediterranean.

[804]Od. iv. 581.

[805]ἐγὼ γοῦν ἀπορούμενος ἀντιγράφων εἰς τὴν ἀντιβολὴν ἐκ θατέρου θάτερον ἀντέβαλον. Casaubon, who narrates a similar circumstance which occurred to himself, thus explains the passage: Our author, being in want of codices to correct imperfections in his own, and to form a complete copy, availed himself of another author whose account was identical, being either, as he says, the original or a transcript from the first.

[806]The words "Sostratus of Cnidus, son of Dexiphanes, to the gods preservers," are rejected by Kramer as being introduced from the margin.

[807]Od. xvii. 266.

[808]Some word, such as κατοικίαι, seems here to be wanting; ὁδοὶ, which some commentators suppose to be here understood, would be unsuitable to the passage, nor would it convey a proper meaning. *Kramer.*

[809]The word ἐραστής must be here understood, and not υἱὸς. *Groskurd.*

[810]The celebrated general of Mithridates.

[811]See b. xii. c. i. § 2.

[812]He was prevented from carrying on this war by the senate. See b. xii. c. iii. § 34.

[813]The elder sister of Cleopatra.

[814]Six months after.

[815]About B. C. 49.

[816]B. ix. c. v. § 6.

[817]I have adopted the reading, ἀπολιτικὸν, "not understanding or ill-adapted for the duties of citizens," suggested by Kramer.

[818]Od. iv. 481.

[819]No longer existing.

[820]Akabet el Kebira or Marsa Sollom.

[821]Baretoun, or Berek-Marsa. "Alexander, after passing 1600 stadia through that part of the desert where water was to be found to Parætonium, then turned inland to visit the oracle of Ammon." *Arrian*, b. iii. § 3.

[822]"Wines which have been very carefully prepared with sea-water never cause head-aches." *Athenæus*, b. i. c. i. 59, p. 54. Bohn's Classical Library.

[823]Cape Deras.

[824]The exact site is not ascertained, but it was not far from Aboukir.

[825]"Hellanicus says that the vine was first discovered in Plinthine, a city of Egypt," and that for those "who, on account of their poverty, could not get wine, there was introduced a custom of drinking beer made of barley." *Athenæus*, b. i. c. i. 61, p. 56. Bohn's Classical Library.

[826]"The Mareotic wine is erroneously stated by Athenæus (p. 55. Bohn's Classical Library) to have obtained its name from a fountain called Marea. The fountain and town derived their name from Maro, who was one of the companions of Bacchus." The wine is praised by Horace, Odes I. xxxvii. 14: Mentemque lymphatam Mareotico Redegit in veros timores. Virgil, Geor. ii. 91, calls a vine by this name: Sunt Thasiæ vites, sunt et Mareotides albæ.

[827]The Papyrus.

[828]"There is also the ciborium. Hegesander the Delphian says that Euphorion the poet, when supping with Prytanis, his host, exhibited to him some ciboria, which appeared to be made in a most exquisite and costly manner. Didymus says that it is a kind of drinking-cup, and perhaps it may be the same as that which is called scyphium, which derives its name from being contracted to a narrow space at the bottom, like the Egyptian ciboria." *Athenæus*, b. xi. § 54, p. 761. Bohn's Classical Library.

[829]The two kinds known at present are the Egyptian and the Syracusan, which, according to Professor Parlatori, have the same general appearance, but differ in the number of flower-lobes.

[830]That is, the juice was extracted for its sugar; see b. xvi. c. ii. § 41, and Pliny, xiii. 12.

[831]Od. iv. 228.

[832]The Canobic mouth was situated in the bay of Aboukir; the Bolbitine is the Rosetta mouth; the Sebennytic is the Burlos mouth; the Phatnitic, the Damietta mouth; the Mendesian is that at Dibeh; the Tanitic, that at Omm. Faregeh; the Pelusiac, that at Terraneh.

[833]The watch-tower of Perseus was at the western end of the Delta, according to Herodotus, ii. 15.

[834]The horned Pan.

[835]The people of Busiris worshipped Isis, and at one epoch, according to Hellenic tradition, sacrificed red men, who came over the sea, i. e. the nomades of Syria and Arabia.

[836]Od. iv. 481.

[837]In this nome tradition affirmed that the Hebrew legislator was born and educated.

[838]καὶ is omitted in the translation, as Groskurd proposes.

[839]Memphis was the residence of the Pharaohs, who succeeded Psammitichus, B. C. 616. The Memphite Nome rose into importance on the decline of the kingdom of Thebaïs, and was itself in turn eclipsed by the Hellenic kingdom of Alexandria. The village of

Mitranieh, half concealed in a grove of palm trees, about ten miles south of Gizeh, marks the site of the ancient Memphis. The successive conquerors of the land, indeed, have used its ruins as a stone quarry, so that its exact situation has been a subject of dispute. Major Rennell, however, brings incontestable evidence of the correspondence of Mitranieh with Memphis. Its remains extend over many hundred acres of ground, which are covered with blocks of granite, broken obelisks, columns, and colossal statues. The principal mound corresponds probably with the area of the great temple of Ptah. *Smith.*

[840]The Egyptians say that the ox Mneyis is sacred to the sun, and that Apis is dedicated to the moon. Ælian de Nat. Animal. ii. 11.

[841]Saïs stood in lat. 30° 4' N., on the right bank of the Canopic arm of the Nile. The site of the ancient city is determined not only by the appellation of the modern town Sa-el-Hadjar, which occupies a portion of its area, but also by mounds of ruin corresponding in extent to the importance of Saïs, at least, under the later Pharaohs. The city was artificially raised high above the level of the Delta to be out of the reach of the inundations of the Nile, and served as a landmark to all who ascended the arms of the river, from the Mediterranean to Memphis. Its ruins have been very imperfectly explored, yet traces have been found of the lake on which the mysteries of Isis were performed, as well as of the temple of Neith (Athene) and the necropolis of the Saïte kings. The wall of unburnt brick which surrounded the principal buildings of the city was 70 feet thick, and probably, therefore, at least 100 feet high. It enclosed an area 2325 feet in length by 1960 in breadth. Beyond this enclosure were also two large cemeteries, one for the citizens generally, and the other reserved for the nobles and priests of the higher orders. Saïs was one of the sacred cities of Egypt: its principal deities were Neith, who gave oracles there, and Isis. The mysteries of the latter were celebrated with unusual pomp on the evening of the Feast of Lamps. Herodotus (ii. 59) terms this festival the third of the great feasts in the Egyptian calendar. It was held by night; and every one intending to be present at the sacrifices was required to light a number of lamps in the open air around his house. At what season of the year the feast of burning lamps was celebrated, Herodotus knew, but deemed it wrong to tell (ii. 62); it was, however, probably at either the vernal or autumnal equinox, since it apparently had reference to one of the capital revolutions in the solar course. An inscription, in the temple of Neith, declared her to be the Mother of the Sun. It ran thus, "I am the things that have been, and that are, and that will be; no one has uncovered my skirts; the fruit which I brought forth became the Sun." It is probable, accordingly, that the kindling of the lamps referred to Neith, as the author of light. On the same night, apparently, were performed what the Egyptians designated as the "Mysteries of Isis." Saïs was one of the supposed places of interment of Osiris, for that is evidently the deity whom Herodotus will not name (ii. 171), when he says that there is a burial-place of *him* at Saïs in the temple of Athene. The mysteries were symbolical representations of the sufferings of Osiris, especially his dismemberment by Typhon. They were exhibited on the lake behind the temple of Neith. Portions of the lake may be still discerned near the hamlet Sa-el-Hadjar. *Smith, Dict. of Greek and Roman Geography, Art.* Saïs.

[842]The evil or destroying genius.

[843]Suez.

[844]Pharaoh Necho, under whom and in the execution of the work 120,000 labourers perished. Herod. ii. 158.

[845]κλειστὸν ἐποίησαν τὸν Εὔριπον, "closed the Euripus." Diodorus Siculus, i. 33, thus speaks of this same work. "Darius the Persian left the canal unfinished, as he was informed by some persons, that by cutting through the isthmus he would be the cause of inundating Egypt; for they pointed out to him that the Red Sea was higher than the level of Egypt. The second Ptolemy afterwards completed the canal, and in the most convenient part constructed an artfully contrived barrier, (διάφραγμα,) which he could open when he liked for the passage of vessels, and quickly close again, the operation being easily performed." The immediate communication therefore between the sea and the canal was cut off by a lock; and as there must have been two, there would be a flux and reflux of water between them on the passage of vessels. This probably suggested to our author the word Euripus,

and is to be understood as applying to that portion of the canal included between the locks. By the word Euripus is generally understood the channel between Negropont and the mainland, which is subject to an ebb and flow of the sea. The storing up of water, and the distribution of it for the purposes of irrigation, was no doubt well known to the Egyptians. Diodorus, b. i. 19, ascribes to Osiris the invention. "Osiris confined the Nile by embankments on both sides, so that at the period of its rising it might not inconveniently spread over the country, but that, by gates (διὰ θυρῶν) adapted for the purpose, the stream might be gently discharged as occasion required.

[846]B. i. c. i. § 20.

[847]Bubastis or Artemis, Diana. Herod. ii. 59, 67, 137.

[848]Among those no doubt now at Rome.

[849]This description is illustrated by the remains of the great temple at Philæ, dedicated to Ammon Osiris.

[850]οὐδὲ γραφικόν. These words have been understood by some writers as signifying that there were no paintings, but Letronne has clearly shown that they do not convey this meaning.

[851]George (Syncellus, or companion of the Patriarch), a writer of the eighth century, and who had the reputation of being well versed in history, says that "Ptolemy Philadelphus collected all the writings of Greeks, Chaldæans, Egyptians, and Romans, and had such of them as were not Greek translated into that language, and deposited 100,000 volumes at Alexandria. M. Letronne is disposed to think that Hipparchus, Eratosthenes, Ptolemy, and others borrowed from these sources.

[852]"Sesoosis (Sesostris) raised two obelisks of hard stone, 120 cubits in height, on which were inscribed the greatness of his power, the amount of his revenue, and the number of the nations which he had conquered. At Memphis, in the temple of Vulcan, he erected monolithe images of himself and his wife, 30 cubits in height, and images of his sons, 20 cubits in height," in memory of his escape from fire when his brother Armais attempted to burn him with his wife and children. Diod. Sic. i. 57.

[853]Probably the statue of Venus bore a crescent on the forehead.

[854]We have reason to be surprised that Strabo, who had seen the pyramids, has said so little concerning them. Herodotus and Diodorus Siculus enter into more particulars, and in general are more exact. Some idea of the immense labour required may be obtained from considerations such as follow:—

The base and height being given, we find for the solid contents—

	cubic yards.
1. of the great pyramid	2,864,000
2. of Chephren	2,056,000
3. of Mycerinus	211,000

So that if a wall of (three metres) about 9-1/4 feet in height, and a foot in thickness, were built with the materials of these pyramids, we should have a wall—

	iles.
1. from the great pyramid in length	626
2. from Chephren or Cheops	167
3. from Mycerinus	

The stones, therefore, of the three pyramids would form such a wall 2910 miles in length, or one sufficient to reach from Alexandreia to the coast of Guinea. *Letronne.*

[855]This is a palpable error, and greater than that of Herodotus, who makes the base equal to the height. The ratio of the height to the base in the great pyramid was as 0·627 to 1; and in the second, as 0·640 to 1. Diodorus approaches nearest of all to the truth, as he makes this ratio to be as 6 to 7 or as 0·817 to 1. Strabo should rather have said, "the sides are rather greater than the height;" but all that he says respecting the pyramids is vague and inexact.

[856]ἐν ὕψει μέσως πως μιᾶς τῶν πλευρῶν· μιᾶς is adopted, although not introduced into the text, by Kramer; μέσως πως is connected with ἐν ὕψει, and not with τῶν πλευρῶν, in the sense of "moderately," in which it is also used in b. xi. c. ii. § 18. "The kings who succeeded to the possession of the country, (μέσως ἔπραττον) were moderately successful." The moveable stone has been taken away, and the aperture is at most at about one-twelfth the whole height of the pyramid from its base.

[857]Chembes the Memphite built the largest of the three pyramids, which are reckoned among the seven most remarkable works in the world. They are situated by the side of Libya, distant 120 stadia from Memphis, and 45 from the Nile. These works, by their size and by the artifice and labour employed in their construction, strike the beholder with astonishment and wonder. The base of the largest, the plan of which is quadrilateral, is seven plethra on each side; the height is more than six plethra; the pyramid gradually contracts towards the top, of which each side measures six cubits, and the whole is built of hard stone. Its construction must have been accompanied with great difficulty, but its permanence will be eternal; for although, it is said, not less than a thousand years have passed away to our day (some even say more than 3400 years) since they were built, yet the stones still remain, preserving their original position, and their whole arrangement uninjured by time. The stone is said to have come from a great distance in Arabia, and the process of building was carried on by raising mounds of earth; for at that period no machines had been invented. But it is most marvellous that although such an immense undertaking has been completed, and the whole country around is composed of sand, not a single trace remains of the mounds raised, nor of the fragments of stone broken off by the workmen: indeed the pyramids do not seem to have been raised by the gradual labour of man, but to have been placed by some divine hand in a mass, perfectly formed, down upon the surrounding sands. Some Egyptians undertake to narrate wondrous stories respecting them, such, for instance, that the mounds above-mentioned were composed of salt and nitre, which melted away upon the rising of the river, and completely disappeared without the intervention of human labour. But this cannot be true, for the same number of hands which constructed the mounds would be able to reduce them again to their former state; and 360,000 men, it is said, were employed in the undertaking. The whole was completed in a little less than twenty years. On the death of this king, he was succeeded by his brother Chephren, who reigned 56 years. According to some writers, it was not a brother, but a son, named Chabryis, who was his successor. But all agree that the successor, whoever he was, desired to imitate his predecessor's conception, and built the second pyramid, which resembled the first in its artificial construction, but was inferior to it in size, the sides of the base being a stadium each in length. On the greater pyramid is an inscription which states the amount expended on herbs and radishes for the workmen, and it informs us that 1600 talents were paid for this purpose. The lesser pyramid bears no inscription, and it has an ascent formed in it through an opening in one of the sides. But although the kings built these pyramids for their own tombs, yet it has so happened that none of them have ever been buried in them. For the population, in consequence of the misery to which these works exposed them, and of the cruelty and tyranny of the kings, were incensed against them as the causes of their sufferings; and moreover threatened to tear their bodies in pieces, and to cast them out with insult from their place of burial. Every king therefore, on the approach of death, enjoined his relations to bury his body secretly in a place undistinguished by marks. These were succeeded by king Mycerinus, (whom some call Mecherinus,) son of the king who built the first pyramid. He designed to build a third, but

died before he accomplished it. Each side of the base of this pyramid was three plethra in length, and fifteen tiers of the building were raised of black stone like the Thebaic stone, but the rest was filled up with a stone resembling that of the other pyramids. This work is inferior to the two former in size, but far surpasses them in artificial construction and in the expensiveness of the stone. On its northern side the name of Mycerinus is inscribed, as the person who caused it to be built. He is said to have held in abhorrence the cruelty of his predecessors, and to have been ambitious of leading a just life, and beneficial to his subjects. He performed many actions by which he called forth the affection of his people towards him; and among others he expended a great sum of money in public causes, rewarding the judges who delivered upright judgments, which was not commonly the case. There are three other pyramids, the sides of which are two plethra in length; in workmanship they entirely resemble the others, except in magnitude. These pyramids, it is said, were built by the three before-mentioned kings in honour of their own wives. These works by universal consent are the most remarkable in Egypt, not only in their ponderous construction, but also in the art displayed. We ought, we are told, to admire more the architects than the kings, who supplied the means, for the architects brought their designs to completion by force of mind and the influence of an honourable ambition, but the kings by the power of that wealth which was their portion, or by injuries inflicted on others. There is no agreement whatever, either between the natives of the country or between authors, respecting the pyramids; for some assert that the kings before mentioned built them, others that they were not the builders, but that Armæus built the first and largest; Amasis, the second; and Inaro, the third: but this last is said by some to be the burial-place of Rhodopis, a courtesan, whose lovers were certain governors of nomes, who from affection towards her undertook this great work, and completed at their common charge. Diodorus Siculus, b. i. 63, 64.

[858]Niebuhr says, that in these stones are found small petrified substances in the form of lentils, which appear to be of the same kind of shell of which he collected several at Bushir. Clarke also says, that at the base of the pyramids a variety of calcareous stone is found in detached masses, exactly such as Strabo has described, and appear to be the petrified remains of some unknown animal. Forskal calls them "testacea fossilia kakiensia." Diodorus, as quoted above, says that there are no vestiges of fragments.

[859]The translation follows Letronne's correction, ἐπέοικε for ἀπέοικε.

[860]In the text λίθου πωρείας, Groskurd reads πωρίνου, which word occurs in Herod. v. 62, and translates it "tufstein."

[861]No passage is to be found in his Geography to this effect, it has either been lost from the text, or existed in his other works.

[862]"It is said that the captives from Babylon revolted from the king (Sesostris), being unable to endure the sufferings to which they were exposed in the public works. They seized upon a strong place on the banks of the river, and maintained for some time a contest with the Egyptians, destroying the neighbouring district. At last, having obtained security from molestation, they made a regular settlement of the place, and called it Babylon, after their native city. Under similar circumstances, it is said, a place received the name of Troy which still exists on the banks of the Nile. For Menelaus, on his return from Troy with captives, came to Egypt. The Trojan captives revolted, took up a position, and carried on a war, until having obtained safety for themselves by treaty, they founded a city bearing the name of their native place." I am aware that Ctesias gives a different account of these cities, and says that some of the soldiers who accompanied Semiramis in her invasion of Egypt founded these cities, and gave to them the names of their native cities. Diod. Sic. i. 56.

[863]This passage presents great difficulties. Kramer expresses himself dissatisfied with any explanation hitherto given. Und so dass der Kanal zwei Mündungen hat, zwischen welche ein Theil der Insel seitwärts anfält. *Groskurd.*

[864]Book i. c. iii. § 4.

[865]Herod. ii. 148; Diod. i. 66. See below, § 42.

[866]The translator adopts Kramer's suggestion, of reading εἰσβλέποντα for ἐκπίπτοντα.

[867]The founder, according to Diodorus Siculus, was Mendes or Marrus. B. i. 61.

[868]Bekneseh.

[869]This fish, a species of sturgeon, received its name from the shape of the head (sharp-pointed), and was said to have been produced from the blood of the wounded Osiris. Ælian. Hist. Animal. x. 46.

[870]Eshmoon.

[871]Babout.

[872]The ruins are supposed to be at the modern hamlet of Mensieh.

[873]ὀλόλιθον, probably an interpolation. *Kramer.*

[874]Il. i. 528.

[875]Hu.

[876]Dendera.

[877]Keft.

[878]The ruins are situated lat. 23° 56' N., and about 35° 34' E.

[879]After σταθμοὺς, in the text, follows ὥσπερ τοῖς ἐμπορίοις ὀδεύμασι καὶ διὰ τῶν καμήλων, which Kramer considers to be an interpolation. Groskurd corrects, and reads σταθμοὺς προσφόρους τοῖς ἐμπόροις ὀδεύουσι καὶ πεζῇ καὶ διὰ τῶν καμήλων, "stations for the service of travellers on foot and on camels."

[880]Near old Kosseir; the "Veneris Portus" of Pliny. It was founded by Ptolemy Philadelphus, B. C. 274. The Greek name may signify, "Harbour of the Mouse," but more probably it means the "Harbour of the Mussel," (μύειν, to close, e. g. the shell,) since on the neighbouring coast the pearl-mussel is collected in large quantities. It is uncertain whether the ruins at the village of Abuschaar, represent the site of the ancient Myos Hormus. See Smith's Dict., art. *Myos Hormus*

[881]Il. ix. 383.

[882]Il. ix. 381.

[883]For θήκαις, "tombs," in the text, Kramer is of opinion that we should read Θήβαις, Thebes, which is also the translation of the passage by Guarini.

[884]The meaning of the passage is clear, and can be understood, as critics have already explained, only as implying the intercalation of a 366th day every fourth year. Some have asserted that Julius Cæsar adopted this method of intercalating a day from the civil practice of the Alexandrines; others, on the contrary, appear disposed to believe that J. Cæsar was the first to give an idea of it, according to the advice of Sosigenes. There is truth and error in both these opinions. On the one hand, it is certain that Strabo, who visited Egypt a short time after the conquest of the country by the Romans, would not have omitted to attribute to them the institution of this year, if it really belonged to them. So far from doing so, he says (above, § 29) distinctly, that this method of intercalation was known and practised by the priests of Heliopolis and Thebes. Diodorus Siculus, who visited Egypt just at the time of the first arrival of the Romans, gives the same account as Strabo. Can we therefore believe that the Egyptians before this period were ignorant of the bissextile intercalation? On the other hand, it is not less certain that this method of intercalation was only introduced into civil use at Alexandria from the time of Julius Cæsar: before this period, the incomplete year of 365 days was adopted throughout the whole of Egypt, as is attested by a host of authorities, and confirmed by the date of the Rosetta stone, which only applies to this method of reckoning. Hence we see (I.) that Julius Cæsar really obtained the idea of a fixed year of 365-1/4 days from the Egyptians, where it was employed for scientific or religious purposes only, whilst the incomplete year was the vulgar and common year; (II.) that he made this fixed year the common year, both among the Romans and Alexandrines, who were a people most readily disposed to adopt foreign innovations. It is, however, probable that the rest of Egypt preserved the ancient use of the incomplete year.

[885]Strabo, I think, is the only author who places Crocodilopolis and Aphroditopolis in this part of Egypt. *Letronne.*

[886]For καὶ τῶν ἡμερῶν of the text, Casaubon reads τεκμηρίων, "signs." Coray proposes καὶ μέτρων, "measures." The expression in the text is obscure, and the translation is a conjecture of the meaning.

[887]This was the general opinion of antiquity, and was reproduced by Eratosthenes, Hipparchus, Ptolemy, and others; in short, by all the Alexandrine school. At the time of Eratosthenes, the obliquity of the ecliptic was 23° 45' 17". Syene was therefore 20' 6" from being exactly under the tropic; for 24° 5' 23" (latitude of Syene)—23° 45' 17" = 20' 6". This would be the distance of the centre of the sun from the zenith of Syene; whence it follows that the northern limb of the sun was about 5' from it. In the time of Strabo, the obliquity was only 23° 42' 22"; the difference between the zenith of Syene and the northern limb of the sun was about 8'. Lastly, about 140 of the vulgar era, the obliquity was reduced to 23° 41' 7". Syene was then 24' 16" from the tropic, and its zenith was about 10' from the northern limb of the sun; when the shadows of gnomons of any tolerable size must have been perceptible, and Syene could not have been any longer considered as lying under the tropic. As regards the well which served to ascertain the instant of the solstice, Pliny and Arrian both mention it. The formation of it no doubt belonged to a very remote period. In the time of Strabo, the rays of the sun could not have reached entirely to the bottom, but the shadow was so small that it was not sufficient to shake the ancient opinion. In fact, the angle being about 8', and supposing the depth to have been 50 feet, the northern side would have projected a shadow of about 18 lines; the rest would have remained in full light, and the reflexion would have caused the whole circumference of the well to appear illuminated. *Letronne.*

[888]Kramer considers the passage between brackets to be an interpolation, as the same sense is conveyed in the passage which immediately follows.

[889]The number here given is nearly twice too great. Kramer quotes G. Parthey (de Philis insula) for correcting the error to 50 stadia, and for perceiving that it arose from the very frequent substitution in manuscripts of the letter P (100) for N (50).

[890]Unhewn stones, with a head of Mercury upon them.

[891]Herod. ii. 28, who, however, seems to doubt the veracity of his informant.

[892]Above, § 8.

[893]B. C. 28.

[894]B. xvi. c. 4, § 23.

[895]The modern hamlet of Dakkeh occupies a portion of the site of ancient Pselchis.

[896]Called Primis by Ptolemy and Pliny. It is placed by the former beyond Napata, and just above Meroë. Hence it is identified with Ibrim.

[897]There is great difficulty in determining the true position of Napata, as our author places it much farther north than Pliny; and there is reason for supposing that it is the designation of a royal residence, which might be moveable, rather than of a fixed locality. Ritter brings Napata as far north as Primis and the ruins at Ipsambul, while Mannert, Ukert, and other geographers, believe it to have been Merawe, on the farthest northern point of the region of Meroë. It is, however, generally placed at the east extremity of that great bend of the Nile which skirts the desert of Bahiouda, and near Mount Birkel. Among the ruins which probably cover the site of the ancient Napata are two lions of red granite, one bearing the name of Amuneph III., the other Amuntuonch. They were brought to England by Lord Prudhoe, and now stand at the entrance of the Gallery of Antiquities in the British Museum. See Smith's Dict., art. *Napata.*

[898]The inhabitants of Biscay. See b. iii. c. iii. § 8.

[899]This name was common to the queens of Ethiopia. Acts viii. 27.

[900]B. xvi. c. iv. § 8 et seqq.

[901]Groskurd corrects the text, and translates, "the inhabitants also are small."

[902]The translation follows the proposed correction of the text by Kramer.

[903]ταῖς συμβολαῖς. The passage presents a great difficulty, because Strabo has before asserted that Meroë is surrounded by these rivers, and that their union takes place below, that is, to the north, and not to the south of the city and island; and this notion corresponds with all the ancients have said on the subject. I declare, without hesitation, that I do not understand my author. *Letronne.* Groskurd attempts to avoid the difficulty by translating, "is within the compass of."

[904]The Tacazze.

[905]Bahr-el-Azrek, or Blue River.

[906]Reading διαπλεκομένων ἢ πλίνθων for διαπλεκόμεναι τοίχων ἢ πλίνθων.

[907]The trees called persiai (or perseai) produce a fruit of great sweetness, which was introduced from Ethiopia by the Persians, when Cambyses conquered that country. Diod. Sic. i. 34.

[908]Tsana.

[909]According to Diod. Sic. iii. 9, this was Jupiter.

[910]Above, c. i. § 15.

[911]The sturgeon.

[912]Cyprinus bynni.

[913]Perea Nilotica. *Cuvier*, Histoire Naturelle des Poissons, xii. 5.

[914]Silurus anguillaris. *Linn.*

[915]Pliny, xxxii. 5. Coracini pisces Nilo quidem peculiares sunt. Athenæus, b. vii. c. 83, p. 484. Bohn's Classical Library.

[916]Called by the Arabs gamor-el-Lelleh, or star of the night. *Cuvier.*

[917]The shad.

[918]The mullet.

[919]About six feet. Nicander is the author of two Greek poems that are still extant, and of several others that have been lost. He may be supposed to have been in reputation for about fifty years, cir. B. C. 185-135. The longest of his poems that remains is named Theriaca. It treats (as the name implies) of venomous animals, and the wounds inflicted by them, and contains some curious and interesting zoological passages, together with numerous absurd fables. The other treats of poisons and their antidotes. His works are only consulted by those who are interested in points of zoological and medical antiquities. He is frequently quoted by Athenæus. See Smith's Dict. of Greek and Roman Biography, art.*Nicander.*

[920]Herod. ii. 36.

[921]Strabo does not appear to have been acquainted with the plant from which these tissues were made. Their true name seems to have been cucina, and were made from a palmtree (the Doum palm), called by Theophrastus (Hist. Plant. 4, 2) κουκιοφόρον, and by Pliny "cuci" (b. xiii. 9): "At e diverso, cuci in magno honore, palmæ similis, quando et ejus foliis utuntur ad textilia."

[922]B. xvi. c. 2, § 34.

[923]B. ii. c. 3, § 4; and c. 4, § 3.

[924]B. i. c. 4, § 2.

[925]Cape Spartel, or Espartel. Ampelusia, vine-clad, was the Greek name,—a translation of the native name.

[926]Groskurd reads Tinx, and also with Letronne observes that our author has mistaken two places for one. Tinx, or Trinx=Tangiers. Lixus = Al-Harâtch, or Laraiche.

[927]Cadiz.

[928]Situated between the town Sala (Salee) and Lixus (El-Harâch).

[929]Tyrwhitt reads Apellas, for Ophellas of the text. Apellas was a Cyrenæan navigator, whose Periplus is mentioned by Marcianus of Heracleia. There was an Ophellas of Cyrene, who advanced at the head of an army along the coast, to unite himself to Agathocles, who was then besieging Carthage, B. C. 310. He was put to death by Agathocles soon after his arrival, and no Periplus of his said to have existed; his course also to Carthage was by land.

[930]A people on the west coast of N. Africa, about the situation of whom Strabo, Pliny, and Ptolemy are in perfect agreement with one another, if the thirty days' journey of Strabo between them and Lixus on the west coast of Morocco, to the south of Cape Spartel, be set aside, as an error either of his information or of the text; which latter is not improbable, as numbers in MSS. are so often corrupt. Nor is this mere conjecture, because Strabo contradicts himself, by asserting in another place (b. xvii. c. 3. § 7) that the Pharusii had a great desert between them and Mauretania. When Ezekiel prophesies the fall of Tyre, it is said, (xxvii. 10,) "The men of Pheres (the common version reads Persia) and Lud and

Phut were in thine armies." These Pheres thus joined with Phut, or Mauretanians, and the Ludim, who were nomads of Africa (the Septuagint and the Vulgate understand the Lydians), may be reasonably supposed to belong to the same region. Without the vowel points, the name will represent the powerful and warlike tribe whom the Greeks call Pharusii. *Smith*, art. Pharusii.

[931]Arum esculentum (snake-weed), and arum dracunculus.

[932]Parsnip (?).

[933]Fennel.

[934]Artichoke.

[935]Groskurd reads Hypsicrates.

[936]The rhinoceros.

[937]About six quarts, according to the lowest value of the (chœnix).

[938]Arzila.

[939]Tiga in the text.

[940]The Septem-Fratres of Pliny.

[941]Jebel-el-Mina, or Ximiera, near Ceuta (a corruption of ἑπτὰ, or septem?).

[942]Ape mountain.

[943]The Muluwi, which now forms the frontier between Morocco and Algeria, as it did anciently between the Mauretanians and Numidians.

[944]Cape Hone, or Ras-el-Harsbak. Groskurd corrects the text, and translates: "Near the river is a large promontory, and a neighbouring settlement called Metagonium." Kramer's proposed correction is followed.

[945]Numidia is the central tract of country on the north coast of Africa, which forms the largest portion of the country now occupied by the French, and called Algeria, or Algérie. The continuous system of highlands which extends along the coast of the Mediterranean was in the earliest period occupied by a race of people consisting of many tribes, of whom the *Berbers* of the Algerine territories, or the *Kabyles* or *Quabaily*, as they are called by the inhabitants of the cities, are the representatives. These people, speaking a language which was once spoken from the Fortunate Islands in the west to the cataracts of the Nile, and which still explains many names in ancient African topography, and embracing tribes of quite different characters, whites as well as blacks (though not negroes), were called by the Romans Numidæ; not a proper name, but a common denomination from the Greek form, νομάδες. Afterwards Numida and Numidia became the name of the nation and the country. Sometimes they were called Maurusii Numidæ, while the later writers always speak of them under the general name of Mauri. The most powerful among these tribes were the Massyli, whose territories extended from the river Ampsaga to Tretum promontory; and the Massæsyli, occupying the country to the west, as far as the river Mulucha. *Smith, Dict.* art. Numidia.

[946]Cartagena.

[947]Marseilles.

[948]The words περιτραχήλια ξύλινα offer some difficulty. Paul Louis Courier, who is of authority on this subject, says that Strabo, having little experience in horses, has mistaken the first word for another, and intended to speak of the horse's nose, and not his neck. Letronne and Groskurd both agree that ξύλινα is rightly to be translated, "of cotton."

[949]Constantine.

[950]The Pharusii, and not the Mauretanians, came with Hercules from the East, according to Pliny, Mela, and Sallust; hence Letronne conjectures that we should read here Pharusii.

[951]A. D. 18 or 19 at latest, but the exact date is uncertain.

[952]Groskurd corrects the text, and translates, "there existed in the Bay Emporicus very many Phœnician cities."

[953]*Plutarch*. Sertorius.

[954]Ebba-Ras.

[955]Probably Tafna.

[956]Jama.

[957]According to Shaw, who however did visit the place, its ruins are still to be seen by the present Tucumbrit; others identify it with Areschkul of the Arabs, at the mouth of the Tafna near Rasgun.

[958]In the text μεγέθει δὲ ἑπτασπονδύλων, scorpions "of seven joints" in the tail; the correction of Letronne, which Kramer supports, is adopted. Groskurd however retains the text, and reads μεγέθει δὲ [ὑπερβαλλόντων καὶ ἐσθ' ὅτε ἑπτασπονδύλων, "of enormous size, and sometimes of seven joints."

[959]Cherchell, a corruption of Cæsarea-Iol.

[960]Ebba Ras (the seven capes) or Bougaron.

[961]Bougie.

[962]Shaw has the merit of having first pointed out the true situation of this celebrated city. Before his time it was sought sometimes at Biserta, sometimes at Farina, but he fixed it near the little miserable "Douar," which has a holy tomb called Boushatter, and with this view many writers have agreed. Adherbal, however, was besieged and captured in Cirta (Constantine), B. C. 109.

[963]An unknown name. Letronne supposes Thisica to be meant, mentioned by Ptolemy, iv. 3.

[964]Vaga or Vacca, now Bayjah.

[965]Shaw takes Ferreanah to have been the ancient Thala or Telepte, but Lapie seeks it at Haouch-el-Khima.

[966]Cafsa.

[967]Jama.

[968]Probably near the ruins of Leptis Parva.

[969]El Aliah.

[970]Karkenah or Ramlah.

[971]Hippo Regius, Bonah; and Hippo Zaritus, Bizerta.

[972]Wady Mejerdah.

[973]Letronne corrects this reading to 2000, which is the number given by Polybius and Arrian.

[974]By the Romans, Numidæ.

[975]Pantellaria.

[976]Marsala.

[977]Kramer is of opinion that this passage from the beginning of the section is an interpolation. Cossura (the island Pantellaria) is nowhere else spelt Corsura; Cossuros is the spelling observed immediately below. Its distance from Aspis is differently stated in b. vi. c. ii. § 11, to be 88 miles from Aspis. Ægimurus is the small island Zembra, near Cape Bon; near it is also another small low rocky island. From the shape and appearance of the former, more especially in some positions, we may attribute the name Aræ (altars), given to them, as in Pliny: "Ægimuræ Aræ, scopuli verius quam insulæ;" and they are the "Aræ" of Virgil, Æn. i. 108.

[978]i. e. sacred to Mercury. Cape Bon.

[979]Cape Aclibia, from the Latin Clypea. B. vi. c. 2, § 11.

[980]Malta.

[981]Sousah.

[982]Demass.

[983]Lampedusa.

[984]Kramer's proposed emendation is followed.

[985]Gulf of Cabes.

[986]Jerba or Zerbi. It produced the "lotus-zizyphus" or the carob, now common in the islands of the Mediterranean and on the continent.

[987]Od. ix. 84.

[988]Sabrata?

[989]Lebida.

[990]Gerace. See b. vi. c. i. § 7, 8.

[991]The Cinifo or Wadi-Quasam.

[992]Cape Canan or Mesrata.

[993]See b. ii. c. v. § 20.

[994]Its position, like that of so many places on the Great Syrtis, can hardly be determined with certainty. A full discussion of these localities will be found in Barth's Wanderungen.

[995]About the middle of the fourth century, B. C., according to a story in Sallust, these monuments commemorated the patriotic sacrifice of two Philæni, Carthaginian envoys.

[996]Gulf of Suez.

[997]Ben Ghazi. Berenice previously bore the name Hesperides, which name seems to have been derived from the fancy which found the fabled Gardens of the Hesperides in the fertile terraces of Cyrenaïca.

[998]Ras-Teyonas.

[999]Cape Catacolo.

[1000]Groskurd justly supposes that the name Chelonatas (Cape Tornese) is here wanting in the text.

[1001]Zante.

[1002]Tochira.

[1003]The name has survived to the present day in that of the district of which it was the capital, the province of Barca, in the regency of Tripoli. The position of Barca is accurately described by Scylax, who places its harbour 500 stadia from Cyrene, and 620 from Hesperides, and the city itself 100 stadia from the sea. It stood on the summit of the terraces which overlook the west coast of the Greater Syrtis, in a plain now called El-Merjeh; and the same name is often given to the ruins which mark the site of Barca, but the Arabs call them El-Medinah. See Smith, art. *Barca.*

[1004]Ras-al-Razat or Ras Sem. Scylax here placed the gardens and lake of the Hesperides.

[1005]Cape Matapan, which is more than a degree and a half more to the east than Phycus.

[1006]In b. viii. c. v. § 1, it is stated to be 3000.

[1007]Santorin.

[1008]Kavo Krio.

[1009]B. C. 631.

[1010]B. C. 330.

[1011]Flourished about B. C. 366. The Cyrenaïc system resembles in most points those of Heracleitus and Protagoras, as given in Plato's Theætetus. The doctrines that a subject only knows objects through the prism of the impression which he receives, and that man is the measure of all things, are stated or implied in the Cyrenaïc system, and lead at once to the consequence, that what we call reality is appearance; so that the whole fabric of human knowledge becomes a fantastic picture. The principle on which it rests, viz. that knowledge is sensation, is the foundation of Locke's Modern Ideology, though he did not perceive its connexion with the consequences to which it led the Cyrenaïcs. To revive these was reserved for Hume. *Smith's Greek and Roman Biography and Mythology.*

[1012]This great astronomer and learned man, whose name so frequently occurs in the course of this work, was born about B. C. 276. He was placed, by Ptolemy Euergetes, over the library of Alexandria. His greatest work, and that which must always make his name conspicuous in scientific history, is the attempt which he made to measure the magnitude of the earth, in which he brought forward and used the method which is employed to this day. See vol. i. page 9, of this translation, note[42].

[1013]Carneades was born about B. C. 213. In the year B. C. 155, when he was fifty-eight years old, he was chosen with Diogenes the Stoic, and Critolaus the Peripatetic, to go as ambassador to Rome, to deprecate the fine of 500 talents, which had been imposed on the Athenians, for the destruction of Oropus. During his stay at Rome, he attracted great notice from his eloquent declamations on philosophical subjects, and it was here that, in the presence of Cato the Elder, he delivered his famous orations on Justice. The first oration was in commendation of virtue; in the second justice was proved not to be a virtue, but a mere matter of compact, for the maintenance of civil society. The honest mind of Cato was shocked at this, and he moved the senate to send the philosopher home to his school, and save the Roman youth from his demoralizing doctrines. He left no writings, and all that is known of his lectures is derived from his intimate friend and pupil, Cleitomachus. See Smith, Dict. of Greek and Roman Biography.

[1014]Marsa-al-Halal or Al Natroun.

[1015]Ras-al-Tyn.

[1016]Grabusa.

[1017]Ras-el-Milhr.

[1018]Marsa Sollom, or Akabet-el-Kebira, the present boundary of Tripoli and Egypt.

[1019]Baretoun or Berek Marsa.

[1020]Kramer's reading of this passage is followed.

[1021]Groskurd has a long note on this passage, and reads τοὺς κατ᾽ αὐτὸν Νασαμῶνας. The words in the original text, τοὺς κατ᾽ αὐτὸ μαλακῶς, present the great difficulty; but Kramer reads τοῦ for τοὺς, and has adopted in the text Falconer's proposed correction, κατ᾽ Αὐτόμαλά πως. The name Augila is wanting in the text; it is supplied by Groskurd, and approved by Kramer, who refers to Herod. iv. 172, 182.

[1022]Aujela, an oasis in the desert of Barca; it still retains its ancient name, and forms one of the chief stations on the caravan route from Cairo to Fezzan.

[1023]Τῆς καθ᾽ ἡμᾶς οἰκουμένης, Groskurd translates as inhabited to our time; but Strabo refers to the then known world, having before, b. i. c. iv. § 6, in a remarkable manner conjectured the existence of other habitable worlds (such as America) in the latitude of Athens. "We call that (part of the temperate zone) the habitable earth (οἰκουμένην) in which we dwell, and with which we are acquainted; but it is possible, that in the same temperate zone there may be two or even more habitable earths, especially near the circle of latitude drawn through Athens and the Atlantic Ocean." The latitudes of Athens and Washington do not differ by one degree.

[1024]B. vi. c. iv. § 2.

[1025]B. ii. c. v. § 31.

[1026]Guadalquiver (Wad-el-Kebir, the Great River).

[1027]B. iv. c. i. § 6.

[1028]B. iii. c. iv. § 20.

INDEX.

182

- Acmon, ii. 191.
- Aconites, people of Sardinia, i. 334.
- Acontia (Acuteia?), t. of Spain, i. 228.
- Acontium, mtn of Bœotia, i. 113.
- Acqui. *See* Aquæ-Statiellæ.
- Acra (*C. Takli*), ii. 222.
- Acræa, Venus, iii. 69.
- Acrææ, t. of Laconia, ii. 15, 41.
- Acræphiæ, Acræphium, t. of Bœotia, ii. 107, 109, 110.
- Acragas, city of Sicily (*Girgenti*), i. 411, 415.
- Acrathos (*C. Monte Santo*), i. 512.
- Acrisius, ii. 118, 211.
- Acritas, prom. of Messenia, i. 36.
- Acrocorinthus, ii. 38, 60, 61, 62, 70.
- Acrolissus. *See* Lissus, i. 485.
- Acropolis, ii. 86.
- Acrothŏï, t. on Mount Athos, i. 512, 513.
- Acta, Lepre, iii. 3.
- Actæon, ii. 87.
- Acte, ii. 77, 79, 174.
- Acte, Actice. *See* Attica, ii. 80.
- Actē. *See* Acarnania.
- Actia, Actian games, i. 499.
- Actian war, i. 36; ii. 330.
- Actis, ii. 375.
- Actium, prom. of Acarnania (*La Punta*), ii. 115, 161.
- ——, t. of Acarnania, ii. 73, 115, 159, 161, 171.
- ——,[Pg 300] battle of, ii. 208; iii. 230, 233.
- Acusilaus, Argive, ii. 189.
- Acyphas, or Pindus, ii. 128.
- ——, in the Œtæan district, ii. 136.
- Ada, dr of Hecatomnus, iii. 35, 36.
- ——, dr of Pixodarus, iii. 35.
- *Ada. See* Patræus.
- Adada, c. of Pisidia, ii. 324.
- Adæ, town of Æolis, iii. 397.
- Adarbal, Adherbal, iii. 284.
- *Adda*, r. of Gaul, i. 287, 304, 312, 317.
- Adeimantus, ii. 350.
- Adiabene, ii. 272; iii. 142, 146, 154.
- Adiatorix, son of Domnecleus, ii. 288.

- ——, f. of Dyteutus, ii. 308.
- Admetus, i. 72; ii. 155.
- Adobogion, ii. 401.
- Adonis, iii. 170.
- ——, river of Phœnicia (*Nahr-Ibrahim*), iii. 170.
- Ador, ii. 270.
- *Adra. See* Abdera.
- Adramyttene, district of Mysia, ii. 370, 400.
- Adramytteni, ii. 383, 384.
- Adramyttium, t. of Mysia (*Adramytti*), ii. 339, 340, 371, 374, 376, 384, 386, 387.
- ——, gulf of, ii. 342, 374, 386, 400.
- Adrapsa, iii. 126.
- Adrasteia, district of Mysia, ii. 317, 332, 337, 348.
- ——, t. of Mysia, ii. 345-348.
- ——, mtn near Cyzicus, ii. 331.
- ——, (Nemesis), ii. 348.
- Adrastus, ii. 59, 97.
- ——, ii. 174, 346, 348.
- Adria, city and naval station of Picenum (*Atri*), i. 358. *See* Atria.
- Adrias, r. i. 487.
- Adriatic (*Gulf of Venice*),

i. 31, 72, 73, 75, 89, 96, 139, 141, 158, 159, 163, 164, 186,193, 291, 303, 307, 308, 314, 315, 3 19, 324-
326, 336, 338, 357, 373, 425, 432,435, 442, 463, 475, 481, 483, 486, 487, 492, 495, 505, 506;
ii. 119, 289, 290, 301,378.

- Adrion, mtn of Dalmatia. *See* Ardium, i. 484.
- Adrumes (*Sousah*), iii. 288.
- *Adshane. See* Canæ.
- Adula, Adulas, mtn, i. 287, 304, 317.
- Æa, city, i. 31, 32, 72-74.
- ——, ftn, i. 509, 510.
- Æacidæ, i. 496; ii. 83.
- Æacus, ii. 57.
- Ææa, i. 6, 32, 73.
- Æaneium, ii. 126.
- Æanes, ii. 126.
- Æanis, ii. 126.
- Æas, r. of Illyria. *See* Aias.
- ——, son of Telamon. *See* Ajax.
- Æclus, i. 493; ii. 152, 154.
- Ædepsus, t. of Eubœa (*Dipso*) i. 94; ii. 125, 152.
- Ædui, i. 278, 286-288.

- Æeta, i. 72, 73.
- Æetes, i. 72-74.
- Æga, prom. of Æolis, ii. 388.
- ——, city of Achaia, ii. 71-73.
- Ægæ, t. of Mysia, ii. 397.
- ——, t. of Eubœa, ii. 72, 98.
- Ægææ, Ægæ, t. of Cilicia (*Ajas*), iii. 60.
- ——, t. of Laconia, ii. 42.
- Ægæan Sea (*Egio-Pelago*), i. 42, 187-190, 195, 487, 496, 505, 512, 518, 519; ii. 72,152, 193, 207, 341, 388.
- Ægæi, ii. 71.
- Ægaleum, mtn of Messenia, ii. 35.
- Ægeira, t. of Achæa, ii. 71, 73.
- Ægeirus, t. in Lesbos, ii. 391.
- Ægesta, t. of Sicily, i. 379, 411, 415; ii. 378.
- Ægestani, i. 401, 411.
- Ægestes, Trojan, i. 378, 411.
- Ægeus, ii. 81.
- Ægialeia, Ægialus, ii. 3, 23, 67, 68, 72, 288.
- [Pg 301]Ægiali (Sicyon), ii. 66.
- Ægialians, ii. 53, 67, 68.
- Ægialus, Ægialeia, in Paphlagonia, ii. 288, 291.
- Ægieis, Ægienses, ii. 73, 157.
- Ægilieis, in Attica, ii. 89.
- Ægilips, ii. 161, 163.
- Ægimius, ii. 128.
- Ægimurus (*Al Djamur*), i. 185, 422; iii. 287.
- Ægina, t. of Argolis, ii. 57, 58.
- ——, island, i. 84, 187; ii. 47, 49, 54, 57, 58, 82, 136.
- ——, city, ii. 55, 58.
- Æginetæ, ii. 57, 58, 83.
- Æginium, t. of Thessaly, i. 501.
- Ægirussa, ii. 84.
- Ægisthus, i. 25.
- Ægium, city of Achæa (*Vostitza*), ii. 5, 6, 8, 59, 71-73, 77.
- Ægletes Apollo. *See* Anaphe.
- Ægospotami, t. and r., i. 438, 518.
- Ægua, t. of Spain, i. 213.
- Ægys, city and district of Laconia, ii. 42, 153.
- Ælana, Aila (*Ailah*), t. of Arabia Petræa, iii. 176, 191.
- Ælanitic Bay, iii. 176, 191, 204, 254.

185

- Ælius Gallus, i. 178; iii. 209-212, 246, 262, 267.
- ——, (Catus), i. 466.
- Æmilian road, i. 323.
- Æmilianus, Q. Fabius Maximus, i. 277, 285.
- ——, Scipio, iii. 51, 286.
- Æmilius, Paulus, i. 495.
- Ænarium. *See* Arnarium, ii. 73.
- Ænea, t. of the Troad, ii. 372.
- ——, t. of Macedonia, i. 509, 510. *See* Enea *and* Nea.
- Æneas, i. 76, 225, 339, 346, 347; ii. 317, 344, 353, 356, 357, 360, 377, 378, 383.
- Ænesippeia, isl., iii. 235.
- Ænesisphyra, prom. of Egypt, iii. 235.
- Æniana (Ænia), ii. 242.
- Æniānes, people of Thessaly, i. 96; ii. 128, 147, 158, 241, 273.
- Æniates, name of the Paphlagonians, ii. 302.
- Ænius, r. of the Troad, ii. 371.
- Ænobarbus (Cnæus), i. 277.
- ——, (Domitius), i. 285; iii. 24.
- Ænus, mtn of Cephallenia (*Monte Nero*), ii. 167.
- ——, city of Thrace, i. 490, 516, 519.
- Æolia, i. 17, 64, 187, 195, 224; ii. 153; iii. 140.
- Æolian nation, ii. 3.
- ——, colony, ii. 94.
- Æolians, i. 96, 328; ii. 2, 3, 154, 366, 374, 382-386; iii. 34.
- Æolic dialect, ii. 2, 3.
- ——, expedition, ii. 93.
- ——, migration, iii. 2.
- Æolis, ii. 339-341, 346, 366, 384, 398.
- Æolus, king, i. 31, 35, 36, 39, 194, 403, 417, 419; ii. 142.
- ——, Islands of (*Lipari Islands*), i. 84, 89, 185, 194, 383, 403, 420.
- ——, Play of Euripides, ii. 32.
- Æpasian plain, region of Triphylia, ii. 22.
- Æpeia, town of Messenia, ii. 35, 37.
- Æpeia Methone, ii. 37.
- Æpy, town of Triphylia, ii. 23, 24.
- Æpytus, son of Neleus, iii. 2.
- Æqui, i. 339, 343, 344, 353.
- Æquum-Faliscum, i. 335.
- Aëria, t. of the Cavari (*Le mont Ventoux*), i. 277.
- Æsar, r. of Etruria, i. 330.
- Æsarus, r. near Crotona (*Esaro*), i. 393.

- Æschines, Athenian, ii. 188.
- ——, Milesian, iii. 5.
- Æschylus, i. 52, 68, 329, 386, 458, 462; ii. 13, 73, 82, 154, 187, 337, 390; iii. 130.
- Æsculapius. *See* Asclepius.
- Æsēpus, r. of Mysia, *Satal-dere*, ii. 300, 316, 317, 330, 332, 337-341, 344-348, 353,357, 369, 371, 372.
- [Pg 302]Æsernia, city of the Samnites (*Isernia*), i. 353, 371.
- Æsis, r. of Umbria (*Fiumesino*), i. 324, 337, 357, 435.
- Æsyetes, tomb of, ii. 361, 364.
- Æthalia, island (*Elba*), i. 185, 332-334.
- Æthaloeïs, r. of Scepsia, ii. 190.
- Æthices, people of Epirus and Thessaly, i. 499, 501; ii. 131, 137, 144.
- Æthicia, i. 501.
- Æthiopia. *See* Ethiopia.
- Ætna, Mount, i. 31, 35, 84, 368, 369, 386, 403, 404, 406, 411, 413-415, 418.
- Ætnæans, i. 405.
- Ætolia, i. 493, 499, 501, 505;
ii. 6, 10, 33, 72, 75, 114, 129, 136, 150, 155, 156, 159,160, 171, 172, 174.
- Ætna, town of Sicily, i. 405, 414.
- Ætōli, Ætolians, i. 381; ii. 2, 6, 12, 30, 33, 121, 122, 127-131, 146, 158-161, 169,172, 175, 176, *passim.*
- Ætolian mountains, ii. 115, 131.
- ——, promontories, i. 93.
- ——, polity, i. 494.
- Ætolicus, Demetrius, ii. 160.
- Ætōlus, ii. 33, 122, 176, 177.
- Æxoneis, vill. of Attica, ii. 89.
- Æxonici, ii. 89.
- *Afium-karahissar. See* Synnada.
- Afranius, legate of Pompey, i. 242.
- *Afreen. See* Oenoparas.
- *Africa. See* Libya, iii. 274-278.
- African coast, i. 76.
- ——, sea, ii. 193, 194, 199, 212.
- Agamedes, ii. 119.
- Agamemnon, i. 17, 63, 499;
ii. 22, 35, 36, 53, 56, 57, 59, 83, 115, 174, 186, 340, 356,368, 374; iii. 10, 15, *passim.*
- Agapenor, iii. 70.
- Agatha, a city of Gaul (*Agde*), i. 269, 272.
- Agatharchides, iii. 34, 208.
- Agathocleia, iii. 231.
- Agathocles, tyrant of Sicily, i. 383, 427; iii. 288.

- Agathocles, father of Lysimachus, ii. 399.
- ——, son of Lysimachus, ii. 400.
- Agathyrnum, i. 401.
- *Agde. See* Agatha.
- Agdistis, the goddess Rhea, ii. 184.
- ——, temple of, ii. 320.
- *Agenois. See* Nitiobriges.
- Agesilaus, i. 427.
- Agidæ, ii. 44.
- Agis, ii. 43, 44.
- *Aglio, L'Osteria dell'. See* Algidum.
- Agnu-Ceras, promontory of Egypt, iii. 239.
- Agoracritus, ii. 87.
- Agra, village of Attica, ii. 91.
- Agradates, iii. 132.
- Agræa, district of Ætolia, ii. 10.
- Agræi, a people of Ætolia, ii. 158, 160, 179.
- ——, people of Arabia, iii. 189.
- Agræus, ii. 77.
- Agri, ii. 223.
- *Agri. See* Aciris.
- Agriades, ii. 8.
- Agriānes, a people of Thrace, i. 488, 514, 515.
- Agrigentini, i. 401.
- Agrigentum. *See* Acragas.
- Agrii, a people of Ethiopia, iii. 196.
- Agrippa, i. 289, 310, 350, 364; ii. 36, 350; iii. 170.
- Agrius, ii. 175, 179.
- Agylla, t. of Etruria, i. 328, 335.
- Agyllæi, i. 328.
- *Aiaghi-dagh. See* Zagrum.
- Aias, *or* Aous, i. 410, 411, 486.
- Aigan, ii. 388.
- Aila. *See* Ælana.
- *Ain-el-Hiyeh. See* Enydra.
- *Aix. See* Sextiæ.
- *Ajas. See* Ægææ.
- Ajax, son of Teucer, iii. 55, 56.
- ——, temple of, ii. 357, 359.
- ——, son of Telamon, ii. 83, 84, 102.
- ——, the Locrian, ii. 126, 367.

- [Pg 303]*Ajazzo, Aias, Bay of. See* Issus.
- *Ak-Su. See* Cestrus.
- *Akaba. See* Aila.
- *Akabel-el-Kebira. See* Catabathmus.
- *Ak-Liman. See* Armene, ii. 291.
- *Ak-Schehr. See* Philomelium.
- *Ala Schehr. See* Philadelphia.
- Alabanda, t. of Caria (*Arab-Nissar*), ii. 347; iii. 34, 37, 38, 40.
- Alæan Minerva, ii. 75.
- Alæis Æxōnici, vill. of Attica, ii. 89.
- Alæsa, t. of Sicily (*I Bagni*), i. 401, 411.
- Alalcŏmĕnæ, t. of Bœotia, i. 501; ii. 107, 110.
- ——, t. of Asteria, ii. 168.
- Alalcomenium, temple of Minerva, ii. 106, 110.
- *Alara. See* Ptolemaïs.
- *Alatri. See* Aletrium.
- Alazia, t. of Mysia, ii. 299.
- Alazōnes, ii. 298, 299.
- Alazonia, Alazonium, t. of the Troad, ii. 300, 371.
- Alazonius, r. of Albania, ii. 230, 231, 234.
- Alba (*Albi*), i. 340, 344, 349, 353, 356.
- Alban wine, i. 347.
- Albania (*Shirvan*), ii. 217, 226, 230-235, 238, 243, 267, 268.
- Albanians, i. 178, 195, 341, 344, 440; ii. 217, 232-235, 248, 260, 267-272, 307.
- Albanus, Mount (*Monte Albano*), i. 340, 351; (*Monte Cavo*), 355, 356.
- *Albi. See* Alba.
- Albia, Alpionia, i. 300.
- Albienses, i. 302.
- Albii, i. 482.
- Albingaunum (*Albinga*), i. 300, 301.
- Albiœci, i. 302.
- Albis, r. (*Elbe*), i. 22, 444-447, 451.
- Albium Intemelium (*Vintimille*), i. 300, 301.
- Albius, Mount, i. 300, 483.
- Albūla, cold waters, i. 354.
- Alcæus, poet of Mitylene, i. 58; ii. 108, 109, 366, 375, 391, 393; iii. 40.
- Alcestis, i. 72.
- Alchædamnus, iii. 166.
- Alcimedon, ii. 42.
- Alcimus, ii. 42.
- Alcmæōn, i. 499; ii. 122, 174.

189

191

- Amarynthia Diana, ii. 155.
- Amarynthium, the, ii. 156.
- Amarynthus, t. of Eubœa, ii. 155.
- Amaseia, city of Pontus (*Amasija*), ii. 295, 306, 311, 312; iii. 252.
- Amasenses, ii. 312.
- Amasias, r. of Germany (*Ems*), i. 444, 445.
- Amasis, ii. 311.
- Amastris, city of Paphlagonia, (*Amassera*), i. 475; ii. 285, 289, 290, 291, 302.
- Amastris, wife of Dionysius, tyrant of Heracleia, ii. 291.
- Amathus, r. of Elis, ii. 7, 11, 26, 38.
- ——, t. of Cyprus (*Limasol*), iii. 69.
- ——, t. of Laconia, ii. 41.
- Amathusii, ii. 13.
- Amazones, plain of the, i. 82, 190.
- Amazonides, ii. 298.
- Amazons, ii. 300, 301, 328, *passim.*
- Ambarvia, i. 341.
- Ambiani, i. 289, 309.
- [Pg 305]Amblada, t. of Pisidia, ii. 324.
- ——, wine, ii. 324.
- Ambracia, t. of Thesprotia (*Arta*), i. 498, 499; ii. 159, 161.
- Ambracian Gulf, Ambracian Sea, (*The Gulf of Arta*), i. 186, 495, 496-498, 501, 505; ii. 4, 129, 158, 161, 171.
- Ambrōnes, i. 274.
- Ambryseis, ii. 123.
- Ambrysus, t. of Phocis (*Distomo*), ii. 122.
- *Amelia. See* Ameria.
- Amĕnanus, r. of Sicily (*Judicello*), i. 356.
- Ameria, t. of Umbria (*Amelia*), i. 338.
- ——, t. of Pontus, ii. 306.
- Amisēne, ii. 290, 294, 296.
- Amiseni, ii. 290, 294, 296, 310, 311.
- Amisus (*Samsun*), i. 106, 107, 109, 113, 114, 190; ii. 227, 256, 289, 294, 296, 302,310; iii. 44, 56, 62, 63.
- Amiternum, city of the Sabines, i. 338, 359.
- Ammon Balithon, prom., iii. 288.
- ——, seat of oracle, i. 504; iii. 226, 253, 258, 283.
- ——, temple of, i. 78, 79, 87, 88; iii. 253, 258, 289, 294, 295.
- Ammonia, iii. 235.
- Amnias, r. of Paphlagonia (*Gok-Irmak*), ii. 313.
- Amnisus, port of Cnossus, ii. 196.

193

- Anabūra, t. of Pisidia, ii. 324.
- Anacharsis, i. 463, 465; iii. 86.
- Anacreon, i. 226; iii. 2, 9, 17, 40.
- Anactorium, t. of Acarnania, ii. 159, 161.
- Anacyndaraxes, iii. 55.
- Anadatus, ii. 246.
- Anæa, Anaïtis, iii. 137, 144. *See* Anaïtis.
- [Pg 306]Anagnia, t. of the Hernici (*Anagni*), i. 353.
- Anagurasii, vill. of Attica, ii. 89.
- Anaïtis, ii. 246.
- ——, temple of, ii. 274, 275, 309.
- Anaphe, isl. (*Nanfio*), i. 73; ii. 206, 207.
- Anaphlystii, ii. 89.
- Anaphlystus, vill. of Attica, ii. 89.
- Anapias, i. 406.
- Anariacæ, ii. 240-242, 248.
- Anariace, ii. 241.
- Anas (*Guadiana*), r. of Spain, i. 208-212, 214, 222, 228, 230, 243.
- Anaurus, r. of Magnesia, ii. 139.
- Anaxagoras, iii. 20.
- Anaxarchus, ii. 356.
- Anaxēnor, iii. 23.
- Anaxicrates, iii. 191.
- Anaxilas, the tyrant of Rhegium, i. 384, 385.
- Anaximander, the Milesian, i. 1, 12; iii. 5.
- Anaximenes of Lampsacus, disciple of Anaximander, ii. 350; iii. 5, 20.
- Ancæus, iii. 2.
- Anchiale, a town of Pontus, i. 490.
- ——, t. of Cilicia, iii. 55, 56.
- Anchialus, ii. 166.
- Anchises, i. 339; ii. 344, 353, 378.
- Anchoë, ii. 100.
- Ancon, Ancona, i. 315, 337, 357, 435, 483.
- Ancus Martius, i. 326, 345, 348.
- Ancyra, t. of Galatia (*Angora*), i. 279; ii. 320.
- ——, t. of Phrygia (*Simau-Gol*), ii. 320, 332.
- ——, t. of Gaul, i. 279.
- Andania, t. of Arcadia, ii. 11, 24, 37, 156.
- Andeira, city of Mysia, ii. 381, 386, 387.
- Andeirene, ii. 387.
- Andetrium, t. of Dalmatia, i. 484.

- Andirus, r. of the Troad, ii. 370.
- Andizetii, i. 483.
- Andræmōn, iii. 2.
- Andriace, t. of Thrace, i. 490.
- Andriclus, mtn of Cilicia (*Kara Gedik*), iii. 52.
- Andrii, i. 511.
- Androclus, iii. 2, 11.
- Andromache, ii. 343, 356, 363, 394.
- Andromeda, i. 68; iii. 175.
- Andron, ii. 81, 167, 195.
- Andronicus, iii. 33.
- Andropompus, iii. 2.
- Andros, isl. (*Andro*), ii. 156, 208, 210.
- Androsthenes, iii. 186.
- Anemurium, prom. of Cilicia (*Inamur*), iii. 52, 54, 68.
- Anemoreia, Anemoleia, t. of Phocis, ii. 123.
- *Angelo-Castron. See* Arsinoe.
- *Angora. See* Ancyra.
- Anias, r. of Arcadia, ii. 76.
- Anigriades, ii. 19, 20.
- Anigrus, r. of Triphylia, ii. 20, 21.
- Anio, r. of Latium (*Teverone*), i. 349.
- Anniceric sect, iii. 293.
- Anniceris, iii. 293.
- Annius, ii. 17.
- Ansander, i. 479.
- Antæus, iii. 281.
- *Antakieh. See* Epidaphne.
- Antalcidas, i. 438.
- Antandria, ii. 375, 384.
- Antandrians, ii. 386.
- Antandros (*San Dimitri*), ii. 186.
- ——, t. of the Troad (*Antandro*), ii. 375, 376, 384.
- Antemnæ, t. of Latium, i. 341.
- Antenor, i. 76, 225, 236, 316; ii. 289, 377.
- Antenoridæ, ii. 360, 377.
- Anthedon, c. of Bœotia, i. 25; ii. 92. 98. 102. 103. 106.
- Antheia, ii. 35, 37.
- Anthemis, ii. 168.
- Anthemus, iii. 8.
- Anthemusia, iii. 158.

- Anthes, ii. 56; iii. 35.
- *Antibes. See* Antipolis.
- Anticasius, mtn of Syria, iii. 164.
- Anticeites, r. of the Mæotæ, ii. 221, 222.
- Anti-Cinolis, t. of Paphlagonia, ii. 291.
- [Pg 307]Anticlides, i. 329.
- Anticragus, iii. 46.
- Anticyra, t. of Phocis (*Aspra-Spitia*), ii. 114, 116, 122, 129.
- ——, t. of Locris, ii. 137.
- ——, t. on the Maliac bay, ii. 116. 137.
- Antigonia, t. of Bithynia, ii. 318.
- ——, t. of the Troad, ii. 355, 361.
- ——, t. of Syria, iii. 162.
- Antigonus, son of Philip, ii. 318, 355, 361; iii. 20, 55, 162.
- ——, king of Macedonia, ii. 70.
- ——, of Apelles, iii. 36.
- Antilibanus, mtn of Syria, iii. 149, 169, 171.
- Antilochus, ii. 359.
- Antimachus, ii. 18, 42, 74, 104, 348.
- Antimenidas, ii. 391.
- Antimnestus, i. 385.
- Antioch, i. 416; ii. 307, 409; iii. 21. 24. 43. 118. 159. 161-164.
- ——, city of Mygdonia, iii. 157.
- Antiocheia, city of Caria, ii. 409; iii. 21, 24, 43.
- ——, city of Phrygia (*Ialobatsch*), ii. 307, 322, 333.
- ——, city of Margiana, ii. 252.
- Antiochis, dr of Achæus, and mother of Attalus, ii. 400.
- Antiochus, the Syracusan, i. 360, 379, 385, 394, 398, 399, 424.
- ——, the Great, i. 439; ii. 269, 273, 285, 355, 400; iii. 46, 153, 176.
- ——, Sōter, ii. 252, 333, 400.
- ——, son of Demetrius, iii. 51.
- ——, son of Epiphanes, iii. 162.
- ——, Ierax, iii. 198.
- ——, philosopher of Ascalon, iii. 175.
- Antiope, ii. 97.
- *Antiparos. See* Oliarus. Antipater, of Macedon, i. 513; ii. 56. 136. 318.
- Antipater, son of Sisis, ii. 304.
- Antipater, Derbētes, the robber, ii. 278. 322; iii. 64.
- ——, of Tarsus, the Stoic, iii. 58.
- ——, of Tyre, iii. 173.
- Antiphanes, the Bergæan, i. 74, 152, 154.

196

- Antiphellus, t. of Lycia, iii. 47.
- Antiphilus, harbour of, iii. 196.
- Antiphræ, iii. 235, 236.
- Antiphus, ii. 149, 403; iii. 31.
- Antipolis, t. of Gaul (*Antibes*), i. 267, 275, 276, 301.
- Antirrhium, prom. of Ætolia (*Castle of Roumelia*), ii. 6, 73, 79, 128, 171.
- ——, cape, ii. 6.
- Antirrhodus, isl. iii. 230.
- Antissa, t. of Lesbos, i. 93; ii. 393.
- Antitaurus (*Dudschik Dagh*), ii. 259. 260.
- Antium, t. of the Volsci (*Capo d' Anzo*), i. 344-346, 355.
- Antonius, Marcus, ii. 166.
- ——, Caius, ii. 166.
- Antony, i. 213, 499; ii. 36, 263, 271, 274, 285, 294, 330, 334, 357, 358; iii. 23, 52,56, 58, 72, 159, 184, 230, 231, 233, 281.
- Antrōn, t. of Thessaly, ii. 24, 135, 138, 139.
- ——, Ass of, ii. 139.
- Anubis, iii. 245, 257.
- *Anzo, Capo d'. See* Antium.
- Aones, i. 493; ii. 88, 93.
- Aonian plain, ii. 108.
- Aornum, castle of India, iii. 6.
- Aornus, bay of Campania, i. 39.
- Aorsi, ii. 219, 239.
- Aōus, r. of Illyria, i. 486. *See* Aias.
- *Aouste. See* Augusta.
- Apæsus, t. of the Troad. *See* Pæsus, ii. 346, 349.
- Apama, wife of Seleucus Nicator, ii. 334; iii. 161.
- ——, wife of Prusias, ii. 315.
- Apameia, city of Syria (*Kulat-el-Mudik*), ii. 250; iii. 33, 161-166, 171.
- [Pg 308]Apameia, city of Phrygia (*Aphiom Kara Hissar*), ii. 322, 323, 332-336, 407, 410; iii.43.
- ——, city of Media, ii. 250, 264.
- ——, city of Bithynia, ii. 315.
- Apameis, ii. 316.
- Aparni, *see* Parni, ii. 245, 246.
- Apasiacæ, ii. 248.
- Apaturum, ii. 223.
- Apellas. *See* Ophelas.
- Apelles, philosopher, i. 23.
- ——, painter, iii. 14.

197

200

- Apollonopolis, city of Egypt, iii. 261. 263.
- Appaïtæ, ii. 296.
- Appia Via, i. 346, 347, 351, 352, 355, 370, 431, 432.
- Apsus, r. of Illyria, i. 486.
- Apsynthis, district of Thrace, i. 519.
- Apsyrtides (islands), i. 484.
- Apsyrtus, i. 484.
- Aptera, t. of Crete, ii. 200.
- Apuli, i. 360, 432, 436.
- Apulia. *See* Daunia, i. 423, 432.
- Aquæ-Statiellæ, t. of Cisalpine Gaul (*Acqui*), i. 323.
- Aquileia, i. 186, 307, 309, 310, 319, 321, 324, 435, 448, 482.
- Aquinum, t. of Latium (*Aquino*), i. 352.
- Aquitani, i. 264, 265, 282, 283, 284.
- Aquitania, Aquitaine, i. 242, 247, 282-285, 296, 310.
- Arabia, i. 63, 197, 458; iii. 86, 88, 89, 132, 148, 149, 171, 176, 185, 186, 189-217, 241, 243, 247, 252, 261, 266.
- ——, Felix, i. 41, 63, 129, 130. 178. 196; iii. 128, 159, 171, 176, 185, 186, 189, 209, 213; iii. 76.
- [Pg 310]Arabia Nabatæa, iii. 241.
- Arabian Gulf, the (*Red Sea*), i. 47, 50, 55, 56, 60, 62, 67, 75, 79, 87, 123, 129, 130, 149, 152, 178, 183, 196, 200, 458; ii. 210; iii. 74, 88, 176, 185, 189, 191, 194-208, 210, 217, 224, 226, 235, 241, 243, 244, 260, 270, 271, 291.
- Arabians, i. 63, 66, 67, 196, 458; ii. 154, et passim.
- ——, Skenitæ, ii. 146, 158, 159.
- ——, Troglodyte, i. 2, 67.
- Arabs, tribes of, i. 440.
- Arabus, daughter of, i. 67.
- Araby the Blest, iii. 76.
- Arachōsia, distr. of Ariana, ii. 252.
- Arachōti, ii. 249.
- ——, people of Ariana, ii. 248; iii. 122, 124, 126.
- Aracynthus, a mtn of Ætolia (*M. Zigos*), ii. 160, 172.
- Aradii, iii. 167, 168, 170, 215.
- Aradus, isl. in the Persian Gulf, (*Arek*), iii. 187.
- ——, city of Phœnicia, iii. 167, 169, 172.
- Aræthyrea, distr. of Argolis, ii. 59, 66.
- Aragus, r. (*Arak*), ii. 230, 231.
- *Arak. See* Aragus.
- Arambi, iii. 216. *See* Arabians.
- Aramæi, Arammæans, i. 66; ii. 404; iii. 216.

- Arar, r. of Gaul (*Saone*), i. 277, 278, 281, 286, 287, 288.
- Ararene, distr. of Arabia, iii. 212.
- Arathus, r. of Epirus, i. 498, 501.
- Aratus, poet, i. 4, 156; ii. 42, 73, 199, 207, 209; iii. 55.
- ——, leader of the Achæi, ii. 66, 70.
- Arausio, t. of the Cavari (*Orange*), i. 277.
- Araxēnæ, distr. of Armenia, i. 113; ii. 242.
- Araxenian plain, ii. 268, 270.
- Araxēnus, ii. 268.
- Araxes, r. of Armenia (*Eraskh*, or *Aras*), i. 96; ii. 217, 232, 263, 268, 270, 272.
- Araxes, r. of Scythia, ii. 247.
- ——, r. of Persis (*Bendamir*), iii. 132.
- Araxus, prom. of Elis (*Cape Papa*), ii. 5, 6, 7, 8, 15, 74, 169.
- Arbaces, iii. 143.
- Arbēla, t. of Assyria (*Erbil*), i. 123; ii. 274; iii. 143, 144, 259.
- Arbēlus, iii. 144.
- Arbies, iii. 120.
- Arbis, r. of Gedrosia (*Purali*), iii. 120.
- Arcadia, i. 94, 343, 416; ii. 5, 6, 7, 8, 10, 11, 14, 22, 28, 32, 33, 37, 52, 74-77, 142,153, 156, 327, 339; iii. 145.
- Arcadian mountains, ii. 40.
- ——, cities, ii. 71.
- Arcadians, Arcades, i. 328, 329, 432; ii. 3, 8, 22, 24, 31, 39, 76.
- Arcadicus, ii. 16.
- Arcesilaus, i. 23; ii. 387.
- Arceuthus, r. of Syria, iii. 164.
- Archæanax, ii. 365.
- Archan, ii. 388.
- Archedēmus, the Stoic, iii. 58.
- Archelaus, king, ii. 277, 278, 282, 284, 285, 305-308; iii. 54, 232.
- ——, priest, ii. 308; iii. 232.
- ——, father of priest, ii. 308; iii. 232.
- ——, son of Penthilus, ii. 340.
- ——, physician, iii. 20.
- ——, play of Euripides, i. 329.
- Archemachus, ii. 178.
- Archias, Corinthian, founder of Syracuse, i. 394, 406, 407; ii. 63.
- ——, general of Antipater, ii. 55.
- Archidamus, i. 427.
- Archilochus, ii. 50, 169, 210, 298; iii. 23.
- Archimedes, i. 85, 87.

- Archytas, i. 427.
- Arconnesus, iii. 16, 35.
- Arctic Circle, i. 4, 5, 144, 200.
- ———, constellation, i. 5.
- Arcton, mtn, ii. 331.
- Arcturus, i. 201; iii. 82, 83.
- Ardania, prom. of Marmara, i. 64.
- [Pg 311]Ardanis, Ardanixis (*Ras-el-Milhr*), iii. 294.
- Ardea, city of the Rutuli, i. 339, 346, 371.
- *Ardgeh. See* Argæus.
- Ardia, distr. of Dalmatia, i. 481.
- Ardiæi, i. 483-485, 487, 488, 489, 505.
- Ardium, mtn of Dalmatia, i. 484.
- Arduenna (*forest of Ardennes*), i. 290.
- Arĕcomisci. *See* Volcæ.
- Arĕgon, ii. 16.
- Areion, ii. 97.
- Areius, iii. 53.
- *Arek. See* Aradus.
- Arelate, t. of Gaul (*Arles*), i. 272.
- Arĕne, t. of Triphylia, ii. 20-23, 27, 38.
- Areopagite code, i. 390.
- Arĕs. *See* Mars.
- Aretas, iii. 212.
- Arĕte, iii. 212, 293.
- Arethusa, castle of Syria, iii. 166, 167.
- ———, ftn of Chalcis, ii. 157.
- ———, ftn of the Island of Ortygia, i. 408, 409.
- ———, city of Macedonia, i. 514.
- *Arezzo. See* Arretium.
- Argæsus, mtn of Cappadocia (*Ardgeh or Edschise Dagh*), i. 113; ii. 282.
- Arganthōnius, king of Tartessus, i. 226.
- ———, mtn of Bithynia, ii. 315.
- Argeadæ, i. 506, 508.
- Argeia. *See* Argia.
- Argennum, prom. of Ionia, iii. 18.
- *Argentière. See* Cimolus.
- Argestes (N.W. wind), i. 45; ii. 80.
- Argia, Argolis, Argeia, i. 187, 416; ii. 6, 51, 66.
- Argian territory, ii. 52, 58.
- Argillæ (under-ground passages), i. 363.

- Argilus, t. of Macedonia, i. 512.
- Arginussæ, islands, ii. 388, 391.
- Argissa, Argūra, t. of Pelasgiotis, ii. 143. 144.
- Argive territory, ii. 51-55, 76, 158.
- Argives, i. 55, 102; ii. 8, 39, 47, 48, 49, 51-55, 58-60, 66, 97, 174, 175.
- Argo, the ship, i. 72, 73, 332; ii. 139. 315.
- Argolic Bay (*Gulf of Napoli*), ii. 6, 47.
- Argolica, ii. 58, 75.
- Argolis (*see* Argia), i. 410, 416; ii. 58.
- Argonautic expedition, i. 31.
- Argonautics, iii. 34.
- Argonauts, i. 71-73, 332; ii. 21, 111, 139, 148, 331.
- Argos, i. 35, 329, 410, 486; ii. 12, 42, 43, 48-56, 58-60, 71, 76, 77, 97, 110, 124, 133,203; iii. 41. 60.
- ——, Pelasgic, i. 328; ii. 50, 52, 132, 133.
- ——, Inachian, ii. 74.
- ——, Achæan, ii. 43, 49, 50.
- ——, castle of Cappadocia, ii. 281.
- ——, Amphilochian, city of Acarnania (*Neochori*), i. 410, 499; ii. 159. 174. 175.
- ——, Hippium, i. 320, 433; ii. 49.
- ——, Jasum, ii. 49, 50, 52.
- ——, Orestic, city of Epirus or Macedonia, i. 500.
- ——, distr. of Peloponnesus, ii. 50-55.
- Argoüs, harbour of Æthalia, i. 332.
- Argūra. *See* Argissa.
- Argyria, city of the Troad, ii. 300, 371.
- Argyrippa. *See* Argos-Hippium.
- Argyrippeni, port of, i. 433.
- *Argyrokastro. See* Phyle.
- Argyrusci, i. 344.
- Aria, distr. of Persia, i. 112-114; ii. 246, 251, 252; iii. 124, 125.
- Ariamazas, rock of, ii. 254.
- Ariana, i. 121, 125, 127-129; ii. 218. 252. 253. 263; iii. 78, 88, 119-129.
- Ariani, i. 66, 104, 196; iii. 125.
- Ariarathes, ii. 277, 283, 401.
- Aricia, t. of Latium (*La Riccia*), i. 344, 355.
- [Pg 312]Aridæus, iii. 229.
- Arii (*Herat*), ii. 245, 248, 249, 254; iii. 124, 125.
- Arima, mtns of Cilicia, ii. 405.
- Arimæi. *See* Aramæi.
- Arimaspi, i. 32; ii. 240.

- Arimaspian poems, ii. 349.
- Arimi, ii. 304, 403-405; iii. 163, 216.
- Ariminum, t. of Umbria (*Rimini*), i. 314, 315, 322, 324, 326, 336, 337, 357.
- Ariminus, r. of Umbria, i. 323.
- Arimus, ii. 406.
- Ariobarzanes, ii. 285.
- Arion, ii. 393.
- Arisba, t. of the Troad, ii. 344, 350, 351; iii. 5.
- Arisbus, in Thrace, ii. 351.
- Aristarcha, i. 268, 269.
- Aristarchus, i. 48, 49, 51, 57, 156; ii. 380; iii. 26.
- Aristeas, of Proconnesus, i. 32; ii. 349; iii. 10.
- Aristeides, painter, ii. 64.
- Aristera, or left coast of Pontus, ii. 286.
- Aristio, ii. 89.
- Aristippus, Socratic philosopher, iii. 293.
- ——, Metrodidactos, iii. 293.
- Aristo, Athenian, i. 23.
- ——, of Ceos, ii. 210; iii. 225, 226.
- ——, of Cos, iii. 36.
- ——, of Rhegium, i. 390, 391.
- Aristobulus, ii. 243, 254; iii. 55, 83-86, 95, 102, 111, 112, 133, 148, 150, 187, 274.
- ——, king of Judæa, iii. 180.
- Aristocles, iii. 34.
- Aristocrates, ii. 39.
- Aristodemus, iii. 26, 27.
- Aristonicus, grammarian, i. 60.
- ——, iii. 20, 21.
- Aristopatra, iii. 96.
- Aristotle, i. 44, 143, 144, 156, 229, 273, 459, 494, 512, 513; ii. 18, 55, 56, 64, 151,154, 156, 363, 378-382, 392, 393; iii. 86, 98, 173, 225.
- Aristoxenus, i. 25.
- Aristus, of Salamis, iii. 69, 134.
- Arius, r. of Aria, ii. 252, 254.
- Ariusia, in Chios, iii. 19.
- *Arles. See* Arelate.
- Armĕne, t. of Paphlagonia (*Ak-Liman*), ii. 291.
- Armĕnia, i. 72, 76, 78, 95, 113, 115, 120, 122, 123, 126, 127, 177, 195; ii. 217, 218,226, 227, 230-235, 238, 245, 259, 260-272, 276, 284-286, 304-306, 309, 310; iii.108, 109, 147, 150, 154, 156, 157.
- ——, Greater, ii. 260, 262, 305; iii. 150.
- ——, the Lesser, ii. 260, 267, 269, 286, 294, 296, 304, 305, 310; iii. 63, 150.

- ——, gates of, i. 123, 124.
- ——, mountains, i. 96, 115, 120, 122, 126, 127; ii. 226, 241.
- Armenians, i. 66, 196, 440; ii. 216, 230, 239, 260, 263-277, 294, 296, 304, 309; iii.216.
- Armenium, t. of Thessaly, ii. 235, 272.
- Armenius, i. 446.
- Armenus, ii. 235, 272.
- *Armyrus. See* Itonus.
- Arnæi, inhabitants of Thessaly, ii. 93.
- Arnæus, ii. 300.
- Arnarium, ii. 70.
- *Arnauti. See* Acamas.
- Arnè, city of Thessaly, i. 92; ii. 107. 110. 138. 143.
- Arnus, r. of Etruria (*Arno*), i. 330.
- Arŏma, t. of Lydia, iii. 26.
- Arotrebæ. *See* Artabri.
- Arotria. *See* Eretria.
- *Arpas-Kalessi. See* Coscinia.
- Arpi, t. of the Daunii. *See* Argos Hippium and Argyrippa, i. 433.
- Arpina, ii. 32.
- Arrechi, ii. 223.
- Arrētium, city of Etruria (*Arezzo*), i. 330, 335, 336.
- Arrhabæus, i. 500.
- Arsacæ, iii. 97, 160.
- Arsaces, a Scythian, ii. 248, 251.
- [Pg 313]Arsaces, son of Pharnaces, ii. 311.
- Arsacia, city of Media, same as Rhaga, ii. 264.
- Arsēne, lake of Armenia (*Thospitis or Van*), ii. 270.
- Arses, iii. 141.
- Arsinoë, t. of Cilicia (*Softa-Kalessi*), iii. 52.
- ——, two towns of Cyprus, iii. 69. 70. 72.
- ——, t. of Ethiopia, iii. 193, 199.
- ——, t. of Cyrene, iii. 291.
- ——, t. of Egypt, iii. 243, 244.
- ——, formerly *Crocodilopolis*, iii. 256. 257.
- ——, t. of Œtolia (*Angelo-Castron*), ii. 172.
- ——, same as Ephesus, iii. 12.
- ——, same as Patara, iii. 47.
- ——, prom. of Cyprus, iii. 70.
- Arsinoite nome, iii. 223, 253, 256.
- Arsinoites, iii. 256.

- Arsīnus. *See* Erasīnus.
- Arsites, satrap of Phrygia, iii. 188.
- *Arsus. See* Rhosus.
- *Arta, Gulf of. See* Ambracic Gulf.
- Artabazus, ii. 334.
- Artabri, Arotrebæ, people of Lusitania, i. 181, 206, 221, 230, 262.
- ——, port of the, i. 230.
- Artacaëna, city of Aria, ii. 252.
- Artace, mtn, ii. 332.
- ——, island, ii. 332; iii. 5.
- ——, t. there situated, ii. 340; iii. 5.
- Artacēne, iii. 144.
- Artagēræ, castle of Armenia, ii. 270.
- *Artaki. See* Cyzicus.
- Artamita, Apollodorus of, ii. 252.
- Artanes, ii. 273.
- Artavasdes, ii. 263, 270, 271, 274.
- Artaxata, city of Armenia, ii. 269, 270.
- Artaxerxes, i. 78; iii. 34.
- Artaxias, ii. 269, 270, 273.
- Artaxiasata, *see* Artaxata, ii. 270.
- Artemidorus, of Ephesus,

i. 207. 208. 223. 235. 236. 239. 246. 251. 255. 258. 274.277. 295. 332. 364. 393. 402. 435. 518; ii. 5, 48, 77; iii. 12, 15, 34, 43-45, 53, 59,62, 118, 192, 201-203, 208, 239, 243, 276, 281, 282.

- Artemidorus, of Cnidus, iii. 34.
- ——, of Tarsus, iii. 59.
- Artemis. *See* Diana.
- Artemisia, iii. 35.
- Artemisium, Dianium, t. of Iberia, i. 239.
- ——, prom. of Caria, with temple of Diana, iii. 28.
- ——, or Grove of Diana, i. 355. 356.
- ——, at Ephesus, ii. 73.
- Artemita, t. of Assyria (*Shirban*), ii. 257, 264; iii. 152.
- ——, one of the Echinades, i. 93.
- Artimachus, ii. 348.
- Artis, iii. 2.
- Aruaci, i. 243.
- Arupenum, t. of the Iapodes (*Auersperg, or the Flecken Mungava*), i. 309. 483.
- Arverni (*inhabitants of Vélai*), i. 281. 284. 285. 289. 291. 293.
- Arxata, city of Armenia, ii. 270.
- *Arzila. See* Zelis.

- Asander, i. 479; ii. 224, 401.
- Asbystæ, i. 198.
- Asca, t. of Arabia, iii. 212.
- Ascæus, ii. 307.
- Ascalon, city of Judæa (*Asculan*), iii. 175, 176.
- Ascalonitæ, iii. 175.
- Ascania, region of Phrygia, ii. 300, 316-318; iii. 66, 67.
- ——, region of Mysia or Bithynia, ii. 316-318; iii. 66, 67.
- ——, vill. of Mysia, iii. 67.
- Ascanius, lake of Bithynia (*Isnik-Gol*), ii. 314, 318; iii. 66, 67.
- ——, son of Æneas, i. 339, 340; ii. 377, 378.
- ——, leader of the Phrygians and Mysians, ii. 316, 317.
- Asclēpiadæ, in Thessaly, ii. 136, 142.
- Asclepiades, of Prusa, ii. 318.
- ——, the Myrlean, i. 235, 249.
- [Pg 314]Asclepieium, iii. 36.
- Asclepius, i. 114; ii. 9, 56, 141; iii. 22.
- ——, temple of, at Carthage, iii. 285.
- ——, Celæni, in the Troad, ii. 371.
- ——, in the Island of Cos, iii. 36.
- ——, of Epidaurus, ii. 56, 321.
- ——, of Gerenia, ii. 36.
- ——, of Olenus, ii. 71.
- ——, of Tricca, ii. 36, 56, 141.
- ——, grove of, between Berytus and Sidon, iii. 171.
- Ascra, t. of Bœotia, ii. 104, 105, 110, 122, 398.
- Asculum Picēnum (*Ascoli*), i. 358.
- Asdrubal, i. 238.
- ——, wife of, iii. 285.
- Asea, a village of Arcadia, ii. 15.
- Asia,

i. 22, 50, 55, 56, 88, 103, 105, 156, 161, 162, 179, 183, 187, 188, 190, 191, 194, 197, 213, 431, 437, 439-441, 453, 465, 466, 477, 478, 490, 510, 517, 518; ii. 2, 32, 60, 62, 68, 93, 145, 154, 209; iii. 38, 39, 98, *et passim.*

- ——, Upper, ii. 244.
- ——, Eastern, ii. 244.
- ——, a Roman province, ii. 401.
- ——, within the Taurus, ii. 333.
- Asiatic coast, i. 491.
- Asii, ii. 245.
- Asinæan Gulf. *See* Messenian Gulf, ii. 35.
- Asinæus, ii. 36.

208

- Asine, city of Messenia, ii. 35-37, 54, 55.
- ——, city of Laconia, ii. 41.
- ——, city of Argolis (*Fornos*), ii. 49. 54. 58.
- ——, Hermionic, ii. 36.
- Asinius, i. 287.
- Asioneis, Esioneis, ii. 405.
- Asisium, i. 338.
- Asius, son of Dymas, ii. 351.
- ——, poet, i. 399.
- ——, Hyrtacides, the Trojan, ii. 344. 345. 350. 351.
- ——, meadow of, iii. 26.
- Asius, temple to, iii. 26.
- Asōpia, vill. of Sicyonia, ii. 103.
- Asōpian district, ii. 66.
- ——, Thebes, ii. 74.
- Asōpus, r. of Sicyon, i. 410; ii. 66, 103.
- ——, r. of Bœotia, ii. 103, 104, 108.
- ——, r. of Phthiotis, ii. 67, 129.
- ——, r. of the isl. of Paros, ii. 66.
- ——, city of Laconia, ii. 41.
- Aspaneus, near Ida, ii. 376.
- Aspasiacæ, *see* Apasiacæ, ii. 248.
- Aspendus, t. of Pamphylia (*Balkesi*), ii. 323; iii. 49.
- Asphalius (name of Neptune), i. 90.
- Aspiōnus, satrapy of, ii. 253.
- Aspis, t. by the Greater Syrtis, iii. 290.
- ——, t. of the Carthaginians, i. 421; iii. 288.
- ——, island, iii. 16.
- Asplēdon, city of Bœotia, ii. 113.
- Aspordēnum, ii. 393.
- Asporēne, Aspordēne, ii. 393.
- *Aspra-Spitia. See* Anticyra.
- *Aspro-potamo*, r. *See* Achelōus.
- *Aspro-vuna. See* Luca.
- Aspurgiani, a nation of Mœotis, ii. 223. 305.
- Assacanus, land of, in India, iii. 82, 90.
- *Assouan. See* Syene.
- Assus, t. of Mysia (*Beramkoi*), ii. 339. 376. 386. 390; iii. 140.
- ——, people of, ii. 375, 381, 382.
- Assyria, iii. 34, 142-160.
- Assyrians, i. 66.

209

- Asta, city of Iberia, i. 211, 213, 215.
- Astaboras, r. of Ethiopia (*Tacazze*), iii. 194, 195, 219, 270.
- Astacēni, iii. 90.
- Astacus, t. of Bithynia, ii. 171.
- ——, t. of Acarnania, ii. 171.
- ——, Gulf of (*Ismid*), ii. 171, 315.
- Astæ, people of Thrace, i. 516.
- Astapus, r. of Ethiopia (*The Blue Nile*), iii. 195, 219, 270.
- Astasobas, r. of Ethiopia, iii. 195, 219, 270.
- Astëeis, iii. 4.
- [Pg 315]Asteria, Asteris, isl. (*Dascaglio*), i. 93; ii. 168.
- Asterium, ii. 142, 143.
- Asteropæus, i. 514.
- Asti, people of Thrace, i. 490, 492.
- Asturian mountains, i. 250.
- Asturians, i. 229, 233, 241, 243, 250.
- Asty, ii. 85, 87, 91.
- Astyages, ii. 264; iii. 134.
- Astygis, t. of Spain (*Ecija*), i. 213.
- Astyochea, ii. 9.
- Astypalæa, prom. of Attica, ii. 89.
- ——, prom. of Caria, iii. 37.
- ——, old city of the Coans, iii. 36.
- ——, one of the Sporades (*Istanpolia or Stanpalia*), ii. 212, 213.
- Astypalæans, inhabitants of Rhætium, ii. 368.
- Astyra, t. of Mysia, ii. 376, 386, 387.
- ——, t. of the Troad, ii. 353; iii. 66.
- Atabyris, mtn of Rhodes (*Abatro*), ii. 164; iii. 33.
- Atagis (*Aude*), r. of Rhætia, i. 308.
- Atalanta, isl. near Attica, ii. 85, 125.
- ——, opposite Eubœa (*Talanta*), i. 95; ii. 85, 125.
- Atargata, iii. 216.
- Atargatis, iii. 158.
- Atarneitæ, ii. 383.
- Atarneus, city of Mysia (*Dikeli-Koi*), ii. 339, 376, 382, 389, 398; iii. 66.
- ——, city of the Troad, ii. 387.
- Atax, r. of Gaul, i. 272, 282.
- Ateas, king of the Bospori, i. 472.
- Ategua, t. of Spain, i. 213.
- Atella, t. of Campania, i. 370.
- Ateporix, ii. 310.

- Atlantis, island of, i. 154.
- Atlas, father of Calypso, i. 39.
- ——, mtn of Mauritania, iii. 276.
- Atmŏni, tribe of the Bastarnæ, i. 470.
- Atrax, city of Pelasgiotis, ii. 142, 146.
- Atrebates, people of Gaul, i. 289, 290.
- Atreus, i. 25, 35; ii. 53.
- *Atri. See* Adria.
- Atria, city of Cisalpine Gaul (*Adria*), i. 319.
- Atropatene, Atropatia, or Atropatian Media, ii. 238, 263, 267, 270.
- Atropateni, Atropatii, ii. 264, 274.
- Atropates, satrap of Media, ii. 262.
- Attaleia, city of Pamphylia (*Adalia*), iii. 49.
- Attalic kings, kings of Pergamus, ii. 320.
- Attalici, ii. 315.
- Attalus, first king of Pergamus, i. 440; ii. 390, 400.
- ——, Philadelphus, ii. 400; iii. 13. 49.
- ——, Philometor, ii. 401; iii. 21.
- ——, brother of Philetærus, ii. 400.
- Attasii, tribe of the Massagetæ, ii. 248.
- Attea, t. of the Troad, ii. 376.
- Atthis, daughter of Cranaus, ii. 87.
- Attica, i. 40, 43, 105, 187; ii. 2, 3, 87.
-         ——, Silver mines of, i. 221.
- Attic dialect, ii. 2.
-         ——, miners, i. 221.
- Attica, i. 493, 506, 507; ii. 57, 62, 67, 78-81, 84, 86-91, 95-97, 99, *passim*.
- *Attock. See* Choaspes.
- Aturia, part of Assyria, iii. 142-144.
- Atys, father of Tyrrhenus, i. 326, 329.
- Auases *or* Oases, in Libya, i. 197; iii. 226, 258.
- Aude. *See* Atagis.
- Aufidus, r. of Apulia (*Ofanto*), i. 346. 433.
- Auge, daughter of Aleus, ii. 389.
- Augeas, king of the Epeii, ii. 10, 13, 27, 30, 31, 170.
- Augeiæ, city of Laconia, ii. 42.
- ——, city of Locris, ii. 42.
- Augila (*Aujela*), iii. 295.
- Augusta, city of the Salassi, (*Aouste*), i. 306.
- ——, Emerita, t. of the Turduli in Spain (*Merida*), i. 227, 250.
- Aulis, city of Bœotia (*Vathi*), i. 16. 457; ii. 58, 92-97, 103, 151.

212

213

- Balearicus, Metellus, i. 252.
- Balithōn. *See* Ammon.
- *Balk. See* Bactra.
- *Balkesi. See* Aspendus.
- *Ballyk. See* Metropolis.
- [Pg 318]Bambyce, t. of Syria, iii. 158, 163.
- Bamōnītis, part of Pontus, ii. 302.
- Bandobēne, distr. of India, iii. 89.
- *Bara. See* Paros.
- Barathra, Berethra, ii. 76; iii. 241.
- Barbarian laws, i. 240.
- Barbarians, i. 17, 18, 77, 104, 114.
- Barbarium, Cape, prom. of Lusitania (*Capo Espichel*), i. 227.
- Barca, city of Cyrene, same as Ptolemais, iii. 291, 292.
- Barcas, Hamilcar, father of Hannibal, i. 226, 238.
- Bards, Celtic poets, i. 294.
- Bardyli, Bardyali, Bardyētæ, Bardyītæ, people of Spain, i. 233, 243.
- *Baretoun. See* Parætonium.
- Bargasa, t. of Caria, iii. 34.
- Bargosa, city of India, iii. 119.
- Bargus, r. of Illyria, i. 488.
- Bargylia, t. of Caria, ii. 383; iii. 37.
- Baris (Zaris), temple of, ii. 273.
- Baris, city, i. 429.
- Barium (*Bari*), i. 432, 433.
- Barnichius, r. of Elis. *See* Enipeus, ii. 32.
- Barnus, city of Macedonia, i. 495.
- Basgœdariza, fortress of Armenia, ii. 304.
- Basileius, r. of Mesopotamia, iii. 158.
- Basilii, i. 470.
- Basoropeda, district of Armenia, ii. 269.
- Bassus, Cecilius, iii. 166.
- Bastarnæ, i. 141, 177, 194, 443, 451-453, 468-471.
- Bastetani, Bastuli, i. 210, 212, 234, 243, 245.
- Bastetania, i. 232, 235.
- Bata, t. of Pontus (*Pschate*), ii. 225.
- Bathynias, i. 518.
- Bathys Limen (Deep Harbour), in Aulis, ii. 95.
- Batiæ, city of the Cassopæi, i. 497.
- Batieia, ii. 328, 361, 399.
- Batōn, leader of the Pannonii, i. 483.

- Batōn, historian, ii. 293.
- Battus, founder of Cyrene, iii. 292.
- *Baubola. See* Bilbilis.
- *Bayjah. See* Vaga.
- Bear, the (constellation), i. 4, 5.
- ——, Greater, i. 21, 117-120.
- ——, Lesser, i. 117-120, 200.
- *Beas. See* Hypanis.
- *Beaucaire. See* Ugernum.
- Bebryces, a people of Thrace, i. 453; ii. 287, 304, 346; iii. 63.
- *Beit-el-ma. See* Daphne.
- Beitylus, ii. 36. *See* Œtylus.
- *Beja. See* Pax Augusta.
- *Beknesch. See* Oxyrynchus.
- Belbina, an island, ii. 57, 89.
- Belgæ, i. 264, 266, 286, 290-293.
- Bĕlio, r. of Lusitania (*see* Limæa), i. 229.
- Bellerophon, ii. 62, 328, 409; iii. 48.
- Bellovaci, a people of Gaul (*inhabitants of the Beauvoisin*), i. 289, 293, 310.
- Bĕlō (*Rio Barbate*), i. 210.
- Belus, i. 67.
- ——, tomb of, iii. 145; temple of, 153.
- Bembina, city of Argolis, ii. 60.
- *Ben-Ghazi. See* Berenice.
- Bĕnacus, lake of Italy, i. 311.
- *Bendamir. See* Araxes.
- *Bender-el-Kebir. See* Berenice.
- Bendidæan rites, ii. 186, 188.
- Beneventum, t. of Samnium (*Benevento*), i. 370, 371, 431.
- *Bengal, Bay of. See* Ocean, Eastern.
- Berecyntes, people of Phrygia, ii. 184, 337; iii. 66, 67.
- Berecyntia, distr. of Phrygia, ii. 337.
- Berecyntian pipes, ii. 187.
- Berenice, dr. of Salome, iii. 184.
- ——, t. of Cyrene, (*Ben Ghazi*), iii. 291, 292.
- ——, t. in the Troglodytic, iii. 197, 260.
- ——, t. of Egypt (*Bender-el-kebir*), ii. 200; iii. 193, 260.
- [Pg 319]Berenice, Hair of (constellation), i. 4.
- Berga, i. 514.
- Bergæan, the. *See* Antiphanes.
- Bĕrisades, king of the Odrysæ, i. 516.

- Bermium, Bermius, mtn of Macedonia (*Buræus*), i. 510, 511; iii. 66.
- *Bernic. See* Hesperides.
- Berœa, t. of Macedonia (*Karafaja*), i. 511.
- ——, t. of Syria, iii. 163.
- Bĕrones, people of Spain, i. 238, 243.
- Bertiscus, mtn of Macedonia, i. 505.
- Bĕrytus (*Beyrout*), city of Phœnicia, iii. 69, 170, 171.
- Bĕsæeis, Bĕsæenses, ii. 127.
- Besbicus, isl. (*Imrali* or *Kalo-limno*), ii. 332.
- Bĕssa, in Locris, ii. 127.
- Bessi, nation of Thrace, i. 489, 516.
- Bĕssus, ii. 248, 255.
- Betarmones, ii. 190.
- Betteres, t. of Spain, i. 240.
- *Bevagna. See* Mevania.
- *Beyrout. See* Berytus.
- *Beziers. See* Bætera.
- Bias, iii. 7.
- Biasas, name of the Paphlagonians, ii. 302.
- Bibracte, fortress of the Ædui, i. 286.
- *Bieda. See* Blera.
- Bilbilis, t. of the Celtiberians, (*Baubola*), i. 244.
- Billarus, sphere of, ii. 293.
- Bion, philosopher, i. 23, 24; ii. 210.
- ——, astronomer, i. 45.
- Bīsa, ftn of Elis. *See* Pīsa?], ii. 32.
- Bisalti, Bisaltæ, people of Macedonia, i. 506, 514.
- Biscay, people of. *See* Cantabrians.
- Bistones, race of Thrace, i. 515.
- Bistonis, lake of Thrace (*Burum*), i. 92, 515.
- Bisurgis. *See* Visurgis.
- Bithynia, ii. 289, 293, 313-318, 329, 356, 402; iii. 297.
- Bithynians, i. 195, 453; ii. 277, 286, 287, 289, 290, 314, 316, 319, 320, 330; iii. 63.
- ——, of Thrace, ii. 287.
- Bithynium (*Boli*), ii. 317.
- Bituitus, son of Luerius, i. 285.
- Bituriges Cubi, i. 283, 284.
- ——, Vivisci, people of Gaul, i. 283.
- Bizōne, t. of Mœsia, i. 84, 490.
- Bizya, t. of Thrace, i. 516.
- Black Forest, the. *See* Hercynia.

- Black Sea, i. 6, n., 457, 468, 469. *See* Euxine.
- Blaēnē, distr. of Paphlagonia, ii. 313.
- Blascōn, isl., i. 271.
- Blaudus, t. of Phrygia (*Suleimanli*), ii. 320.
- Bleminātis, in Laconia, ii. 15.
- Blemmyes, a people of Ethiopia, iii. 219, 266.
- Blera, t. of Etruria (*Bieda*), i. 335.
- Blēsino, t. of Corsica, i. 333.
- Blessed, Isles of the (*Canary Islands*), i. 3, 226.
- Blucium (Luceium ?), fortress of the Tolistobogii, ii. 320.
- Boagrius, r. of Locris (*Boagrio*), i. 95; ii. 126.
- Bōcalia (?), Bōcarus, r. of Salamis, ii. 83.
- Bocchus, king of Mauritania, iii. 280.
- Bœa, t. of Laconia, ii. 41.
- Bœbē, t. of Magnesia, ii. 139, 142, 272.
- Bœbēis or Bœbias, lake of Thessaly, ii. 131, 139, 142, 146-148, 235; iii. 22.
- Bœi, people of Gaul, i. 291-485.
- Bœōnōa, t. of Elis, ii. 9.
- Bœoti, t. of Laconia, ii. 2.
- ——, (Athenians), ii. 81.
- Bœotia, i. 6, n., 16, 94, 493, 494, 506, 507; ii. 4, 6, 36, 48, 62, 78, 79, 82, 90-115,122-125, 136, 138, 142, 151, 154, *et passim*; iii. 31.
- Bœotian coast, ii. 98.
- [Pg 320]Bœotians, i. 102, 493; ii. 98, 101, 102, 105, 134, 175, *et passim*.
- Bœōtus, son of Melanippe, i. 399.
- Bœrebistas, king of the Getæ. *See* Byrebistas.
- Boēthus, Sidonian, iii. 173.
- ——, of Tarsus, iii. 58.
- Bœum, city of the Dorians, i. 505; ii. 128, 195.
- *Bog. See* Hypanis.
- *Bogdana. See* Hyampolis.
- Bogodiatarus (? Deïotarus), ii. 320.
- Bogus, king of Mauritania, i. 151, 153, 154; ii. 36; iii. 278, 280.
- Boïanum, city of the Samnites (*Bojano*), i. 371.
- Boii, i. 291, 306, 307, 317, 321, 322, 448, 450, 454, 466, 482, 485.
- Bolbe, lake of Macedonia, i. 514.
- Bolbitine mouth of the Nile, iii. 239.
- *Boli. See* Bithynium.
- *Bologna. See* Bononia.
- *Bolsena. See* Volsinii.
- Bōmianes, a nation of Ætolia, ii. 160.

- *Bondoniza. See* Scarpheia.
- Bonōnes, son of Phraates, iii. 160.
- Bonōnia, city of Italy (*Bologna*), i. 322, 324.
- Boos-Aule, cave of Eubœa, ii. 152.
- Boosura, city of Cyprus (*Bisur*), iii. 70.
- *Bordeaux. See* Burdegala.
- Boreas, Borras, i. 42, 44, 97.
- Boreion, prom. of Cyrene (*Ras-Teyonas*), iii. 291.
- Borrhama, iii. 170.
- Borsippa, t. of Babylonia, iii. 146.
- Borsippeni, Chaldæans, iii. 146.
- Borus, ii. 110.
- Borysthenes (*Dnieper*), i. 98-100, 110, 111, 113, 114, 116, 162, 172-175, 188, 190,191, 202, 203, 442, 443, 451, 457, 470-472, 475, 478; ii. 222, 298.
- Bosporani, Asian, ii. 223.
- ——, European, ii. 223.
- ——, Bosporiani, Bosporians, ii. 223, 224; iii. 180.
- Bosporii, i. 476.
- Bosporus, Cimmerian (*Straits of Kertch or Zabache, Azof*), i. 8, 31, 114, 164, 189, 223,441, 450, 463, 472, 475-478, 480; ii. 216, 219-222, 224, 225, 239, 294, 302, 305,318, 401, 402.
- ——, Thracian, i. 138, 189; ii. 318.
- ——, Mysian, ii. 318.
- Botrys, fortress of Syria, iii. 170.
- Bottiæa, distr. of Macedonia, i. 430, 508, 509.
- Bottiæi, people of Macedonia, i. 425, 506, 508.
- Bottōn, i. 506.
- *Bougie. See* Salda.
- *Bouz Dagh. See* Tmolus.
- Boxos, Boxes, iii. 208.
- *Bracchiano, Lago di. See* Sabatus.
- Brachmānes, philosophers of India, iii. 109-111, 114, 117.
- Branchidæ, priests of Apollo, ii. 254; iii. 4, 259.
- ——, their city in Sogdiana, ii. 254.
- Branchus, ii. 120; iii. 4.
- Braurōn, t. of Attica, ii. 52, 88, 89.
- Breasts, the (*Stethè*), i. 79, 82.
- Brĕnæ, people of Thrace, i. 516.
- Brennus, i. 280.
- Brentĕsium, t. of Iapygia (*Brindisi*), i. 347, 370, 423, 428-435, 497.
- Brescia, i. 317.
- Brettii, i. 315-441.

219

- Breuci, people of Hungary, i. 483.
- Breuni, nation of Illyria, i. 306.
- Brigantii, a people of the Vindelici, i. 307.
- Brigantium (*Briançon*), i. 268, 307.
- Briges, people of Thrace, i. 453, 510.
- Brilessus, mtn of Attica, ii. 90.
- *Brindes. See* Brundusium.
- *Brindisi. See* Brentesium.
- Briseïs, ii. 343, 384.
- Britain, i. 99, 100, 111, 116, 117, 141, 157, 172-175, 181, 193, 263, 264, 281, 283,288-290, 295-298.
- Britannic Islands, British Islands, i. 172, 173, 194, 196, 221.
- [Pg 321]British Channel, i. 192.
- Britomartis, ii. 199, 200.
- Britons, i. 116, 177, 298, 299.
- Briula, iii. 26.
- Brixia, t. of the Insubri, i. 317.
- Brothers, Seven, monuments of the, iii. 278.
- Bructeri, a people of Germany, i. 444, 445, 447.
- Brundusians, i. 430.
- Bruttii, i. 315, 339, 374, 377-383, 391, 431.
- Brutus, the Gallician, i. 228, 230, 233.
- ——, Decimus, vanquished at Philippi, i. 305, 515.
- Bryanium, t. of Macedonia, i. 501.
- Bryges, Brygi, Phryges, ii. 298.
- Brygi, people of Epirus, i. 500, 501.
- Bubastite nome, iii. 245.
- Bubastus, t. of Egypt, iii. 245.
- Bubōn, t. of Lycia (*Ebedschek-Dirmil*), ii. 410.
- Buca, t. of the Frentani, i. 359, 436.
- Bucephālia, city of India, iii. 91.
- Bucephalus, the horse of Alexander, iii. 91, 92.
- Buchetium, city of the Cassopæi, i. 497.
- Bucolopolis, t. of Judæa, iii. 175.
- Budŏrus, r. of Eubœa, ii. 153.
- ——, mtn of Salamis, ii. 153.
- *Budrun. See* Teos.
- Bujæmum, i. 444.
- Bulliones, people of Illyria, i. 500.
- Buprasian district, ii. 18.
- Buprasii, Buprasians, ii. 12, 13, 27.

221

- *Cadiz.* See Gades, Gadeira.
- Cadmē, same as Priene, iii. 7.
- Cadmeia, citadel of Thebes, ii. 108, 109.
- [Pg 322]Cadmeian victory, i. 224.
- ——, territory, i. 493; ii. 93.
- Cadmus, founder of Cadmeia, i. 493, 500; ii. 93, 154.
- ——, Melesian, i. 281; iii. 66.
- ——, r. of Phrygia, ii. 334.
- ——, mtn of Phrygia, ii. 334.
- Cadurci, a people of Gaul (*Querci*), i. 284.
- Cadusii, a people of Asia, ii. 240-242, 245, 248, 249, 263, 264.
- Cæcias, name of a wind (N.E.), i. 45.
- Cæcilius Bassus, iii. 165.
- Cæcubum, distr. of Latium, i. 345, 347.
- Cælius, Mount, i. 348.
- Cænepolis, ii. 36.
- Cæni, people of Thrace, ii. 401.
- Cænys, prom. of Italy, i. 385, 400.
- Cæpio, Q. S., a Roman general, i. 280.
- ——, tower of, i. 211.
- Cæratus, same as Cnossus, ii. 190.
- Cærea, c. of Etruria, i. 328.
- Cæretana, hot-springs, i. 328.
- Cæretani, i. 327, 335.
- Cæsar, i. 213, 241, 242, 270, 271, 285, 290, 305; ii. 44, 270, 274, 278, 297.
- ——, Augusta, c. of the Celtiberi (*Saragossa*), i. 227, 242, 244.
- ——, Augustus, i. 234, 265, 275, 286, 298, 304, 306, 308, 349-351, 369, 388, 404,408, 411, 439, 441, 444, 446, 448, 467, 483, 484; ii. 294, 309, 334, 356-358, 392,402; iii. 36, 53, 54, 58, 59, 74, 118, 159, 184, 209, 231, 233, 281, 296, 297.
- ——, Julius or divus, i. 265, 285, 288, 297, 298, 317, 350, 439, 457, 497-499; ii. 65,294, 308; iii. 20, 24, 227, 284, 287.
- Cæsarea, c. of Numidia, iii. 284.
- Cæsarium, temple of Alexandria, iii. 230.
- Cæsēna, c. of Italy, i. 322.
- *Cafsa.* See Capsa.
- Caiata, gulf of (*Gaëta*), i. 347.
- ——, promontory of, i. 347.
- Caicus, r. of Mysia (*Bakyr-Tschai*), ii. 326, 327, 339, 376, 383, 387-390, 395, 397,401.
- ——, plain of, ii. 332, 388-390, 401; iii. 82.
- Caieta, nurse of Æneas, i. 347.
- Cainochorion, fortress of Pontus, ii. 306.

- Calabri, i. 422, 423.
- Calabria, i. 430.
- Calachene, distr. of Assyria, ii. 235, 272; iii. 142.
- Calaguris, t. of Spain (*Calahorra*), i. 242.
- Calamis, i. 490.
- Calanus, iii. 74, 112, 113, 115, 116.
- Calasarna, t. of Lucania, i. 379.
- Calatia, c. of Campania (*Le Galazze*), i. 431.
- Calauria, island (*Poros*), i. 187; ii. 49, 55.
- Calbis, r. of Caria (*Doloman Ischai*), iii. 28.
- Calchas the prophet, iii. 15, 50, 59, 60.
- ——, shrine of, i. 434; ii. 324.
- Calche, the, ii. 271.
- Cale-Peuce, ii. 371.
- Calenian wine, i. 361.
- Calēs, c. of Campania (*Calvi*), i. 352, 370.
- Caleti, people of Gaul, i. 281, 289.
- Callaïci, people of Spain, i. 222-251.
- Callanian plain, ii. 407.
- Callas, r. of Eubœa, ii. 152.
- Callateria (*Galazze*), i. 370.
- Callatis, t. of Mœsia (*Mangalia*), i. 489, 490; ii. 288.
- Calliarus, t. of Locris, ii. 127.
- Callias, ii. 393.
- Callicolōnē, ii. 362.
- Callidromus, part of Œta, ii. 129.
- Callimachus, i. 70-72, 321, 459; ii. 21, 29, 87, 141, 199, 206; iii. 9, 35, 245, 292.
- Callinicus, iii. 162, 168. *See* Seleucus.
- Callīnus, ii. 373, 405; iii. 3, 22, 23, 50.
- [Pg 323]Calliŏpe, ii. 189.
- Callipidæ, nation of Scythia, ii. 298.
- Callipolis, t. of Sicily, i. 412.
- ——, t. of the Thracian Chersonesus (*Gallipoli*), i. 518; ii. 349.
- ——, t. of Macedonia, i. 514.
- *Calliste. See* Thēra.
- Callisthenes, ii. 39, 254, 288, 356, 383, 405; iii. 5, 49, 66, 225, 245, 261, *passim*.
- Callydium, fortress of Mysia, ii. 330.
- Calpas, r. of Bithynia, ii. 288.
- Calpè, t. of Spain, i. 81, 210, 212.
- ——, rock of Spain (*Gibraltar*), i. 164, 234, 235, 253, 255.
- *Calvi. See* Cales.

- Calybe, t. of the Asti, i. 492.
- Calycadnus, r. of Cilicia (*Kelikdni*), ii. 405; iii. 53-55.
- Calydna, same as Tenedos, ii. 214, 372.
- Calydnæ, islands, ii. 212-214, 372.
- Calydōn, c. of Ætolia, ii. 127, 155, 159, 160, 171, 172, 175, 179.
- Calymna, Calymnæ, isl. (*Calimno*), ii. 214.
- Calynda, c. of Caria, iii. 28.
- Calypso, island of, i. 459.
- Camarina, c. of Sicily (*Torre di Camarana*), i. 401, 411.
- *Camasch. See* Commagene.
- Cambysene, distr. of Armenia, ii. 232, 234, 269.
- Cambyses, ii. 190; iii. 141, 224, 245, 261.
- Cameirus, t. of Rhodes (*Camiro*), iii. 31, 33.
- Camertes, t. of Umbria, i. 338.
- Camici, t. of Sicily, i. 413, 425.
- Camillus, son of Vulcan, ii. 189.
- Camisa, fortress of Pontus, ii. 310.
- Camisene, distr. of Cappadocia, ii. 293, 310.
- *Campanella, Puntadella. See* Athenæum and Sirenussæ.
- Campani, Campanians, i. 352, 357, 361, 366, 369, 371, 373, 377, 387, 404.
- Campania, i. 326, 344, 346, 360, 361, 369-371, 373, 379, 429, 431, 432.
- Campodunum, t. of the Vindelici, i. 307.
- Campsiani, people of Germany, i. 445.
- Campus Martius, i. 350, 371.
- Camuni, people of the Rhæti, i. 306.
- Canæ, c. of Æolia, ii. 153, 388.
- ——, mtns (*Adschane*), ii. 339, 342, 376, 388, 390, 391.
- *Canary Islands. See* Blessed, Islands of the.
- *Canan, Cape. See* Cephalæ.
- Canastræum, prom. of Macedonia (*Cape Pailuri*), i. 510.
- Canastrum, prom. of Pallene (*Cape Pailuri*), i. 511, 512.
- Candace, queen of Ethiopia, iii. 268, 269.
- Candavia, mtns of Illyria, i. 495, 500.
- *Candia. See* Crete.
- Canēthus, hill of Eubœa, ii. 154.
- Canidius, ii. 231.
- Cannæ, t. of Apulia, i. 436.
- Canopic mouth of the Nile, i. 101; iii. 237, 238. *See* Nile.
- ——, gate of Alexandria, iii. 231, 237.
- ——, canal, iii. 231, 237, 239.
- Canōpus, constellation, i. 4, 180.

- ——, c. of Egypt (*Aboukir*), i. 130; iii. 48, 222, 237, 238.
- Cantabria, i. 236, 247.
- Cantabrian mtns, i. 250.
- Cantabrians, i. 230, 233, 234, 239, 241, 243, 246-248, 250, 439.
- ——, Conish, the, i. 243.
- Cantharius, prom. of Samos, iii. 10.
- Cantharōlĕthron, i. 511.
- Cantium. *See* Kent.
- Canusitæ, emporium of the, i. 433.
- Canusium, t. of Apulia (*Canosa*), i. 431, 433.
- Capedunum, t. of the Scordisci, i. 488.
- Caphareus, prom. of Eubœa, ii. 48.
- Caphyeis, t. of Arcadia, ii. 75.
- [Pg 324]Capitol, the, i. 298, 342, 348, 351, 424, 490; iii. 8.
- Capitŭlum, t. of Latium, i. 353.
- Capnobatæ, i. 454, 455.
- *Capo Boeo. See* Lilybæum.
- Cappadocia, i. 113, 195, 262, 279; ii. 216, 218, 246, 259-261, 265, 273, 276-286, 301,307, 310, 314, 319-322; iii. 35, 44, 54, 63-65, 137, 150, 232.
- ——, the Great, ii. 278, 293, 294, 307, 321.
- ——, Upper, ii. 259.
- ——, on Pontus, ii. 278.
- Cappadocians, i. 440; ii. 273-286, 290, 301, 320, 322.
- Capreæ, Capriæ, Capria, isl. (*Capri*), i. 34, 93, 185, 368, 369, 387.
- Capria, l. of Paphlagonia, iii. 49.
- Caprus, port of *Chalcidia*, i. 512, 513.
- ——, island, i. 512, 513.
- ——, r. of Phrygia, ii. 334.
- ——, r. of Assyria (*The Little Zab*), iii. 144.
- Capsa, t. of Numidia (*Cafsa*), iii. 284.
- Capua, (*S. Maria di Capoa*), i. 351, 360, 370, 431.
- Capyæ, t. of Arcadia, ii. 378.
- Capys, ii. 378.
- Caracoma, i. 516.
- Caralis, t. of Sardinia (*Cagliari*), i. 333.
- *Caraman. See* Laranda.
- Carambis, prom. of Paphlagonia (*Kerempi-Burun*), i. 188, 476; ii. 225, 291, 293.
- Carana, t. of Pontus, ii. 310.
- Caranītis, ii. 310.
- Carcathiocerta, t. of Armenia (*Kharput*), ii. 268.
- *Carchi. See* Chalcia.

- Carcinites Gulf, i. 471, 473, 474, 478.
- Carcoras, r. of Noricus, i. 482.
- Cardaces, iii. 138.
- Cardamylæ, t. of Messenia (*Scardamula*), ii. 35-37.
- Cardia, t. of the Thracian Chersonesus, i. 517.
- *Cardiana. See* Lagusa.
- Cardūchi, people of Asia, iii. 157 Carēnitis, distr. of Armenia, ii. 269.
- Carēsēnē, distr. of the Troad, ii. 371.
- Carēsus, t. of the Troad, ii. 304, 371.
- ———, r. of the Troad, ii. 357, 371.
- Caria, i. 8, 102, 103, 133, 140, 172, 187, 190, 195, 202, 493;

ii. 56, 68, 259, 298, 313,329, 333, 334, 383, 407, 409; iii. 1, 2, 6, 27-44, 59.

- ———, coast, iii. 34.
- Carians, i. 96, 103, 493, 494; ii. 50, 56, 88, 277, 327-329, 383; iii. 2, 35, 38-43, 63.
- Cariatæ, ii. 254.
- Carmalas, r. of Cataonia, ii. 280-283.
- Carmania (*Kerman*), i. 121-126, 129, 131, 132, 135, 196, 201;

iii. 109, 120, 122, 124,125, 127-133, 146, 152, 186, 187.

- Carmanians, iii. 120.
- Carmēl, mtn of Judæa, iii. 175.
- Carmentis, mother of Evander, i. 343.
- Carmō, t. of Spain (*Carmona*), i. 213.
- Carmylessus, t. of Lycia, iii. 46.
- Carna, Carnana, c. of Arabia, iii. 190.
- Carneades, iii. 293.
- Carneates, mtn of Sicyonia, ii. 66.
- Carni, i. 307-309, 321, 448, 482, 483.
- Carnus, t. of Syria (*Carnoon*), iii. 167.
- Carnutes (*people of the Chartrain*), i. 284, 289.
- Carpasia, t. of Cyprus, iii. 69.
- Carpasian islands, iii. 69.
- Carpathian Sea, i. 187; ii. 212; iii. 68.
- Carpathus, ii. 212, 213.
- Carpetani, i. 209, 212, 228, 229, 243.
- Carpetania, distr. of Spain, i. 214.
- Carrhæ, c. of Mesopotamia, iii. 157.
- Carseoli, t. of Latium (*Carsoli*), i. 353.
- Carsūli, t. of Umbria, i. 337.
- [Pg 325]Carta, t. of Hyrcania, ii. 242.
- Cartalia, t. of Spain, i. 239.
- Carteïa, c. of Spain, i. 210, 213, 218, 226.

226

- Cartera, Comè, vill. of Thrace, i. 515.
- *Cartero. See* Heracleium.
- Carthæ, ii. 210.
- Carthage, in Africa, i. 101, 140, 180, 184, 197, 198, 201, 403, 411, 439; ii. 331; iii.51, 282, 284-291.
  - Carthagena, in Spain, i. 222, 234, 238, 239, 245, 251, 262, 334; iii. 279.
  - Carthaginian Bay, iii. 285, 287.
  - ——, wars, iii. 284.
- Carthaginians, i. 104, 226, 238, 334, 377, 403, 404, 408, 424, 438, 439; ii. 71, 73; iii.240, 275, 285.
  - Carura, t. of Phrygia, ii. 334, 336, 409; iii. 43.
  - Caryanda, t. and isl. of Caria, ii. 318, 340; iii. 37.
  - Caryandians, iii. 37.
  - Carystian marble, ii. 140, 153.
  - Carystus (*Castel Rosso*), t. of Eubœa, ii. 153.
  - ——, in Laconia, ii. 153.
  - Casiana, fortress of Syria, iii. 165.
  - Casii, ii. 213, 214.
  - Casilīnum, t. of Campania (*Nova Capua*), i. 351-353, 370, 431.
  - Casīnum, t. of Latium, i. 352.
  - Casium, mtn of Egypt (*El Kas*), i. 62, 79, 87, 91; iii. 149, 233.
  - ——, mtn of Syria (*Ras el Kasaroun*), iii. 162, 164, 174-177.
  - *Caslona. See* Castulōn.
  - Caspian Sea, i. 54, 102, 109, 115, 122, 123, 132, 135-138, 183, 194, 451, 471; ii. 216-218, 226, 227, 230, 232, 235, 239, 240-246, 249, 255, 256, 260, 267, 270, 272.
  - Caspian Gates (*Firouz-Koh*), i. 94, 100, 121, 124, 125, 127, 130-132, 136-139, 202; ii.218, 237, 242, 249, 250, 259-265; iii. 120, 124, 125, 130, 153.
  - ——, tribes, ii. 234.
  - Caspiana, distr. of Albania, ii. 234, 269.
  - Caspii, ii. 226, 248, 253, 258.
  - Caspius, mtn of the Caucasus, i. 137-139; ii. 226.
  - Cassander, king of Macedonia, i. 509-511; ii. 88, 89.
  - Cassandra, i. 398, 511; ii. 367.
  - Cassandria, i. 511.
  - Cassiŏpē, port of Epirus (*Cassiopo*), i. 497.
  - Cassiopeia, constellation, i. 202.
  - Cassiterides (*Scilly Islands*), i. 181, 194, 221, 262.
  - Cassius, i. 515; iii. 164.
  - Cassōpæi, people of Epirus, i. 493, 496-498.
  - Castabala, t. of Cilicia, ii. 278, 281.
  - Castalian fountain, ii. 116.
  - *Castel Franco. See* Phœnix.

- *Castel Rosso. See* Carystus.
- Castellum, port of Firmum Picenum (*Porto di Fermo*), i. 357.
- *Castezzio. See* Clastidium.
- Casthanæa, t. of Magnesia, ii. 148.
- Castor, father of Deiotarus, ii. 314.
- ——, son of Saocondarus, ii. 321.
- Castor and Pollux. *See* Dioscuri.
- Castrum, Castrum Novum, t. of Picenum (*Giulia Nova*), i. 357, 358.
- Castulōn, Castlōn (*Caslona*), t. of Spain, i. 214, 222, 228, 241, 250.
- Casus, ii. 212-214.
- Casystes, iii. 17.
- Catabathmus, mtn and t. of Egypt, *Akabet-el-Kebira*, iii. 226, 235, 275, 294.
- Catacecaumene, distr. of Mysia, or Lydia, ii. 332, 335, 336, 403, 404, 406; iii. 8, 43.
- Catacecaumene, wine of, ii. 406; iii. 8.
- *Catacolo, Cape. See* Ichthys.
- Catana, c. of Sicily (*Catania*), i. 356, 367, 402, 403-405, 411, 415.
- Catanæa, i. 405, 411.
- Catanæi, Catanæans, i. 405, 406, 412.
- [Pg 326]Cataones, Cataonians, people inhabiting the Taurus, ii. 269, 276, 277; iii. 64.
- Cataonia, part of Cappadocia, i. 82, 202; ii. 259, 276-279, 280; iii. 59, 65.
- Cataractes, r. of Pamphylia, iii. 49.
- Cataracts, of Teverone, i. 353.
- ——, of the Euphrates, iii. 147.
- ——, of the Nile, iii. 217, 265.
- Catennenses, mtn of Pisidia, ii. 324.
- Cathæa (? Cathay), distr. of India, iii. 92.
- Cathæi, iii. 93.
- Cathylci, people of Germany. *See* Caulci, i. 447.
- Cato, Marcus, ii. 250; iii. 58, 72, 291.
- Catocas. *See* Menippus.
- Catopterius, near Parnassus, ii. 123.
- Catoriges, an Alpine nation, i. 303.
- Cattabaneis, people of Arabia, iii. 190.
- Cattabania, iii. 191.
- Caucasian mtns, i. 106, 115-117, 130, 131, 162, 177, 195, 273; ii. 219, 220, 230-232, 235, 258, 269; iii. 79.
- ——, tribes, ii. 227.
- Caucasus, ii. 224-226, 229-235, 238, 239, 245, 267; iii. 77, 78, 80, 107, 125.
- Cauci, a people of Germany, i. 445.
- Caucon, r. i. 14, 15; ii. 74.

- ——, monument of a, ii. 18.
- Caucōnes, in Elis, i. 493, 494; ii. 8, 14, 15, 18, 19, 28, 31, 74, 286-288, 290, 327, 383,394.
- Cauconia, ancient name of Elis, ii. 18.
- Cauconiatæ, Cauconītæ, in Paphlagonia, ii. 18, 286-288; iii. 63, 65.
- Cauconis, Cauconitis, ii. 14, 74.
- Caudium, t. of the Samnites (*S. Maria di Goti, Paolisi*), i. 370, 431.
- Caulci, people of Germany, i. 445.
- Caulōnia, t. of the Bruttii, i. 392. *See* Aulonia.
- Caunians, iii. 28.
- Caunus, t. of Caria (*Dalian*), iii. 28.
- *Cavaillon. See* Caballio.
- *Cavaliere. See* Zephyrium.
- Cavari, people of Gaul, i. 276-278.
- *Cavo, Monte. See* Albanus.
- Caÿster, r. of Ionia, ii. 145, 396, 397, 402-407; iii. 10, 14, 26.
- ——, plain of, ii. 397; iii. 82.
- ——, Larisæans in the, ii. 397.
- Caÿstrius, iii. 26.
- Ceans, ii. 210.
- Cĕbrēn, Cebrēnē, t. of the Troad, ii. 373, 375, 376.
- Cĕbrēni, in the Troad, ii. 361, 375.
- ——, in Thrace, ii. 351.
- Cĕbrēnia, a part of the Troad, ii. 360, 362.
- Cĕbriŏnes, ii. 360.
- Cecrŏpia, citadel of Athens, ii. 88.
- Cĕcrops, i. 493; ii. 87, 88, 101.
- Ceii, inhabitants of Ceus, ii. 253.
- Cĕladōn, r. of Elis, ii. 15, 22.
- Celænæ, hill of the Troad, ii. 333, 390.
- ——, t. of Phrygia, ii. 333, 335, 407.
- Celæno, one of the Danaids, ii. 335.
- Celænus, son of Neptune, ii. 335.
- Celenderis, t. of Cilicia (*Kilandria*), iii. 52, 177.
- Cĕlia, t. of Apulia (*Ceglie*), i. 431.
- Celmis, one of the Dactyls, ii. 191.
- Cĕlōssa, mtn of Sicyonia, ii. 66.
- Celsa, t. of Spain (*Xelsa*), i. 241, 242.
- Celtica. *See* Keltica.
- Cemmenus, mtn (*the Cevennes*), i. 193, 264-267, 272, 276, 277, 279, 282, 283, 285,310.

- Cēnæum, prom. of Eubœa (*C. Lithada*), i. 94; ii. 126, 130, 137, 150.
- Cenchreæ, port of the Corinthians, (*Kankri*), i. 85, 88; ii. 49, 62, 63.
- ——, t. of Argolis, ii. 58.
- Cencrius, r. near Ephesus, iii. 11.
- Cenomani, people of Cisalpine Gaul, i. 321.
- [Pg 327]Centauri, Centaurs, ii. 20.
- Centoripa, t. of Sicily (*Centorbe*), i. 411, 414.
- Centrones, Alpine ntn, i. 303, 305, 309.
- Ceōs, island (*Zia*), ii. 156, 208, 210.
- *Ceperano. See* Fregellæ.
- Cephalæ, prom. (*Cape Canan*), iii. 289, 290, 291.
- Cephallēnes, Cephallenians, ii. 83, 161, 162, 166, 167, 173.
- Cephallenia (*Cephalonia*), i. 187; ii. 5, 9, 15, 25, 161-169; iii. 8.
- Cephalœdium, t. of Sicily (*Cefalu*), i. 401, 411.
- Cephalōn, ii. 350.
- Cephalus, son of Deïonius, ii. 162, 166, 170, 173.
- Cēphēnes, i. 67.
- Cephisia, t. of Attica, ii. 88.
- Cēphissis, lake of Bœotia, ii. 102, 107.
- Cēphissus, r. of Phocis and Bœotia (*Mauropotamos*), i. 25; ii. 91, 98, 100-102, 123,124, 128.
- ——, r. of Attica, ii. 91, 124.
- ——, r. of Salamis, ii. 124.
- ——, r. of Sicyonia, Scyrus, Argolis, ii. 124.
- ——, ftn of Apollonia, ii. 124.
- Cephisus, r. ii. 351.
- Cēpi, t. of the Cimmerian Bosporus, ii. 223.
- Ceramietæ, iii. 40.
- Ceramus, t. of Caria, iii. 34.
- Cerasus, t. of Pontus, ii. 296.
- Cerata, mtns of Attica, ii. 84.
- Ceraunia, part of the Caucasus, ii. 232, 235.
- Ceraunian mtns, on the coast of Albania, i. 31, 159, 429, 432, 435, 486, 487, 489, 497,500; ii. 78, 79.
- Cerberus, ii. 40.
- Cerbesii, people of Phrygia, ii. 337.
- Cercaphus, father of Ormenus, ii. 142; iii. 32.
- Cercesura, t. of Egypt, iii. 247.
- Cercetæ, people of Asia, ii. 219, 225.
- Cerceteus, mtn of Icaria, ii. 212.
- Cercinna, isl. and town (*Karkenah*), i. 185; iii. 285, 288.

- Cercinnītis, island, iii. 288.
- Cercītæ, people of Pontus, ii. 296.
- Cercyra. *See* Corcyra.
- Cereate, t. of Latium (*Cerretano*), i. 353.
- Ceres. *See* Demeter, i. 95, 295, 516; ii. 66, 118, 130, 138, 139, 183.
- ——, Eleusinian, temple of the, ii. 84; iii. 2.
- ——, temples of, i. 411; ii. 17, 138, 139.
- ——, grove of, ii. 17.
- Cēreus, r. of Eubœa, ii. 137.
- *Cerigo, isl. See* Cythera.
- *Cerretano. See* Cereate.
- Cērilli, t. of the Bruttii (*Cirella*), i. 380.
- Cērinthus, t. of Eubœa, ii. 152, 153.
- Cerne, island. *See* Kerne.
- Cersobleptes, king of the Odrysæ, i. 516.
- Ceryneia, t. of Achæa, ii. 73.
- Cestrus, r. of Pamphylia (*Ak-su*), ii. 325; iii. 49.
- Cētæi, ii. 389, 395; iii. 63, 65.
- Cēteium, r. of Mysia, ii. 389.
- *Cevennes, the. See* Cemmenus.
- *Ceylon. See* Taprobane.
- Chaa, c. of Triphylia, ii. 22.
- Chaalla, c. of Arabia, iii. 212.
- Chaarene, distr. of Ariana, iii. 126.
- Chabaca, c. of Pontus, ii. 296.
- Chabrias, iii. 241.
- ——, rampart of, vill. in the Delta of Egypt, iii. 177.
- Chabum, c. of the Tauric Chersonesus, i. 479.
- Chæanœtæ, Chamæeunæ, Chamæcœtæ, people of the Caucasus, ii. 239.
- Chærēmōn, iii. 246.
- Chærōneia, c. of Bœotia (*Kapurna*), ii. 101, 110, 111, 123.
- Chalcēdōn, c. of Bithynia, i. 491; ii. 286, 289, 314, 315, 318, 380.
- ——, temple at, i. 491; ii. 289, 315.
- Chalcedonian shore, i. 491.
- Chalcēdonians, i. 491, 492.
- [Pg 328]Chalcētŏres, Chalcētŏr, c. of Caria, iii. 6, 37.
- Chalcia, Chalcis, mtn of Ætolia (*Varassova*), ii. 160, 171, 172.
- ——, one of the Sporades (*Carchi*), ii. 212, 213.
- Chalcideis, in Eubœa, iii. 17.
- Chalcidenses, Chalcidians, in Eubœa, i. 361, 365, 369, 385, 404, 506; ii. 154, 157, 158.

- ——, in Ionia, iii. 17.
- ——, in Thrace, i. 506.
- Chalcidic cities, i. 513.
- Chalcidica, distr. of Syria, iii. 166.
- Chalcis, iii. 33.
- ——, c. of Eubœa, i. 65, 86, 90; ii. 96, 151-156, 160, 162, 178, 188.
- ——, or Hypochalcis, c. of Ætolia, ii. 127, 155, 160, 172.
- ——, c. of Triphylia, ii. 16, 25, 26.
- ——, c. of Syria (*Balbek and Kalkos*), iii. 166, 170.
- ——, r. of Triphylia, ii. 16.
- ——, or Chalcia, mtn of Ætolia (*Varassova*), ii. 160, 171, 172.
- Chaldæans, i. 35; iii. 185.
- Chaldæi, people of Pontus, ii. 296, 297, 300, 304, 305.
- Chalestra, c. of Macedonia, i. 508, 509, 510.
- *Châlons-sur-Saone. See* Cabyllinum.
- Chalybē, c. of Pontus, ii. 297.
- Chalybes, people of Pontus, ii. 269, 297, 298, 300; iii. 63, 64.
- Chalybonian wine, iii. 140.
- Chalonītis, distr. of Assyria, ii. 271; iii. 142.
- Chamæcœtæ, ii. 239.
- Chamanēnē, prefecture of Cappadocia, ii. 278, 284, 285.
- Chanes, r. of Albania, ii. 230.
- Chaones, nation of Epirus, i. 496, 497.
- Charadra, c. of Messenia, ii. 36.
- Charadrūs, fortress of Cilicia (*Charadran*), iii. 52.
- Charax, t. of Corsica, i. 333.
- Charax, place near the Greater Syrtis, iii. 290.
- ——, Chabriou. *See* Chabrias.
- ——, Patrŏclou. *See* Patrŏclus.
- Charaxus, iii. 250.
- Chares, of Lindus, iii. 29.
- ——, r. of Colchis, ii. 229.
- Charilaus, king of Sparta, ii. 204.
- Charimortus, altar of, on the coast of Egypt, iii. 201.
- Charmides, father of Phidias, ii. 29.
- Charmŏlĕo, i. 247.
- Charmŏthas, c. and port of Arabia, iii. 205.
- Charon, of Lampsacus, ii. 340, 350.
- Charondas, ii. 284.
- Charonia, ii. 385. *See* Plutonium.
- Charonitis, ii. 271.

232

- Charonium, sacred cave, iii. 6, 25.
- Charybdis, in the frith of Sicily (*Garafalo*), i. 31, 32, 35, 37, 39, 69, 404, 416.
- Chatramōtītæ, people of Arabia Felix, iii. 190.
- Chatramōtītis, iii. 191.
- Chatti, people of Germany, i. 445-447.
- Chattuarii, people of Germany, 445-447.
- Chaubi, i. 445.
- Chaulotæi, people of Arabia, iii. 189.
- Chazēnē, distr. of Assyria, iii. 142.
- Cheimĕrium, prom. of Epirus, i. 497.
- Cheirocrates (*leg.* Deinocrates), architect, iii. 12.
- Chelidoniæ, islands, near the coast of Pamphylia (*Schelidan Adassi*), ii. 259; iii. 27, 43,47, 48, 61, 62, 68.
- Chĕlōnatas, prom. of Elis (*Cape Tornese*), ii. 5, 9, 15, 22, 167; iii. 291.
- Chĕlōnophagi, iii. 199.
- *Chenab. See* Acesines.
- Cherronesus, t. of Spain (*Peniscola*), i. 239.
- Chersicrates, i. 407.
- Chersiphron, iii. 12.
- [Pg 329]*Cherso and Ossero. See* Absyrtides.
- Chersonesus, c. of the Tauric Chersonese, i. 474-480; ii. 288.
- ———, same as Apamea in Syria, iii. 165.
- ———, port of Lyctus in the isl. of Crete, ii. 199, 200; iii. 294.
- ———, fortress of Egypt, iii. 236.
- ———, prom. and port of Cyrenæa (*Ras-el-Tyn*), iii. 294.
- ———, Thracian, by the Hellespont (*Peninsula of Gallipoli*), i. 140, 506, 517; ii. 171,291, 349, 357, 358.
- ———, Tauric or Scythian, by the Palus Mæotis, i. 474-480; ii. 291; iii. 61.
- ———, Greater, i. 471, 474, 475, 478.
- ———, Smaller, i. 475.
- Chersonitæ, i. 475-480.
- Cherūsci, people of Germany, i. 445-447.
- Chian pottery, i. 487.
- Chiana, i. 349.
- Chians, ii. 396; iii. 19.
- Chieti. *See* Teatea.
- Chiliocōmon, ii. 312.
- Chimæra, monster, iii. 46.
- ———, valley of Lycia, iii. 46, 47.
- ———, mtns of, in Albania. *See* Ceraunian mountains.
- Chimerium, promontory, i. 497.

- Chios, isl. (*Skio*), i. 187; ii. 204, 213, 349, 368, 394; iii. 2, 3, 8, 19.
- ——, wine of, iii. 36.
- *Chiusi. See* Clusium.
- *Chlomos. See* Cnemis.
- Chlōris, mother of Nestor, ii. 20.
- Choaspes, r. of India (*Attock*), iii. 89.
- ——, r. of Persia (*Ab-Zal*), i. 75; iii. 131, 132.
- Chœnicides, ii. 292.
- Chœrilus, poet, i. 465; iii. 55.
- Chōne, c. of Lucania, i. 378, 380.
- Chōnes, inhabitants of Lucania, i. 377, 378, 380.
- Chōnia, iii. 33.
- Chorasmii, people of the Sacæ or Massagetæ, ii. 248.
- Chordiraza, c. of the Mygdones in Mesopotamia (*Racca*), iii. 157.
- Chorene, ii. 250.
- Chorzēne, distr. of Armenia (*Kars*), ii. 269.
- Chrysa, c. of the Troad, ii. 373, 374, 384-386.
- ——, Cilician, ii. 385.
- Chrysaoreōn, Chrysaoric body, in Caria, iii. 39, 40.
- Chryseïs, ii. 343, 384, 385.
- Chryses, ii. 385.
- Chrysippus, Stoic, i. 463; ii. 382; iii. 55.
- *Chryso. See* Crisa.
- Chrysopolis, vill. in Bithynia, ii. 315.
- Chrysorrhoas, r. of Syria, iii. 169.
- *Chun. See* Mallus.
- Chytrium, place near Clazomene, iii. 20.
- Cibotus, port of Alexandria, iii. 230.
- Cibyra, Great, city of Phrygia (*Chorsum*), ii. 409, 410; iii. 27, 45.
- Cibyratæ, ii. 409, 410; iii. 50.
- ——, the Little, in Pamphylia, iii. 50.
- Cibyrātis, Cibyratica, ii. 408, 410; iii. 27.
- Cicero, ii. 166; iii. 40, 234.
- Cichyrus, i. 497; ii. 10.
- Cicōnes, people of Thrace, i. 508, 515, 519.
- Cicynēthus, isl. (*Trikeri*), ii. 140.
- Cicysium, ii. 32.
- Cidēnas, iii. 146.
- Cierus, t. of Thessaly, ii. 138.
- Cilbianum, plain, in Lydia, ii. 407.

- Cilicia, i. 75, 76, 82, 96, 105, 107, 109, 110, 130, 189, 190; ii. 74, 115, 244, 259, 276,278-281, 285, 404; iii. 28, 44, 50-64, 73, 160, 162, 177, 216.
- [Pg 330]Cilicia, Tracheia, ii. 276-278, 281, 285, 322; iii. 44, 45, 50, 54, 68.
- ——, Lyrnessian, ii. 345.
- ——, Pedias, iii. 50.
- ——, sea of, i. 129; ii. 218, 281.
- Cilician Gates, ii. 281, 283; iii. 53, 61.
- Cilicians, i. 196; ii. 197, 216, 345, 322, 327, 329; iii. 1, 50-64.
- ——, in the Troad, ii. 375, 383, 385, 389, 394, 395; iii. 49, 63.
- Cilla, t. of the Troad, ii. 384, 385.
- Cillæum, mtn of the Troad, ii. 384.
- ——, mtn of Lesbos, ii. 384.
- Cillæan Apollo, ii. 384, 385.
- Cillanian plain, in Phrygia, ii. 407.
- Cillus, r. near Cilla, ii. 385.
- ——, charioteer of Pelops, ii. 385.
- Cimarus, prom. of Crete, ii. 193, 195.
- Cimbri, nation of Germany, i. 154, 288, 292, 319, 445, 448-451.
- Cimiata, fortress of Paphlagonia, ii. 314.
- Cimiatēnē, distr. of Paphlagonia, ii. 314.
- Ciminius, lake, in Etruria (*Lago di Vico or di Ronciglione*), i. 336.
- Cimmerian Bosporus. *See* Bosporus.
- ——, village, ii. 222.
- Cimmerians, Cimmerii, Kimmerii, i. 8, 31, 96, 223, 224, 363, 364, 476; ii. 221, 246,301, 329, 405.
- ——, Cimbri Cimmerii, i. 450.
- Cimmericum, city of the Cimmerian Bosporus, ii. 221.
- Cimmeris, i. 459.
- Cimmerium, hill in the Tauric Chersonesus, i. 476.
- Cimōlus, isl. *Argentière*, ii. 207, 208.
- Cindya, vill. of Caria, iii. 37.
- Cindyas Artemis, iii. 37.
- Cineas, historian, i. 503.
- Cingulum, Mount, i. 337.
- *Cinifo*, r. *See* Cinyps.
- Cinnamon country, i. 99, 111, 115, 144, 171, 179-181, 199, 200.
- Cinōlis, t. of Paphlagonia (*Kinoli*), ii. 291.
- Cinōlis, Anti, ii. 291.
- Cinyras, tyrant of Byblus, i. 63; iii. 170.
- Circæum, prom. of Latium (*Monte Circello*), i. 35, 344, 346.
- Circe, i. 31, 69, 70, 73, 332, 346; ii. 85.

- *Circello, Monte. See* Circæum.
- *Cirella. See* Cerilli.
- Cirphis, t. of Phocis, ii. 114.
- ——, mtn of Phocis, ii. 116.
- Cirra, t. of Phocis, ii. 114, 116.
- Cirrha, ii. 77.
- Cirta, c. of Numidia (*Constantine*), iii. 280, 285.
- Cisamus, t. of Crete (*Kisamos*), ii. 200.
- Cispadana, i. 316, 321, 322, 323.
- Cisseus, i. 509, 510.
- Cissia, mother of Memnon, iii. 130.
- Cissii, same as Susii, iii. 130.
- Cissus, father of Althæmenes, ii. 77, 203.
- Cissus, t. of Macedonia, i. 509, 510.
- Cisthēnē, t. of Mysia, ii. 376.
- ——, isl. and t. near Lycia, iii. 47.
- Cithærōn, i. 40; ii. 62, 82, 97, 99, 103, 107, 108.
- Citium, c. of Cyprus, i. 24; ii. 382; iii. 69.
- Citrum, t. of Macedonia, i. 509.
- Cius, friend of Hercules, ii. 315.
- ——, c. of Bithynia, ii. 314.
- *Civita Lavinia. See* Lanuvium.
- Cizari, citadel of Phazemonitis, ii. 311.
- Clanis, r. in the Norican Alps, i. 308.
- ——, r. of Latium, i. 347.
- Clarus, c. of Ionia, iii. 15, 50.
- Clastidium, t. of Cisalpine Gaul (*Castezzio*), i. 323.
- Claterna (*Quaderna*), i. 322.
- Clautinatii, people of the Vindelici, i. 307.
- Clazomenæ, c. of Ionia (*Kelisman*), i. 91; iii. 3, 20.
- Clazomenians, i. 517; ii. 221; iii. 17.
- [Pg 331]Cleanactidæ, tyrants of Mitylene, ii. 391.
- Cleandria, t. of the Troad, ii. 371.
- Cleandridas, leader of the Thurii, i. 398.
- Cleanthes, Stoic, ii. 382.
- ——, painter, ii. 16.
- Cleides, islands, iii. 68-70.
- Cleitor, t. of Arcadia, ii. 75.
- Cleobūlus, iii. 33.
- Cleomachus, iii. 23.
- Cleombrotus, founder of Heræa, ii. 8.

- Cleōn, ii. 330.
- Cleōnæ, t. on Mt Athos, i. 512, 513.
- ——, city of Argolis, ii. 59, 60, 66.
- Cleōnæi, ii. 60.
- Cleonymus, i. 427; ii. 8.
- Cleopatra, daughter of Auletes, i. 440, 499; iii. 52-56, 71, 72, 231-234, 281.
- ——, wife of Euergetes II., i. 149, 150.
- ——, Selene, iii. 161.
- Cleopatris, t. of Egypt, iii. 210, 243, 244.
- Cleophanes, rhetorician, ii. 318.
- Cleuas, leader of the Æolians, ii. 340.
- Climax, mtn of Lycia, iii. 48.
- ——, mtn of Cœle-Syria, iii. 170.
- Clitarchus, i. 332, 449; ii. 217, 237; iii. 117.
- Clusium (Chiusi), city of Etruria, i. 327, 336, 349.
- Clymĕnē, i. 52.
- Clypea, city of the Carthaginians. See Aspis.
- Clytemnestra, i. 25.
- Cnemīdes, t. of the Locrians, ii. 126.
- Cnemis, mtn in Locris (Chlomos), ii. 114, 125.
- Cnidian wine, iii. 8.
- ——, territory, ii. 213.
- Cnidii, Cnidians, i. 417, 484; iii. 30.
- Cnidus, city of Caria (Crio), i. 180, 187; iii. 8, 31, 34, 227, 247.
- Cnōpia, Thebaïc, vill. of Bœotia, ii. 96.
- Cnōpus, son of Codrus, iii. 2.
- Cnossus, city of Crete (Makro Teichos), i. 430; ii. 195-197, 200, 202.
- Cnuphis, god of the Egyptians, iii. 263.
- Coa, same as Cos.
- Coans, iii. 31, 36.
- Cōbialus, vill. of Paphlagonia, ii. 291.
- Cōbus, of Trerus, i. 96.
- Cōcalus, i. 413, 425.
- Coccēius, i. 364.
- Coccēs. See Ptolemy.
- Cōdridæ, ii. 68.
- Cōdrus, i. 493; ii. 68, 82; iii. 2, 3, 30.
- Cœle-Syria, i. 201.
- ——, -Elis, ii. 7, 8.
- Cœlius, Roman historian, i. 343.
- Cœus, ii. 208.

239

- Corinthians, i. 486, 511; ii. 49, 63, 64, 78, 82, 111.
- Coriscus, ii. 378.
- Cornelius Gallus, prefect of Egypt, iii. 267.
- Corœbus, ii. 30.
- [Pg 333]Corocondamē, t. of the Cimmerian Bosporus (*Taman*), ii. 222, 225.
- Corocondamitis, lake, ii. 222.
- Corōne, city of Messenia, ii. 37.
- Corōneia, city of Bœotia, ii. 101, 107, 108, 111, 136.
- ——, city of Thessaly, ii. 136.
- ——, Messenian, ii. 108, 136.
- Corōnii, Coronenses, ii. 108.
- Coropassus, t. of Lycaonia, ii. 322; iii. 43.
- Corpīli, people of Thrace, i. 516.
- Corpilice, in the Hellespont, i. 519.
- *Corsica. See* Cyrnus.
- Corsiæ, the Furni Islands, ii. 212; iii. 7.
- Corsūra, island (*Pantalaria*), iii. 287.
- Corus, r. of Iberia, same as Cyrus, ii. 230.
- Corybantes, i. 516; ii. 180, 184, 188, 191.
- Corybantium, ii. 190.
- Cŏrybissa, near Scepsis, ii. 190.
- Corybus, ii. 188.
- Cŏrycæans, pirates of Ionia, iii. 18.
- Cŏrycian cave, ii. 405; iii. 54.
- Corycium, ii. 115.
- Cŏrycus, mtn and prom. of Ionia, iii. 17, 18.
- ——, prom. of Crete (*Grabusa*), ii. 41; iii. 294.
- ——, prom. of Cilicia, iii. 54, 70.
- ——, coast of Lycia, iii. 48.
- ——, city of Lycia, iii. 49, 55.
- Corydalleis, ii. 85.
- Corydallus, mtn of Attica (*San Giorgio*), ii. 85, 90.
- Coryphantis, t. of the Mitylenæans, ii. 376.
- Coryphasium, mtn and prom. of Messenia (*Mount St. Nicolas*), ii. 11, 21, 22, 26, 28,35.
- ——, t. of Messenia, ii. 211.
- Cos, island (*Stanko*), i. 187, 519; ii. 212-214; iii. 8, 30, 36, 74, 94.
- ——, city, ii. 56, 328; iii. 74, 94.
- Cŏssa, r. of Latium, i. 352.
- Coscinia, t. of Caria (*Arpas-Kalessi*), iii. 26.
- Coscinii, ii. 347.

- Cosentia (*Cosenza*), i. 382.
- Cossa, Cossæ, city of Etruria, i. 330, 334, 335.
- Cossæa, distr. of Asia, iii. 153.
- Cossæan mtns, iii. 150.
- Cossæi, ii. 261, 264; iii. 148, 153.
- Cossūra, island and town (*Pantalaria*), i. 185, 421; iii. 288.
- Cŏteis, prom. of Mauritania (*Cape Espartel*), iii. 276, 279.
- Cŏthŏn, island and port of the Carthaginians, iii. 285, 286.
- Cŏthus, i. 493; ii. 152, 154.
- Cŏtiaeium, t. of Phrygia (*Kiutaha*), ii. 332.
- Cotinæ, t. of Bætica, i. 214.
- Cottius, country of, in the Alps, i. 268, 303, 323.
- Cŏtuantii, i. 307.
- Cotyliæ, waters at, i. 338.
- Cŏtylus, summit of Mt. Ida, ii. 369.
- Cŏtys, prince of the Sapæi, ii. 305.
- ——, king of the Odrysæ, i. 516.
- ——, goddess of the Edoni, ii. 187, 189.
- Cotytia, rites of, ii. 186.
- Cragus, c. of Lycia, iii. 46.
- ——, mtn and prom. of Lycia, iii. 46.
- ——, rocks of Cilicia, iii. 52.
- Crambūsa, t. of Lycia (*Garabusa*), iii. 48.
- ——, isl. of Cilicia, iii. 54.
- Cranaë, island, ii. 90.
- Cranaï, ii. 87.
- Cranaüs, king of the Athenians, ii. 87.
- Cranes, battles of the, i. 109.
- Cranii, t. of Cephallenia, ii. 166, 167.
- Crannŏn, t. of Thessaly, i. 507; ii. 146, 147.
- Crannonii, i. 507; ii. 10, 147.
- Crapathus or Carpathus, ii. 212-214.
- Crassus, Publius, i. 263; iii. 21.
- ——, triumvir, iii. 157, 159.
- Crater (*Bay of Naples*), i. 360, 369.
- Craterus, iii. 96, 121, 127.
- [Pg 334]Crates, the miner of Chalcis, ii. 101.
- ——, of Mallos, i. 4, 6, 48, 49, 57, 60, 155, 156, 176, 237; ii. 143, 380; iii. 60.
- Crāthis, r. of Achæa, ii. 72.
- ——, r. of Italy (*Crati*), i. 396; ii. 72, 157.
- Cratippus, iii. 25.

243

- Cyanæan rocks, same as the Symplēgades, i. 32, 137, 138, 224, 490, 491, 518; ii. 292.
- Cyané, lake. *See* Mantianē.
- Cyaxares, king of the Medes, iii. 239.
- Cybĕbe, same as Cybĕle.
- Cybĕla, mtn of Phrygia, ii. 321.
- Cybĕle, or Cybĕbe, name of Rhea, ii. 184-186, 321.
- Cybĕlia, t. of Ionia, iii. 18.
- Cybiosactes, king of the Egyptians, iii. 232.
- Cybistra, t. of Cataonia (*Eregli*), ii. 278, 281, 284.
- Cybrene, ii. 360.
- Cychreia, same as Salamis, ii. 82.
- Cychreus, ii. 83.
- Cychrides, serpent, ii. 83.
- Cyclades, islands, i. 90, 187; ii. 47, 192, 207-214; iii. 7.
- Cyclopæ, Cyclops, i. 31-33, 64; ii. 54, 354.
- Cyclopean mode of life, ii. 233.
- Cyclopeia, ii. 48.
- Cycnus, king of the Colonæ, ii. 64, 350, 373.
- Cydippe, wife of Cercaphus, iii. 32.
- Cydnus, r. of Cilicia (*Kara-sui*), i. 75; iii. 56, 57, 59.
- Cydonia, city of Crete, ii. 58, 195, 198, 200.
- Cydonians, people of Crete, i. 328; ii. 195.
- Cydoniatæ, ii. 199, 200.
- Cydrēlus, son of Codrus, iii. 2.
- Cydriæ, t. of Epirus, i. 501.
- Cyinda, fortress of Cilicia, iii. 55.
- Cyllēnē, city of Elis, ii. 9, 13.
- ——, mtn of Arcadia, ii. 75, 76.
- Cynætha, t. of Arcadia, ii. 75.
- Cynamolgi, people of Ethiopia, iii. 196.
- Cynia, lake, in Ætolia, ii. 171.
- Cynocephali, people of Ethiopia, i. 68, 458; iii. 200.
- ——, in Thessaly, ii. 146.
- Cynōpolis, city of Egypt, iii. 240, 257.
- Cynopolite nome, iii. 257.
- Cynossema, ii. 357; iii. 34, 236.
- Cynthus, Cythnus, mtn of Delos, ii. 208.
- Cynthus (*Thermia*), ii. 207.
- Cynūria, distr. of Argolis, ii. 51, 58.
- Cȳnus, t. and prom. of Locris (*Kyno*), i. 95; ii. 125, 126, 153, 388.

- Cythnus, island, ii. 208.
- Cytinium, t. of Locris, ii. 128, 195.
- Cytōrum, t. of Paphlagonia, ii. 288, 291.
- Cytōrus, t. of Pontus, ii. 296.
- ——, son of Phrixus, ii. 291.
- Cyzicene, ii. 317, 338-341, 347.
- Cyziceni, i. 189; ii. 299, 331, 332, 340, 341, 347, 349; iii. 5.
- Cyzicus, island and city (*Artaki*), i. 71, 152, 189, 518; ii. 316, 330-332, 346, 348, 349,402; iii. 5, 30, 34, 67.
- Daci, Dacians, Daæ, i. 309, 317, 467, 468, 481.
- Dactyli, Idæan, ii. 30, 180, 191.
- Dædala, t. of Caria, iii. 28, 45, 46.
- ——, mtn of Lycia, iii. 45, 46.
- Dædalus, father of Iapyx, i. 425; ii. 197 ; iii. 10.
- Daēs, of Colonæ, ii. 384.
- Dahæ, ii. 241, 245, 257.
- Daisitiatæ, nation of Hungary, i. 483.
- *Dalian. See* Caunus.
- Daliōn, r. of Triphylia, ii. 17.
- Dalmatæ, Dalmatians, i. 484, 487.
- Dalmatia, Dalmatice, i. 483, 484.
- Dalmatium, city of the Dalmatæ, i. 484.
- Damascus, city of Syria, iii. 169-171.
- Damasia, t. of the Licattii, i. 307.
- Damastes, historian, i. 74, 75; ii. 340; iii. 70, 71.
- Damastium, in Epirus, i. 500.
- Damasus the Athenian, iii. 2.
- ——, Scombrus, iii. 25.
- Damnamenus, one of the Idæan Dactyli, ii. 191.
- Danaë, mother of Perseus, ii. 211.
- ——, play of Æschylus, i. 329.
- Danai, i. 329; ii. 49, 52, 133, 329.
- Danaïdes, ii. 52, 335; iii. 33.
- Danala, fortress of the Galatæ, ii. 320.
- Danaus, i. 35, 329, 493; ii. 52, 53; iii. 51.
- Dandarii, ii. 223, 224.
- Danthēlētæ, people of Thrace, i. 489.
- *Danube. See* Ister.
- [Pg 337]Daorizi, a nation of Dalmatia, i. 484.
- Daphitas, the grammarian, iii. 22.
- Daphne, city of Syria (*Beit-el-mà*), iii. 118, 162.

- Daphnus, t. of Phocis or Locris, (?) i. 95; ii. 114, 124-126.
- ——, port of Ethiopia, iii. 200.
- Darada, city of, iii. 197.
- Darapsa, city of Bactriana, ii. 253.
- *Dardanelles, Strait of the. See* Hellespont.
- Dardani, Dardanii, Dardanians, i. 485, 489; ii. 77, 162, 353, 375; iii. 41.
- Dardania, distr. of the Troad, i. 481, 516; ii. 317, 353, 354, 360, 369, 371, 375.
- Dardaniatæ, Dardanii, a people of Illyria, i. 485, 505; ii. 3.
- Dardanica, a region of Illyria, i. 485.
- Dardanis, Dardanian prom. ii. 357.
- Dardanium, i. 347.
- Dardanus, t. of the Troad, ii. 347, 352, 357, 366.
- ——, brother of Jasion, i. 516; ii. 19, 353, 354.
- Darieces, name of Darius, iii. 216.
- Darius, i. 148, 152, 462, 463, 465, 468, 469; ii. 347; iii. 60, 89, 133, 134, 141, 144,188, 216, 244, 259.
- ——, son of Hystaspes, iii. 5, 9, 163.
- ——, father of Xerxes, ii. 352.
- ——, conquered by Alexander, ii. 291.
- ——, Longimanus, iii. 140.
- Dasarētii, a people of Illyria, i. 485, 489.
- Dascylītis, lake of Mysia (*Jaskili*), ii. 329-332, 346.
- Dascylium, t. of Mysia, ii. 331, 340.
- *Daskalio. See* Asteria.
- Dasmenda, fortress of Cappadocia, ii. 284.
- Dastarcum, a fortress of Cataonia, ii. 280.
- Dasteira, city of Armenia, ii. 305.
- Dateni, people of Macedonia, i. 513.
- Datis, ii. 90.
- Datum, city of Thrace, i. 512-514.
- Daulia, Daulis, city of Phocis, ii. 114, 122.
- Daulieis, i. 493; ii. 123.
- Daulius, king of Crissa, i. 399.
- Daunia, *see* Apulia, i. 425, 434; iii. 32.
- Daunii, i. 320, 360, 422, 428, 431-433, 436.
- Davi, i. 467.
- Dazimonitis, distr. of Pontus (*Kas Owa*), ii. 295.
- Debæ, people of Arabia, iii. 206.
- Dēcæneus, Getæan bard, i. 457, 467; iii. 180.
- Deceleia, t. of Attica, ii. 88.
- Deciētæ, a people of the Ligurians, i. 301.

- ——, of Skepsis, i. 71, 74, 90, 502, 513, 518; ii. 10, 11, 17, 56, 142, 143, 168, 189,190, 298-300, 355, 360, 364, 375, 377, 380, 383, 404, 405; iii. 66.
- ——, son of Euthydemus, ii. 253.
- Demi, ii. 90.
- Dēmocles, historian, i. 91.
- Dēmŏcŏŏn, son of Priam, ii. 344.
- Dēmocritus, i. 95, 102, 103; iii. 98.
- Demosthenes, i. 182; ii. 55, 56, 123, 152, 188.
- Dēmus, i. 460; ii. 374.
- *Denia. See* Dianium.
- *Deras, Cape. See* Derhis.
- Derbe, t. of Lycaonia, ii. 278, 322; iii. 64.
- Derbices, people of Margiana, i. 249, 258.
- *Dercĕto. See* Atargatis.
- Derdæ, iii. 101.
- *Derekoi. See* Myus.
- Derhis, a port of Marmara (*Deras*), iii. 236.
- Derrhis, prom. of Macedonia, i. 511, 512.
- Derthon (*Tortona*), i. 323.
- Dertōssa, t. of Spain (*Tortosa*), i. 239, 241.
- *Descura. See* Sitacene.
- Deucalion, king of Thessaly, i. 494; ii. 67, 125, 134, 139, 140, 149.
- ——, island, ii. 139.
- Deudorix, the Sicambrian, i. 446.
- Deuriŏpes, people of Macedonia, i. 501.
- Deuriŏpus, district of Macedonia, i. 500.
- Dexia, or the right of Pontus, ii. 286.
- *Dhiles. See* Delos.
- Dïa, temple of, at Sicyon, ii. 66.
- ——, isl. near Crete (*Standia*), ii. 207.
- ——, in the Arabian Gulf, iii. 205.
- Diacŏpēne, district of Pontus, ii. 312.
- Diades. *See* Athenæ Diades.
- Diagesbes, people of Sardinia, i. 333.
- *Diakopton. See* Bura.
- Diana (Artemis), i. 270, 385; ii. 16, 73, 208, 348; iii. 146, 153, 162.
- ——, of Ephesus, i. 268, 269.
- ——, of Ephesus, temple of (the Ephesium), i. 238-240, 268, 275; iii. 11.
- ——, Brauronia, ii. 90.
- ——, Perasia, ii. 281.

249

250

251

- Doria Riparia, r. *See* Durias.
- Dorian Tetrapolis, ii. 195.
- Dorians, i. 96, 328, 404, 407;
ii. 2, 3, 43, 58, 67, 81, 82, 114, 115, 125, 128, 131, 147; iii. 30, 31, 40, 43.
- Doric dialect, ii. 2, 3.
- Doricha, courtesan, iii. 250.
- Dōris, at Parnassus, ii. 55, 136.
- Doris, or Histiæotis, in Thessaly, ii. 141, 195.
- Doriscus, t. of Thrace, i. 5, 6.
- Dorium, in Messenia, ii. 23, 24.
- Dōrus, son of Hellen, ii. 67, 340.
- Dorylæum, t. of Phrygia (*Eski-Schehr*), ii. 332.
- Dorylaüs, the tactician, great-great-grandfather of Strabo, ii. 198, 307.
- ———, son of Philetærus, ii. 198, 307.
- Dosci, a Mæotic race, ii. 223.
- Dotium, c. and plain of Thessaly, i. 96; ii. 147.
- *Doubs*, r. of Gaul, i. 278, 281, 286. *See* Dubis.
- *Douro*, r. *See* Durius.
- Drabēscus, t. of Macedonia, i. 512.
- Drabus, t. of Thrace, i. 517.
- [Pg 341]Drabus (*Drave*), r. of Noricus and Hungary, i. 483.
- Dracanum, t. of Icaria, iii. 36.
- Draco, companion of Ulysses, tomb of, i. 376.
- ———, Python, ii. 120.
- *Dragomestre. See* Crithote.
- *Dragone, Monte. See* Sinuessa.
- *Dramesi. See* Delium.
- Drangæ, people of Ariana, iii. 122, 124-126.
- Drangē, Drangianē, district of Ariana, (*Sigistan*), ii. 249; iii. 142.
- *Drave*, r. *See* Drabus.
- Drecanum, in the island of Cos, iii. 10, 36.
- Drepanum, prom. of Achaia, ii. 6.
- ———, of Icaria, iii. 8, 10.
- ———, of Marmara, iii. 235.
- Drilon, r. of Dalmatia (*Drin*), i. 485.
- Drium, hill in Daunia, i. 434.
- ———, c. of Macedonia, i. 509.
- Dromi, iii. 245.
- Dromichætes, king of the Getæ, i. 464, 469.
- Dromos, iii. 245, 248.
- Druentia, i. 268.

- Echedōrus, r. of Macedonia (*Gallico*), i. 509.
- Echeiæ, t. of Laconia, ii. 37.
- Echinades, islands (*Curzolari*), i. 93, 187; ii. 5, 12, 25, 162, 167, 169-171.
- Echīnus, t. of Phthiotis (*Echino*), i. 94; ii. 136, 138, 147.
- *Ecija. See* Astygis.
- Ecrēgma (mouth of the lake Sirbonis), i. 102; iii. 176.
- Edessa, city of Macedonia (*Vodina*), i. 495; ii. 157.
- [Pg 342]Edessa, city of Syria, iii. 158. *See* Bambyce.
- Edōtani, people of Spain, i. 234, 235, 243. *See* Sidētani.
- Edōni, people of Thrace, i. 506, 514.
- *Edschise-Dagh. See* Argæus.
- Eētiōn, king of Thebes, ii. 343, 384, 394.
- Egelastæ, t. of Spain (*Yniesta*), i. 241.
- Egĕria, ftn, i. 356.
- Egertius, founder of Chios, iii. 3.
- Egnatia, city and port of Apulia (*Torre d' Agnazzo*), i. 431, 432.
- Egnatian Way, i. 495, 500, 506, 507, 509.
- Egra, city of Arabia, iii. 212.
- *Egripo. See* Eubœa.
- Egypt, i. 8, 15, 25, 46, 47, 49, 50, 52, 55-64, 67, 68, 79, 87, 88, 90, 91, 103, 129,130, 134, 136, 143, 149, 150, 178, 183, 189, 197, 198, 201, 262, 274, 458, 467,493; ii. 89, 92, 280; iii. 51, 67, 74, 81-84, 88, 90, 95, 102, 103, 190, 210, 211, 217-270, 272, 273, 275, 292-294.
- ——, name of the Nile, i. 46, 56.
- ——, Lower, i. 47, 103, 316; iii. 177.
- Egyptian screws, i. 221.
- ——, exiles, island of the, i. 179.
- ——, Sea, same as Mediterranean, i. 56, 91, 185, 189, 458; iii. 68, 142, 160, 224, 228,266.
- Egyptians, i. 41, 49, 63-65, 155, 197, 233, 440, 456, 463; ii. 304, 308.
- ——, priests of the, i. 35, 96, 154, 180, 196.
- ——, island of the, i. 99.
- Eidomene, t. of Macedonia. *See* Idomene.
- Eileithyia, city of Egypt, iii. 263.
- Eilesium, ii. 106.
- Eïones, vill. of Argolis, ii. 54, 55, 58.
- Eisadici, ii. 239.
- *Eksemil. See* Lysimachia.
- *Eksenide. See* Xanthus.
- *El-Aliah. See* Acholla.
- *El-Arish. See* Rhinocolura.
- *El-Asi. See* Orontes.

256

- Elimiotæ, ii. 137.
- Elis, i. 502; ii. 5, 7, 8-10, 12-15, 17-19, 25, 27, 28, 31-33, 45, 73, 77, 122, 126, 156,162, 167, 169, 170, 176, 177.
- ——, Cœlē, *or* Hollow, ii. 7-9, 12, 18, 23, 25, 30.
- Elisa, modern name of Eleüssa.
- Elisson, or Elissa, r., ii. 9.
- Elixus, ii. 210.
- Ellopia, ii. 152, 153.
- Ellopians, ii. 152, 153.
- Ellops, ii. 152.
- Elōne, t. of Thessaly, ii. 143, 145.
- Elpiæ, city of the Daunii, iii. 32.
- Elui, people of Gaul (*inhabitants of Vivarais*), i. 284.
- Elymæa, Elymaïs, district of Persis, ii. 264; iii. 153, 154.
- Elymæi, ii. 261, 264; iii. 135, 142, 146.
- Elymus, Trojan, ii. 378.
- Elysian Fields, in Spain, i. 3, 62, 225.
- Emathia, district of Macedonia, i. 41, 506.
- ——, city of Macedonia, i. 506.
- Emathoeis, Emathois, same as Pylus, ii. 7, 11, 16.
- *Emboli. See* Amphipolis.
- *Embrun. See* Ebrodunum.
- Emerita. *See* Augusta.
- Emesēni, people of Syria, iii. 166.
- Emōdi mtns, ii. 245; iii. 91, 118.
- Emodus, iii. 78.
- Empedocles, philosopher, i. 414, 418; ii. 42.
- Emporicus, bay, on the Mauritanian shore, iii. 276, 277.
- Emporītæ, in Spain, i. 240.
- Emporium, t. of Spain (*Ampurias*), i. 239.
- ——, of Alexandria, iii. 230.
- ——, of Medma, i. 383.
- ——, of the Segestani (*Castel à Mare*), i. 401, 411.
- *Ems. See* Amasias.
- Enchelii, people of Epirus, i. 500.
- Endĕra, city of Ethiopia, iii. 196.
- Endymiōn, father of Ætolus, ii. 176; iii. 6.
- Enea (*see* Ænea), t. of the Troad, ii. 300.
- Eneta. *See* Heneta.
- Eneti, people of Paphlagonia, i. 316. *See* Heneti.
- *Engia, Gulf of. See* Saronic Sea.

- Enicŏniæ, t. of Corsica, i. 333.
- Enienes, ii. 145.
- Enipeus, r. of Pisatis, ii. 32.
- ——, r. of Thessaly (*Vlacho*), ii. 32, 134.
- Enispe, t. of Arcadia, ii. 75.
- Enna, t. of Sicily (*Castro Johanni*), i. 411, 413.
- Ennea-Hodoi, t. of Macedonia, i. 513.
- Ennius, the poet, i. 429.
- Enŏpe, t. of Messenia, ii. 35, 37.
- Enops, ii. 394.
- Enotocoitæ, iii. 107.
- Enydra, t. of Syria (*Ain-el-Hiyeh*), iii. 167.
- Enyus (Bellona), temple of, ii. 279.
- Eordi, people of Macedonia, i. 495, 500.
- *Eoube. See* Olbia.
- Epacria, t. of Attica, ii. 88.
- Epaminondas, ii. 75, 92, 111.
- Epaphus, ii. 152.
- [Pg 344]Epeius, i. 397; ii. 122.
- Ephesians, ii. 284; iii. 3, 10.
- Ephesium, the, *See* Diana.
- Ephesus, city of Ionia, i. 268; ii. 73, 237, 298, 299, 333, 396; iii. 1-

4, 10, 11, 14, 15,21, 22, 43.
- Ephialtes, traitor, i. 17.
- ——, i. 29.
- Ephorus,

i. 1, 51, 52, 207, 296, 328, 329, 363, 388, 390, 394, 399, 402, 403, 407, 425,449, 464, 465, 469, 499, 501;
ii. 1, 3, 33, 38, 42, 44, 55, 58, 77, 92, 93, 120, 127,162, 174, 176, 177, 196, 197, 200, 201, 204, 291, 298, 299, 341, 366, 398, 399; iii.4, 62-65.
- Ephyra, t. of Elis, i. 502; ii. 9, 10, 52.
- ——, t. of Epirus, i. 497.
- ——, t. of Thesprotia, i. 502; ii. 9, 10, 149.
- ——, t. of Thessaly, ii. 9, 10.
- ——, t. of Perrhæbia, ii. 10.
- ——, vill. of Ætolia, ii. 9, 10. *See* Corinth, Crannŏn.
- Ephyri, i. 507; ii. 10, 147.
- ——, Thesprotic, ii. 10.
- Epicarus. *See* Epidaurus.
- Epicharmus, poet, ii. 42.
- Epicnemidii. *See* Locri.
- Epicteti, in Phrygia, ii. 314, 330, 402.

258

- Epictetus, Ætolia, ii. 159, 172.
- ——, Phrygia, ii. 277, 289, 315, 316, 332.
- Epicurus, philosopher, ii. 350; iii. 9.
- Epidamnus, city of Illyria (*Durazzo*), i. 140, 161, 432, 485, 495, 500; ii. 33, 134.
- Epidanus (*the Jura*), ii. 134.
- Epidaphne (*Antakieh*), iii. 161.
- Epidaurian territory, ii. 47, 75.
- Epidaurii, Epidaurians, ii. 58.
- Epidaurus, city of Argolis (*Pidauro*), ii. 54-57, 321.
- ——, Limēra, t. of Laconia, ii. 48.
- Epigoni, expedition of the, i. 499; ii. 93, 109, 111, 174.
- Epii, people of Elis, ii. 7-10, 12, 13, 15, 18, 25-28, 30, 33, 167, 176.
- Epimenides, of Crete, ii. 200 Epirōtæ;, i. 493, 495, 498-500, 506; ii. 2, 128, 131, 137; iii. 297.
- Epirotic nations, i. 495, 496, 516; ii. 2, 3, 114, 131.
- Epirus, i. 186, 187, 194, 329, 429, 432, 481, 496, 497, 501, 506; ii. 161, 163, 164,174, 183; iii. 297.
- Episarosis, a religious rite, ii. 82.
- Epistrophus, leader of the Halizoni, ii. 297-299, 343, 384, 394.
- Epitalium, t. of Triphylia, ii. 16, 23, 24.
- Epitimæus. *See* Timæus.
- Epizephyrii, ii. 128.
- Epōmeus, Mount, i. 369.
- Eporĕdia (*Ivrea*), i. 306.
- Eræ, t. of Ionia (*Sighadschik*), iii. 17.
- Erana, t. of Messenia, ii. 22, 37.
- Erannoboas (*Hiranjavahu*), iii. 97.
- Erasīnus, Arsīnus, r. of Argolis, i. 416; ii. 52, 76.
- Erasistratus, physician of Ceos, ii. 210, 337.
- Erastus, the Scepsian, ii. 378.
- Eratosthenes, i. 1, 9, 12, 13, 22-26, 28, 29, 33-36, 38, 39, 42, 43, 45, 61, 70, 74, 77,80, 84-88, 97, 98, 100, 103-110, 114, 117-120, 122-135, 138-142, 144, 147, 157,158, 161, 163, 164, 171, 185, 189, 190, 199, 202, 203, 239, 255, 256, 332, 4 57,460, 462, 487; ii. 70, 76, 195, 240, 243, 244, 248, 261, 271; iii. 44, 70, 71, 75, 78,79, 84, 124, 130, 149-151, 156, 183, 186, 188, 189, 192, 208, 220, 276, 281, 293.
- Eratyra, t. of Macedonia, i. 50.
- *Erbil. See* Arbela *and* Lycus.
- Erechtheus, ii. 67.
- *Eregli. See* Cibistra.
- *Erekli. See* Heracleia.
- Erembi, i. 2, 41, 46, 60, 66, 67; iii. 215.

- Eremni, iii. 216.
- [Pg 345]Eressus, t. of Lesbos (*Eresso*), ii. 392.
- Eretria, city of Eubœa (*Vathy*), i. 65; ii. 95, 152, 154-156, 162.
- ——, t. of Thessaly, ii. 136, 154.
- ——, vill. of Attica (*Paleocastro*), ii. 95, 152, 154.
- Eretrici, a sect of philosophers, ii. 82, 156.
- Eretrieis, Eretrians, i. 368; ii. 152, 155.
- Eretrieus, founder of Eretria, ii. 155.
- Erētum, t. of the Sabines, i. 338, 339, 354.
- Erginus, king of the Orchomenii, ii. 112.
- Ericthonius, ii. 374.
- Ericūssa, Ericōdes, one of the Æolian islands, i. 419, 421.
- Eridanus, r. of Attica, i. 320; ii. 88.
- Erigōn, r. of Macedonia, i. 501, 506, 508, 509.
- ——, r. of Thrace, i. 516.
- Erineum, ii. 39, 195.
- Erineus, t. of Doris, ii. 128, 361, 363.
- ——, of Phthiotis, ii. 136.
- Erōs, a work of Praxiteles, ii. 105.
- Erymanthus, r. of Arcadia, ii. 15, 33.
- Erymnæ, t. of Magnesia, ii. 148.
- Eryschæi, people of Ætolia, ii. 172.
- Erythia, isl., i. 222, 223, 254, 406.
- Erythīni, Erythrīni rocks, ii. 288, 291.
- Erythræ, t. of Bœotia, ii. 97, 104, 106.
- ——, of Ionia (*Ritri*), ii. 97, 259; iii. 2, 17, 18.
- Erythræan Sea (*Red Sea*), i. 52, 68, 87, 88, 91, 102, 261; iii. 186.
- ——, Gulf, i. 87.
- Erythræans, ii. 349, 350, 386; iii. 17, 18.
- Erythras, iii. 187, 208.
- Eryx, mtn of Sicily, i. 378, 412, 413, 378.
- ——, t. of Sicily, i. 412; ii. 378.
- *Esaro. See* Æsar.
- *Esdod. See* Gadaris.
- *Eshinoon. See* Hermopolis.
- *Esino. See* Æsis.
- Esioneis, Asioneis, ii. 405.
- *Eski-Hissar. See* Stratoniceis and Laodicea.
- *Eski-Scheur. See* Dorylæum.
- *Eski-Stamboul. See* Alexandria Troad, in the.
- Esōpis, mtn (*Monte Esope*), i. 389.

- Euergetes. *See* Mithridates, Ptolemy.
- *Eugubbio. See* Iguvium.
- Euhēmerus, Messenian, i. 74, 154, 157, 158, 459.
- Eulæus, r. of Susiana, iii. 131, 140.
- Eumæus, ii. 364.
- Eumēdes, founder of Ptolemaïs, iii. 194.
- Eumēlus, son of Admētus, i. 72; ii. 139, 143, 146, 148.
- Eumeneia, city of Phrygia (*Ischekli*), ii. 332.
- Eumenes, brother of Philetærus, ii. 400.
- ——, son of Eumenes, ii. 400.
- ——, son of Attalus, ii. 281, 333, 400; iii. 46, 55.
- ——, grove of, iii. 197.
- ——, harbour of, iii. 198.
- Eumolpus, Thracian, i. 493; ii. 67, 187.
- Eunēos, son of Jason, i. 66, 71, 73.
- Eunomia, elegy of Tyrtæus, ii. 39.
- Eunomus, i. 390, 391.
- Eunostus, harbour of, near Alexandria, iii. 227, 230.
- Eunus, i. 412, 413.
- Euōnymus, one of the Lipari islands, i. 420.
- Eupalium, ii. 128, 159.
- Eupator. *See* Mithridates.
- Eupatŏria, t. of Pontus, *see* Magnopolis, ii. 306.
- Eupatŏrium, t. of the Tauric Chersonnesus, i. 479.
- Euphŏriōn, poet, ii. 42, 318; iii. 67.
- Euphrantas, tower, iii. 290.
- Euphrates (*the Forat, Ferat, or Frat*), i. 75, 100, 101, 122-
124, 126, 127, 129, 134,135, 137, 196, 440; ii. 251, 259-
263, 267, 268, 270, 274, 278, 283, 343, 345; iii.44, 52, 63, 108, 109, 131, 132, 142, 145-
151, 156-163, 166, 185, 186-188.
- Euphrŏnius, poet, ii. 66.
- Eureïs, r. of Mysia, ii. 190.
- Euripides, tragic poet, i. 52, 274, 329; ii. 32, 45, 52, 60, 62, 185, 189, 389, 390;
iii.20, 53, 75.
- Euripus, ii. 92, 96.
- ——, Chalcidian, i. 17, 57, 94; ii. 96, 130, 148, 151, 154.
- ——, Pyrrhæan, ii. 391.
- Eurōmus, t. of Caria, iii. 6, 37.
- Europe, i. 22, 52, 78, 88, 103, 140, 157-164, 183, 188, 191-
194, 205, 206, 303, 442,453, 464, 477, 480, 490, 505, 517; ii. 1, 4, *passim*.
- Europeans, ii. 240.
- Eurōpus, city of Media, ii. 264.

- ——, same as Rhaga, ii. 284.
- ——, city of Macedonia, i. 501.
- ——, r. of Thessaly, i. 501, 507.
- Eurōtas, r. of Laconia (*the Iri or Vasili Potamo*), i. 417, 507; ii. 15, 41, 42, 76, 145.
- Eurus (south-east wind), i. 45.
- Eurycleia, iii. 13.
- Eurycles, leader of the Lacedæmonians, ii. 41, 44.
- Eurycydeium, grove, in Elis, ii. 19.
- Eurydice, mother of Philip, i. 500.
- Eurylochus, ii. 83.
- ——, Thessalian, ii. 116, 120.
- Eurymachus, ii. 173.
- [Pg 347]Eurymedōn, leader of the Athenians, ii. 35.
- ——, r. of Pamphylia (*Kopru-su*), ii. 325; iii. 49.
- Eurypōn, son of Procles, ii. 44.
- Eurypōntidæ, ii. 44.
- Eurypylus, son of Euæmon, ii. 134, 136, 138, 142, 143.
- ——, son of Telephus, ii. 343, 345, 389, 395.
- Eurysthenes, brother of Procles, ii. 42-44, 77.
- Eurysthĕnidæ, ii. 44.
- Eurystheus, king of Mycenæ, ii. 59.
- Eurystheus's-head, ii. 59.
- Eurytănes, people of Ætolia, ii. 156, 160, 179.
- Eurytus, ii. 10, 11, 23, 24, 142; iii. 10.
- Eusebeia, ii. 281, 282. *See* Tyana *and* Mazaca.
- Euthydĕmus, king of the Bactrians, ii. 251, 253.
- ——, orator, iii. 38, 39.
- Euthymus, i. 381.
- Eutresis, ii. 106.
- Euxine, i. 8, 31, 32, 68, 75, 76, 78-81, 84, 86, 89, 95, 96, 102, 106, 113, 139, 163,177, 183, 188-190, 193-195, 202, 245, 440, 442, 443, 451, 452, 467, 474, 476, 481,491, 492, 496, 506; ii. 216, 218, 221, 226, 227, 231, 238, 240, 243, 246, 270, 273,276, 277, 282, 286, 290, 295; iii. 1, 61, 63, 64, 142, 186. *See* Pontus.
- Euxynthetus, ii. 199.
- Evander, i. 343.
- Evenus, r. of Ætolia (*Fidari*), i. 501; ii. 6, 160, 171.
- ——, r. of Mysia, ii. 387.
- Exitani, city of the, in Bætica, i. 235, 255.
- Exterior Sea. *See* Atlantic.
-
- Fabius, the historian, i. 339.

264

- [Pg 348]*Frejus. See* Forum Julium.
- Frentani, people of Italy, i. 358-360, 432, 436.
- Frozen Sea, i. 99.
- Frūsino, c. of Latium (*Frusinone*), i. 352.
- Fucinus, Lake, i. 356.
- Fugitives, t. of, i. 73.
- Fundi, t. of Latium, i. 347.
- Furies, the, i. 262.
- *Furni Islands. See* Corsiæ.
- *Fusaro, Lago di. See* Acherusian Lake.
-
- Gabæ, city of Persis, iii. 131.
- Gabala, city of Syria, iii. 167.
- Gabales, a people of Aquitania, i. 284.
- Gabianē, a province of Elymais, iii. 154.
- Gabii, t. of Latium (*L'Osteria del Pantano*), i. 353, 354.
- Gabinius, historian, iii. 281.
- ——, consul, ii. 308; iii. 232.
- Gabreta, forest of Germany, i. 448.
- Gadara, t. of Judæa, iii. 175.
- Gadaris (*Esdod*), iii. 175, 183.
- Gades, Gadeira (*Cadiz*), i. 60, 150, 152, 153, 157, 161, 164, 180, 208, 210-212, 222,223, 226, 235, 236, 241, 253-262, 296; iii. 276, 278.
- ——, Gates of, i. 256, 258.
- Gadilōn (*Wesir Kopti*), ii. 294.
- Gadilonītis, ii. 294.
- Gaditanians, i. 212, 213, 255, 260.
- Gæsatæ, people of Cisalpine Gaul, i. 317, 322.
- *Gaëta. See* Caiata.
- *Gaeta, Mola di. See* Formiæ.
- Gætuli, people of Libya, i. 198; iii. 276, 282, 289, 294.
- Galabrii, people of Illyria, i. 485.
- Galactophagi, i. 453, 458, 461, 465, 479; ii. 304.
- Galatæ (*see* Celtæ), in Europe, i. 96, 161, 219, 264, 270, 271, 282, 286, 327, 482, 485; ii. 71.
- ——, Cisalpine, i. 313.
- Galatæ, Scordisci, i. 482.
- ——, Alabroges, iii. 184.
- Galatia, part of Phrygia, i. 195; ii. 310, 319-321.
- Galatians, ii. 282-284, 286, 290, 293, 294, 310, 319, 320, 329, 355; iii. 297.
- Galatic or Gallic race, i. 282, 283, 291, 443.

- ——, or Celtic Gulf (*Gulf of Lyons*), i. 160, 174, 184, 192, 206, 249, 271, 283.
- ——, (*Gulf of Aquitaine*), i. 192, 249.
- Galēpsus, t. of Macedonia, i. 512, 513, 515.
- *Galazze. See* Callateria.
- Galilee, district of Judæa, iii. 177, 181.
- Gallesius, mtn of Ionia, iii. 15.
- Gallia Cispadana. *See* Keltica.
- Gallicians, the, i. 228, 229, 233, 243, 246, 250; iii. 63, 65.
- Gallinarian Wood, in Campania (*Pineta di Castel Volturno*), i. 362.
- *Gallipoli. See* Chersonesus, Thracian, Callipolis.
- Gallo-Græcia, i. 195.
- Gallus, r. of Phrygia, ii. 289. *See* Ælius *and* Cornelius.
- Gamabrivi, people of Germany, i. 445.
- Gambarus, prince of Syria, iii. 167.
- Gandaris, district of India, iii. 92.
- Gandarītis, district of India, iii. 89.
- Ganges, r. of India, iii. 74, 79, 80, 90, 96, 97, 108, 117, 118.
- Gangitis, iii. 157.
- Gangra, ii. 314.
- Ganymede, ii. 347.
- Garamantes, a people of Libya, i. 198; iii. 289, 294, 295.
- Garescus, t. of Macedonia, i. 509, 514.
- Gargara, t. of the Troad, ii. 342, 375, 376, 382, 384.
- Garganum, mtn of Italy (*Punta di Viesti*), i. 434-436.
- [Pg 349]Gargareis, Gargarenses, inhabitants of the Gargari, ii.
- ——, people of the Caucasus, ii. 235, 236.
- Gargaris, ii. 381.
- Gargarum, peak of Mount Ida, i. 64; ii. 342.
- Gargasus, son of Cypselus. *See* Gorgus.
- Gargettus, vill. of Attica, ii. 59.
- *Garigliano. See* Liris.
- Garindæi, a people of Arabia, iii. 204.
- Garmānes, philosophers of India, iii. 109, 110.
- *Garonne*, r. *See* Garuna.
- Garsaurītis, province of Cappadocia, ii. 278.
- Garsavira, vill. of Cappadocia (*Mekran*), ii. 281, 284; iii. 74, 121, 124, 125, 128, 156.
- Garuna (*Garonne*), r., i. 265, 282-284, 288, 297.
- Gasterocheires, ii. 54.
- Gasys, ii. 302.
- *Gata. See* Curias.

- Gaudus, island (*Gozo*), i. 71, 421, 459.
- Gaugamēla, village of Aturia (*Karmelis*), i. 123; iii. 144.
- Gaul, i. 192, 264-296, 439. *See* Keltica.
- Gaul, Cisalpine, i. 287, 324, 357.
- ——, Transalpine, i. 264.
- Gauls, the, i. 292-294.
- Gaza, city of Judæa, iii. 171, 176, 191.
- Gazaka, city of Media, ii. 263.
- Gazacene, district of Pontus, ii. 302.
- Gazæans, iii. 160.
- Gazaluïtis, district of Pontus, ii. 302.
- Gazelonītis, ii. 311.
- Gaziūra, t. of Pontus (*Turchal*), ii. 295.
- *Gedis. See* Cadi.
- *Gedis-Tschai. See* Hermus.
- Gedrosia, i. 196, 197; iii. 74, 121, 128, 156, 190.
- Gedrosia, Upper, i. 201.
- Gedrosii, Gedroseni, people of Ariana, iii. 124, 125.
- *Geihun. See* Pyramus.
- Geira, *see* Aphrodisias, ii. 332.
- Gĕla, city of Sicily, i. 412.
- Gēlæ, ii. 235, 241, 245.
- Geloi, i. 411.
- Gelōn, tyrant of Syracuse, i. 149; ii. 158.
- Genabum (*Orleans*), i. 284.
- Genauni, people of Illyria, i. 306.
- Genētes, prom. and river of Pontus (*C. Vona*), ii. 296.
- Gennesarītis, lake and district of Judæa, iii. 169.
- Genoa, i. 300-302, 314, 322, 323.
- *Genoa, Gulf of. See* Liguria.
- Georgi, i. 479; ii. 219.
- Gephyra, Gephyrismi, in Attica, ii. 91.
- Gephyræans, ii. 96. *See* Tanagræi.
- Geræstus, t. and prom. of Eubœa (*C. Mantelo*), ii. 150, 151, 153, 154.
- Geranius, r. of Elis, ii. 11.
- Gerēna, Gerēnia, city of Messenia, i. 459; ii. 12, 28, 36, 37.
- Gerenius, epithet of Nestor, ii. 11, 36.
- Gerēnus, a place in Elis, ii. 11.
- Geres, a Bœotian, iii. 2.
- Gergitha, t. of the Troad, ii. 350, 390.
- ——, vill. near the sources of the Caïcus, ii. 390.

- Gergitheis, t. of Cymæa, ii. 350.
- Gergithium, a place near Lampsacus, ii. 350.
- ——, in Cymæa, ii. 350.
- Gergithius, Cephalon, the, ii. 350.
- Gergovia, city of the Arverni, i. 285.
- German tribes, i. 445.
- German war, i. 289.
- Germanicus, son of Tiberius, i. 441, 446.
- Germans, i. 18, 177, 287, 288, 292, 307, 439, 443, 451, 468, 470, 481.
- Germany, i. 22, 110, 141, 193, 292, 442, 443, 451, 452, 467, 471, 481.
- [Pg 350]Gerræi, iii. 191, 204, 207.
- Gerræidæ, port of the Teii, iii. 17.
- Gerrha, t. of Egypt, i. 79, 87; iii. 177.
- ——, t. of Arabia, iii. 186, 187.
- Gerōn, r. of Elis, ii. 11.
- Gēryon, i. 33, 225, 254, 255, 343, 364.
- Gezatorix, prince of Paphlagonia, ii. 314.
- *Ghela. See* Acila.
- *Giaretta. See* Symæthus.
- *Gibraleon. See* Onoba.
- Getæ, i. 141, 177, 193, 445, 452-457, 461, 463, 464, 466-470, 481; iii. 180.
- ——, desert of the, i. 468, 469.
- *Giaur-Kalessi. See* Balbura.
- Gibraltar, Strait of, i. 62.
- ——, *and Ceuta*, rocks of. *See* Pillars of Hercules.
- Gigartus, a fortress of Syria, iii. 170.
- Gindarus, t. of Syria, iii. 163.
- *Gira-petra. See* Therapytna.
- *Giulia Nova. See* Castrum.
- Glaucias, tyrant, ii. 368.
- Glaucōpium, citadel of Athens, i. 460.
- Glaucus, the Anthedonian, ii. 98.
- ——, Pontius, play of Æschylus, ii. 155.
- ——, of Potniæ, ii. 103.
- ——, r. of Colchis (*Tschorocsu*), ii. 227, 231.
- ——, bay of Caria, iii. 28.
- Glechon, ii. 124.
- Glissas, t. of Bœotia, ii. 107, 108.
- Glycera, courtesan, ii. 105.
- Glycys-Limen, bay and port of Epirus, i. 497.
- Gōgarene, distr. of Armenia, ii. 268, 269.

- *Gok-Irmak. See* Amnias.
- Gomphi, t. of Thessaly, ii. 141.
- Gonnus, t. of Thessaly, ii. 145.
- Gonoessa, ii. 59.
- Gorbeüs, t. of Phrygia, ii. 321.
- Gordium, t. of Phrygia (*Juliopolis*), ii. 321, 330.
- Gordius, king of Phrygia, ii. 321.
- Gordus, place in the Troad, ii. 371.
- Gordyæa, Gordyene, a province of Armenia, i. 123; ii. 268; iii. 146, 156, 157, 162.
- Gordyæan mountains, i. 124; ii. 261.
- Gordyæi, people of Mesopotamia (*the Kurds*), ii. 271, 274; iii. 142, 157.
- Gordyæus, prince of the Gordyæi, ii. 274.
- Gordys, son of Triptolemus, iii. 153, 162.
- Gorgipia, city of the Sindi, ii. 223, 224.
- Gorgons, Gorgo, i. 29, 33, 459; ii. 211.
- ——, Gorgon's Head, the, ii. 62, 211.
- Gorgus, son of Cypselus, i. 498; ii. 161. *See* Gargasus.
- ——, the miner, iii. 93.
- Gortyna, city of Crete (*Hagius Dheka*), ii. 195, 196, 198, 200; iii. 22.
- Gortynii, ii. 197, 202.
- Gortynium, city of Macedonia, i. 504.
- Gorys, t. of India, iii. 89.
- *Goti, S. Maria di. See* Caudium.
- *Gozo. See* Gaudus.
- *Grabusa. See* Corycus.
- Gracchus, Tiberius, i. 244.
- Graces, temple of the, ii. 112.
- Græcia, Magna, i. 377.
- Græa, Graia, t. of Bœotia, ii. 58, 96, 106.
- Granicus, r. of Mysia (*Kodscha-Tschai*), ii. 338, 340, 347, 349, 371.
- Gras, son of Penthilus, ii. 340.
- Gravisci, t. of Etruria, i. 335.
- Grecian cities, i. 350.
- ——, nations, i. 372; ii. 3.
- ——, shore, the, i. 9.
- ——, territories, i. 43.
- Grecians, i. 256, 282.
- Greece,
i. 17, 24, 28, 40, 77, 90, 94, 96, 103, 164, 188, 194, 311, 316, 328, 329, 345,366, 431, 432, 437, 442, 443, 457, 461, 481, 492-494, 496, 501, 505;
ii. 1, 3, 4, 12,28, 29, 49, 50, 71, 78, 158, 159, 177, 178, 185, 193; iii. 41, 42, *et passim.*
- [Pg 351]*Greego. See* Theoprosopon.

269

- Greek language, i. 149.
- ——, tribes, ii. 2.
- ——, cities, i. 393.
- ——, states, i. 427.
- ——, laws, i. 240.
- ——, dialects, ii. 2.
- ——, literature, i. 271.
- Greeks,

i. 16, 49, 51, 54, 57, 67, 70, 73, 77, 102, 104, 191, 192, 194, 224, 232, 233,237, 240, 249, 274, 296, 302, 317, 326, 328, 330, 345, 350, 360, 372, 377, 378,392, 394-396, 403, 407, 408, 411, 422, 427, 439, 450, 453, 462, 463, 468, 478,492, 496, 498, 505, 514; ii. 33, 43, 44, 50, 54, 55, 132, 134, 158, 169, 172, 174,182-184; iii. 40-43, 110, 114.

- ——, Italian, i. 376, 377, 433; ii. 68.
- *Grego. See* Throni.
- Grium, mtn of Caria, iii. 6.
- *Grotta di Pausilipo. See* Cumæ.
- Grūmentum, t. of Lucania, i. 379.
- Gryllus, ii. 95.
- Gryneus, name of Apollo, ii. 393.
- Grynium, city of Æolis, ii. 397.
- *Guadalquiver*, r. *See* Bætis.
- *Guadiana. See* Anas.
- *Gumusch-dagh. See* Thorax.
- *Gura. See* Othrys.
- *Gura*, r. *See* Epidanus.
- Guranii, a people of Armenia, ii. 273.
- Gutōnes, i. 444.
- Gyarus, island (*Jura*), ii. 208.
- Gygæa, a lake of Lydia, afterwards Coloe, ii. 403.
- Gygas, prom. of the Troad, ii. 352.
- Gyges, king of the Lydians, ii. 119, 351; iii. 66.
- Gymnesian or Balearic islands (*Majorca and Minorca*),

i. 185, 194, 216, 217, 239,251; iii. 32.

- Gymnetæ, iii. 117.
- Gymnosophists, Indian philosophers, iii. 180.
- Gynæcopolis, t. of Egypt, iii. 241.
- Gynæcopolite nome, iii. 241.
- Gyrtōn, Gyrtōne, city of Thessaly (*Tcheritchiano*), i. 507; ii. 143-148.
- Gyrtōnii, Gyrtonians, i. 507; ii. 147.
- Gythium, t. of Laconia, ii. 15, 41.
-
- Hades, i. 31, 33, 223-225; ii. 17, 41, 51; iii. 110, 111.

- Hadylium, ii. 123, 124.
- Hæmon. *See* Hæmus.
- Hæmōn, father of Thessalus, ii. 149.
- ——, father of Oxylus, ii. 176.
- Hæmōnia, ancient name of Thessaly, i. 73; ii. 149.
- Hæmus, mtn of Thrace (*Velikidagh*), i. 311, 463, 481, 489, 490, 496, 506, 514; ii. 145.
- *Hagius Dheka. See* Gortyna.
- Halæ, t. of Bœotia, ii. 98, 125.
- ——, in Attica, ii. 98.
- ——, Araphenides, ii. 90, 153.
- ——, Æxoneis, ii. 89.
- Halesian plain, ii. 374.
- Halex, r. (*Alece*), i. 390.
- Haliacmon, r. in Macedonia (*Indesche Karasu*), i. 505-509.
- Haliartia, ii. 107.
- Haliartus, city of Bœotia, i. 25, 457; ii. 101, 106-109.
- Halicarnassus, ii. 56, 374; iii. 5, 30, 34, 35.
- Halieis, ii. 54.
- *Halikes. See* Zoster.
- Halimusii, ii. 89.
- Halisarna, iii. 36.
- Halius, ii. 135.
- Halizoni, Halizones, ii. 297, 299, 300, 371; iii. 63-66.
- Halonnesus, ii. 140, 393; iii. 18.
- Halys (*Kizil-Ermak*), i. 190, 195, 439, 457; ii. 135, 139, 218, 276, 277, 283, 285, 286,290, 293, 294, 301, 302, 311 313, 327; iii. 61, 141, 297.
- [Pg 352]Halys, Phthiotic, ii. 135.
- Hamaxitus, ii. 145, 373-375, 385, 395.
- Hamaxœci, i. 191, 453, 461; ii. 219.
- *Hamedan. See* Ecbatana.
- Hannibal, i. 238, 239, 311, 321, 323, 336, 364, 370, 373, 374, 381, 382, 428, 436,439.
- *Haran. See* Niciphorium.
- Harma, vill. of Bœotia, ii. 97, 99, 106.
- ——, t. of Attica, ii. 96, 97.
- Harmatus, prom. of Æolia, ii. 397.
- Harmōnia, i. 73, 500.
- Harmozi, prom. of Carmania, iii. 186.
- Harpagīa, t. of Mysia, ii. 347.
- Harpagus, general of Cyrus, i. 376.

271

- Harpalus, iii. 292.
- Harpies, the, i. 465.
- Harpina, t. of Pisatis, ii. 32.
- Hebe, Dia, ii. 66.
- Hĕbrus, r. of Thrace (*Maritza*), i. 495, 505, 516, 518; ii. 351.
- Hecatæus, the Milesian, i. 1, 12, 13, 28, 410, 459, 486, 492; ii. 13, 299, 300, 302; iii.5, 6.
- ——, of Teïos or Abdera, iii. 17.
- Hecate, ii. 183, 189; iii. 39.
- Hecaterus, ii. 188.
- Hecatomnus, king of Caria, iii. 35, 38.
- Hecatompolis, ii. 40.
- Hecatompylos, city of the Parthians, ii. 249, 250.
- Hecatonnesi, *see* Apollononnesi, ii. 393.
- Hecatos, name of Apollo, ii. 393.
- Hector, i. 64; ii. 344, 356, 357, 360, 363, 365, 394, 395.
- Hecuba, ii. 168, 351.
- Hecuba's monument, i. 517.
- Hĕdylus, poet, iii. 69.
- Hēdyphōn, r. of Babylon, iii. 154.
- Hegesianax, historian, ii. 355.
- Hegesias, orator, ii. 86; iii. 23.
- Heilĕsium, t. of Bœotia, ii. 100.
- Heilotæ. *See* Helots.
- Heleii, ii. 43; iii. 195.
- Helen, i. 65, 274; ii. 52, 86, 90, 360; iii. 238.
- ——, Claimed, play of Sophocles, iii. 15.
- Helena (*Isola Longa* or *Macronisi*), ii. 90, 208.
- Helēne, isl. *See* Cranæ.
- Heleōn, vill. of Tanagria, ii. 98-100, 143.
- Heliadæ, sons of the Sun, iii. 32.
- Heliades, drs of the Sun, i. 320.
- Helice, city of Achæa, i. 92; ii. 59, 69-73.
- ——, t. of Thessaly, ii. 71.
- Helicon, mtn of Bœotia (*Zagaro Voreni*), i. 40; ii. 62, 99, 101, 104, 105, 107, 109,122, 187, 398.
- Helius, son of Perseus, ii. 41.
- Heliopolis, city of Syria, iii. 166.
- ——, city of Egypt, iii. 241, 245-247.
- Heliopolĭtæ, iii. 21.
- Heliopolite nome, iii. 245.

- Hella, strait, i. 519.
- *Hellada. See* Spercheius.
- Hellanicus, historian of Lesbos, i. 69; ii. 44, 127, 167, 241, 298, 368, 382, 393.
- Hellas. *See* Greece.
- ——, city of Phthiotis, ii. 133, 134.
- ——, Southern Thessaly, ii. 149.
- Hellen, son of Deucalion, ii. 67, 134, 149.
- Hellenes. *See* Greeks.
- ——, t. of Spain, i. 236.
- Hellespont (*Strait of the Dardanelles*), i. 72, 78, 99, 106, 107, 164, 187, 188, 195, 453,481, 496, 517-519; ii. 92, 289, 319, 326, 341, 346, 350, 352.
- ——, mouth of, ii. 352.
- Hellespontia, ii. 277.
- Hellespontiac Phrygia. *See* Phrygia.
- Helli, inhabitants of Dodona, i. 502.
- Hellŏpia, district adjacent to Dodona, i. 502.
- ——, same as Eubœa, ii. 152.
- ——, t. of Eubœa, ii. 152.
- Helos, t. of Laconia, ii. 15, 23, 24, 41, 43, 100.
- [Pg 353]Hĕlos, in Triphylia or Messenia, ii. 23, 24, 100.
- Helots, ii. 43, 44, 287; iii. 96.
- Helvetii (*the Swiss*), i. 287, 288, 293, 306, 310, 447, 448, 450, 482.
- Hĕmeroscopium, city of Spain (? *Denia* or *Artemus*), i. 238, 242.
- Hemicynes, i. 68, 458.
- Heneta, ii. 289, 302.
- Henĕti, people of Italy (*Venetians*), i. 76, 96, 225, 313-316, 319-322, 433, 434; ii. 288,301, 378.
- Henetian horses, i. 316.
- Henetica, the Venetian territory, i. 483; ii. 378.
- Hĕniochi, people of Asia, i. 195; ii. 219, 224, 225, 238; iii. 296.
- Hĕniochia, ii. 224.
- Heorta, t. of the Scordisci, i. 488.
- Hephæsteium, iii. 248.
- Hĕphæstus. *See* Vulcan.
- Heptacōmĕtæ, people of Pontus, ii. 296, 297.
- Heptaporus, r. of Mysia, ii. 304, 341, 347, 357, 371.
- Heptastadium, on the Hellespont, ii. 352.
- ——, between Alexandria and Pharus, iii. 227, 230.
- Hĕra. *See* Juno.
- Heraclæa, city of Magna Grecia, i. 397, 398, 427, 428.
- ——, city of Elis, ii. 32.

273

- Hercules,

i. 3, 15, 76, 207, 210, 224, 236, 256, 257, 273, 274, 277, 326, 333, 343,364, 429, 511, 515, 519; ii. 9, 13, 26-28, 30, 34, 40, 52, 55, 59, 64, 238, 315, 359,380, 386, 389; iii. 31, 74, 76-78, 259, 271, 277, 280, 294.

- ——, Ipoctonus, ii. 386.
- ——, Cornōpiōn, ii. 386.
- ——, Macistian, ii. 22.
- ——, work of Lysippus, i. 424; ii. 171.
- ——, of Myron, iii. 8.
- ——, picture of Aristides, ii. 64.
- ——, labours of, i. 30, 40, 254; ii. 171; iii. 172.
- ——, expedition of, i. 255, 256.
- ——, children of, i. 333; ii. 59.
- ——, descendants of, i. 326.
- ——, companions of, ii. 315.
- ——, Pillars of. *See* Pillars.
- ——, temple, i. 254, 256, 258, 261, 353; iii. 238.
- ——, island, i. 255, 239.
- ——, harbour and grove of, ii. 171.
- ——, Colossus of, i. 424.
- ——, altar, iii. 277.
- ——, warm-baths, ii. 125, 129.
- ——, city, iii. 256.
- ——, port of, Herculis Portus (*Porto Ercole, Formicole*), i. 334, 383.
- Herculeum Promontorium, i. 388.
- Hercynia, forest of (*The Black Forest*), i. 308, 444, 447, 448, 450, 452.
- Herdōnia, t. of Apulia (*Ordona*), i. 131.
- *Hergan Kaleh. See* Amorium.
- Hermæa, t. on the Carthaginian coast, iii. 288.
- ——, prom. (*Cape Bon*) iii. 285, 287.
- Hermagoras, rhetorician, ii. 397.
- Hermeia, images of Mercury, ii. 16.
- Hermeias, tyrant of the Atarnitæ, ii. 382, 387.
- Hermes, i. 67; iii. 119.
- Hermion, ii. 71.
- Hermione, city of Argolis (*Castri*), ii. 49, 54-56, 58.
- Hermionenses, ii. 54.
- Hermionic Gulf (*Gulf of Castri*), i. 92; ii. 6, 47, 49, 63, 79.
- ——, promontory, ii. 207.
- Hermocreōn, architect, ii. 348.
- Hermodorus, Ephesian, iii. 14.
- Hermon, city of, ii. 55.

- Hermonassa, t. of Pontus (*Platana*), ii. 296.
- ——, on the lake Corocondametis, ii. 223.
- Hermōnax, vill. of (*Akkerman*), i. 469.
- Hermonduri, people of Germany, i. 445.
- Hermōnthis, city of Egypt, iii. 263.
- Hermopolis, in Egypt, iii. 239, 241, 257.
- Hermopolite castle, iii. 258.
- Hermus, r. of Lydia, (*Godis-Tschia*), ii. 303, 339, 342, 346, 397, 402, 403; iii. 2.
- ——, plain of, ii. 402. 403; iii. 82.
- Hernici, people of Latium, i. 339, 343, 344, 353.
- Hero, tower of, ii. 352.
- Herod, king of Judæa, iii. 177, 184.
- Herodotus of Halicarnassus, i. 47, 56, 69, 97, 148, 152, 430, 462, 517; ii. 155, 190,241, 273, 275, 277, 280, 290, 298, 328, 393, 403, 405; iii. 35, 82.
- Heroopolis, city of Egypt, near Suez, i. 130, 131; iii. 176, 189, 191, 193, 203, 291.
- Herophilian school of medicine, ii. 336.
- Herostratus, of Ephesus, iii. 12.
- Hērpa, Hērphæ, city of Cappadocia, ii. 281, 283; iii. 44.
- Hesiod, i. 35, 45, 67, 68, 93, 329, 458, 462, 465, 494, 501, 502; ii. 14, 42, 50, 70, 83,104, 110, 188, 241, 348; iii. 22.
- Hesione, daughter of Laomedon, ii. 359.
- [Pg 355]Hesperides, city of Cyrenæa (*Bernic* or *Bengazi*), i. 186; ii. 169; iii. 291.
- ——, of Nympha, i. 226, 273, 459.
- Hesperii. *See* Locri.
- Hesperitæ, Libyans, iii. 22.
- Hestia, goddess. *See* Vesta.
- Hestiæa, ii. 364.
- Hestiæōtis, Histiæōtis, part of Thessaly, ii. 132, 137, 141, 142, 145, 152, 153, 195.
- ——, in Eubœa, ii. 141, 153.
- Hicĕsius, physician, ii. 337.
- Hicetaōn, Trojan, ii. 344.
- Hidrieus, son of Hecatomnus, iii. 35.
- Hiera, *see* Thermessa, isl. sacred to Vulcan, i. 418, 420.
- ——, Sacra, Sacred Promontory, prom. of Lycia, iii. 48.
- Hieracōnnēsos, or island of Hawks, in the Arabian Gulf, iii. 199.
- Hieracōnpolis, city of Egypt, iii. 263.
- Hierapolis, city of Syria, iii. 158.
- ——, city of Phrygia (*Pambuk-Kalessi*), ii. 140, 335, 408, 409.
- Hierapytna, t. of Crete, ii. 144, 188, 189, 194, 199.
- Hierapytnii, ii. 199.
- Hiericus, in Judæa. *See* Jericho.

- Homer, i. 1, 2, 5-9, 11, 16, 19, 25-27, 32, *passim.*
- ——, native land of, ii. 399; iii. 16, 19, 20.
- Homēreium, iii. 20.
- [Pg 356]Homēridæ, in the island of Chios, iii. 19.
- Hŏmŏlē, Hŏmŏlium, t. of Magnesia, ii. 147, 148.
- Hŏmŏnadeis, people of Pisidia, ii. 323, 324; iii. 50, 64.
- Hormiæ, i. 347.
- Hormina, Hyrmina, prom. of Elis, ii. 13.
- Hortēnsius, ii. 250.
- *Hu. See* Diospolis.
- *Huesca. See* Osca.
- Hya, same as Hyampolis.
- Hyacinthine games, i. 424.
- Hyameitis (Hyameia?), distr. or t. of Laconia, ii. 38.
- Hyampea, ii. 123.
- Hyampolis, c. of Bœotia (*Bogdana*), ii. 93, 116, 123.
- ——, c. of Phocis, ii. 93, 123.
- Hyantes, i. 493; ii. 93, 123, 177.
- Hyarōtis, r. of India (*Ravee*), iii. 85, 86, 90, 92.
- Hybla (the Lesser), c. of Sicily, afterwards named Megara, i. 404; ii. 73.
- ——, (the Greater), c. of Sicily, i. 405.
- Hyblæan honey, i. 404.
- Hyblæei Megarenses, ii. 73.
- Hybreas, ii. 409; iii. 38, 39.
- Hybriānes, an Illyrian race, i. 489.
- Hyda, c. of Lydia, ii. 102, 403, 404.
- Hydara, fortress of Armenia, ii. 304.
- Hydarnes, ii. 273.
- Hydaspēs, r. of India (*Jelum*), iii. 74, 82, 84, 88, 90-94, 122.
- Hydatopotami, iii. 164.
- Hydra, prom. of Æolis, ii. 397.
- ——, lake of Ætolia, afterwards Lysimachia, ii. 172.
- Hydracæ, al. Oxydracæ, people of India, iii. 75. *See* Sydracæ.
- Hydrēlus, iii. 26.
- Hydromanteis, iii. 180.
- Hydrūs, c. of Calabria (*Otranto*), i. 429.
- Hydrūssa, isl. near Attica, ii. 89.
- Hyĕla, c. of Lucania, i. 375.
- Hyla, ii. 102.
- Hylæ, c. of Bœotia, ii. 102, 106.
- Hylas, companion of Hercules, ii. 315, 316.

- Hylicus, lake in Bœotia (*Makaris*), ii. 102.
- Hyllus, son of Hercules, ii. 128.
- ——, r. of Lydia, ii. 303, 403.
- Hylobii, iii. 110, 111.
- Hymettus, mtn of Attica, ii. 90, 93.
- Hypæpa, c. of Lydia, ii. 405.
- Hypæsia, distr. of Triphylia, ii. 21.
- Hypæthrum, iii. 7.
- Hypana, c. of Triphylia, ii. 17.
- Hypanis, r. (*Bog*), i. 162, 457, 470.
- ——, r. of Sarmatia (*Kuban*), ii. 222-224.
- ——, r. of India (*Beas*), ii. 252; iii. 74, 82, 90, 94, 97.
- ——, same as Anticeites, ii. 222, 224.
- Hypasii, people of India, iii. 82, 90.
- Hypatus, mtn of Bœotia, ii. 108.
- Hypelæum, iii. 11.
- Hypelæus, ftn near Ephesus, iii. 3.
- Hyperboreans, i. 97, 452; ii. 240; iii. 108.
- Hypereia, ftn in Pharsalia, ii. 134.
- ——, ftn in the city of the Pheræi, ii. 142, 143.
- Hyperēsia, c. of Achæa, ii. 59, 67.
- Hypernotii, i. 97.
- Hyphanteium, mtn near Orchomenus, ii. 124.
- Hyphochalcis, c. of Ætolia, ii. 160.
- Hypocrēmnus, vill. of Ionia, iii. 18, 20.
- Hypsicrates, i. 479; ii. 235.
- Hypsoeis, t. of Elis, ii. 24.
- Hyrcania (*Corcan*), i. 22, 112, 113, 141, 178, 202, 467; ii. 237, 241-246, 252-257, 407; iii. 152.
- Hyrcanian Sea, same as the Caspian, i. 106, 107, 113, 115, 142, 180, 183, 194, 195; ii. 218, 239, 244, 245, 256, 257, 262.
- ——, plain, ii. 407.
- ——, Gulf, ii. 247.
- Hyrcanians, i. 195; ii. 240, 245, 248-250.
- Hyrcanium, fortress of Judæa, iii. 181.
- [Pg 357]Hyrcanus, king of Judæa, iii. 180, 184.
- Hyria, c. of Iapygia, i. 430.
- ——, c. of Bœotia, i. 16; ii. 58, 97, 103.
- Hyriæ, ii. 97.
- Hyrienses, ii. 97.
- Hyrieus, father of Orion, ii. 97.

- Hyrmina, Hormina, prom. of Elis, ii. 13.
- Hyrmine, c. of Elis, ii. 12, 13.
- Hyrtacus, ii. 344, 350.
- Hysiæ, c. of Bœotia, ii. 97.
- ——, c. of Argolis, ii. 58.
- Hysiātæ, ii. 97.
- Hyspirātis, distr. of Armenia, ii. 271.
- Hystaspes, father of Darius, i. 468.
-
- Jaccetania, Jaccetani, in Spain, i. 242.
- *Jaffa. See* Joppa.
- *Ialea. See* Elæa.
- Ialmenus, leader of the Orchomenii, ii. 113.
- Ialysii, iii. 33.
- Ialysus, Iēlysus, city of Rhodes, iii. 33.
- ——, painting of Protogenes, iii. 29, 31.
- *Jama. See* Zama.
- Iamblicus, prince of the Emiseni, iii. 166.
- Iamneia, t. of Judæa (*Jebna*), iii. 175.
- Iaones, ii. 134.
- Iapodes, i. 300, 308, 482-484.
- Iapyges, Iapygians, i. 394, 425, 428.
- Iapygia, i. 159, 164, 187, 314, 315, 388, 399, 400, 422, 428, 430, 435; ii. 98.
- Iapygian promontory (*Cape Leuca* or *Finisterre*), i. 186. 314. 393. 423.
- Iapygum tria Promontoria (*Capo della Castella, Capo Rizzuto,* and *Capo della Nave*), i.393.
- Iapyx, son of Dædalus, i. 425, 430.
- Iardanes, r. of Pisatis, ii. 15, 21.
- Iardanus, tomb of, ii. 22.
- Ias. *See* Attica, ii. 81.
- Iasidæ, ii. 52.
- Iasiōn, brother of Dardanus, founder of Samothracia, i. 516.
- *Iaskili. See* Dascylitis.
- Jasōn, i. 8, 18, 31, 32, 71, 72-74, 76, 89, 224, 332, 333, 375; ii. 139, 224, 235, 266,272, 273, 293.
- Jasonia, Jasonian Shrines, i. 72.
- ——, monuments in Armenia, ii. 235, 266, 272.
- Jasonium, mtn of Media, ii. 266.
- ——, prom. of Pontus (*Jasun*), ii. 296.
- *Jasun. See* Jasonium.
- Iasus, city and island of Caria, iii. 37.

280

- Idrieis, people of Caria, iii. 63.
- Idubeda, mtns of Spain, i. 241, 243.
- Idumæans, people of Judæa, iii. 160, 177.
- *Jebna. See* Iamneia.
- *Jekil-Irmak. See* Iris.
- *Jelum. See* Hydaspes.
- Iēlysus. *See* Ialysus.
- *Ienischer. See* Sigeium.
- *Jerba. See* Meninx.
- Jericho, iii. 177, 181, 209.
- Ierna, (*Ireland*), i. 99, 100, 111, 115-117, 173, 174, 179, 180, 199, 298.
- *Jeroskipo. See* Hierocepia.
- Jerusalem, capital of Judæa, iii. 175, 177, 178, 180.
- *Jeschil Irmak. See* Iris.
- Jews, iii. 142, 160, 175-185, 190, 210, 237, 274.
- Iglētes, i. 249.
- Iguvium, city of Umbria (*Engubbio* or *Gubbio*), i. 338.
- *Ijan Kalessi. See* Sagylium.
- *Ilan-Adassi*, isl. *See* Leuca.
- Ilasarus, iii. 212.
- Ilerda, t. of Spain (*Lerida*), i. 242.
- Ilergetes, nation of Spain, i. 242.
- Ilethyia, i. 335.
- *Ilgun. See* Holmi.
- Iliad of Homer, ii. 364.
- *Ilias. See* Pelinæum.
- Ilibirris, t. and r. of Gaul, i. 272.
- Ilieis, Ilienses, ii. 354-356, 359-362, 366-368.
- Iliocolōne, ii. 350.
- Ilipa, t. of Turditania (*Alcolea*), i. 213, 214, 261.
- Ilissus, r. of Attica, ii. 91.
- Ilium. *See* Troy.
- Illyria,

i. 110, 159, 164, 186, 194, 308, 309, 317, 432, 435, 439, 443, 466, 481, 483,487, 489, 495, 501; iii. 297.

- Illyrian nations, i. 482, 483, 489, 500; ii. 2.
- ——, mountains, i. 492, 495, 499, 501.
- ——, Sea (*Gulf of Venice*), i. 73.
- ——, coast, i. 483, 489.
- Illyrians, Illyrii, i. 306, 308, 319, 466, 468, 481, 482, 485, 488, 493, 506; ii. 2, 157.
- *Ilori. See* Hippus.

282

283

- Io, mother of Epaphus, ii. 152; iii. 57, 162.
- Iōl, t. of the Masæsylii, iii. 284.
- Iolaenses, people of Sardinia, i. 333.
- Iolaus, i. 333; ii. 59.
- Iolcius, same as Jason.
- Iolcus, c. of Magnesia (*Volo*), i. 71, 72, 111, 139-142.
- Ioleia, iii. 10.
- Iōn, son of Xuthus, ii. 67, 87, 152.
- ——, poet, i. 42, 94; iii. 19.
- ——, tragedy of Euripides, ii. 32.
- ——, river of Thessaly, i. 501.
- Ionæum, ii. 19.
- Iones, ii. 2, 5, 13, 53. *See* Ionians.
- Ionia, in Asia, i. 9, 17, 91, 96, 172, 187, 190, 195, 224; ii. 42, 221, 339; iii. 1-9, 12-27, 43, 202.
- ——, same as Attica, i. 257; ii. 67, 68, 81, 87.
- Ioniades, nymphs, ii. 32.
- Ionian colony, ii. 68.
- ——, colonists, ii. 68.
- ——, Gulf, Ionian Sea, i. 186, 388, 429, 486, 487, 495, 499, 500, 501, 507, 518.
- Ionians, i. 96, 102, 224, 256, 269, 397, 404, 458, 493; ii. 3, 43, 56, 67-71, 80-82, 181,298, 303; iii. 34, 40, 41, 43.
- Ionius, i. 487.
- Joppa (*Jaffa*), i. 68; iii. 175, 177.
- Ioras, mtn *See* Jura.
- Jordan, r. of Judæa, iii. 169, 170.
- *Iorghan-Ladik. See* Laodiceia.
- Ios (*Nio*), ii. 207.
- Ioza. *See* Julia.
- Iphicrates, ii. 76; iii. 278.
- Iphidamas, son of Antenor, i. 509, 510.
- Iphigeneia, ii. 279.
- ——, play of Euripides, ii. 60.
- Iphitus, ii. 34.
- ——, Eurytides, ii. 46.
- Ipnus, t. of Magnesia, ii. 148.
- Ira, t. of Messenia, ii. 37.
- *Ireland. See* Ierne.
- Iris, r. of Pontus (*Jekil-Irmak*), i. 82; ii. 295, 300, 311.
- Irra, daughter of Arrhabæus, i. 500.
- Isamus, r. of India, ii. 252.

- Isar, r. of Gaul, i. 276, 277, 288, 303.
- ——, r. of Vindelicia, i. 308.
- Isaura, t. of Isauria, ii. 322; iii. 46, 55.
- [Pg 360]Isauria Palæa, t. of Isauria, ii. 322.
- Isaurica, part of Lycaonia, ii. 322.
- *Ischekli. See* Eumeneia.
- *Ischia. See* Pithecussa.
- Ischopolis, t. of Pontus, ii. 296.
- *Isère*, r. of Gaul. *See* Isar.
- Isinda, t. of Pisidia, ii. 410.
- Isis, iii. 242, 260, 271.
- ——, temple of, iii. 70.
- ——, river, iii. 200.
- *Iskuriah. See* Dioscurias.
- *Islote. See* Scombraria.
- Ismandes. *See* Imandes.
- Ismaris, lake of Thrace, i. 515.
- Ismarus, Ismara, t. of the Ciconi, i. 515.
- Ismēnus, r. by Thebes, ii. 103.
- *Ismid. See* Astacus and Nicomedia.
- *Isnik. See* Nicæa.
- *Isnik-gol. See* Ascanius.
- Isocrates, ii. 398.
- Isodroma Mater, temple of, ii. 145.
- *Isola Longa. See* Helena.
- ——, *Plana. See* Planesia.
- Issa, isl. of the Liburni (*Lissa*), i. 186, 484, 487.
- ——, same as Lesbos, i. 93.
- Isseans, i. 484.
- Issus, iii. 60, 62, 160, 164.
- ——, Sea of, ii. 219; iii. 1.
- ——, Gulf of (*Bay of Ajazzo*, or *Aïas*), i. 75, 105, 106, 160, 179, 183, 189, 190; ii.256, 277, 279, 282; iii. 44, 45, 50, 55-57, 60, 61, 63, 68, 142, 160.
- *Istanpolin. See* Astypalæa.
- Ister, r. (*Danube*), i. 9, 22, 73, 79, 82, 89, 162, 177, 193, 264, 303, 308, 309, 317,319, 439, 440, 442, 443, 447, 450, 452-454, 457, 463, 467-470, 478, 480-483, 487-489, 492; ii. 77, 220, 240, 302.
- ——, sacred mouth of, i. 481, 489.
- ——, town of Mœsia, i. 489, 490.
- Isthmian games, ii. 60, 63.
- Isthmus. *See Suez.*
- Istri, i. 321, 483.

285

- Istria, distr. of Italy, i. <u>89</u>, <u>313</u>, <u>321</u>, <u>483</u>.
- Isus, distr. of Bœotia, ii. <u>98</u>, <u>99</u>.
- Italian cities, i. <u>276</u>.
- ——, revolt, 371.
- ——, headlands, i. <u>139</u>.
- ——, coast, i. <u>184</u>, <u>487</u>.
- Italians, Italiotæ, i. <u>250</u>, <u>302</u>, <u>310</u>, <u>313</u>, <u>358</u>, <u>379</u>; ii. <u>118</u>.
- Italica, c. of Spain, i. <u>213</u>.
- ——, c. of the Peligni, i. <u>358</u> Italy, i. <u>9</u>, <u>31</u>, <u>33</u>-
36, <u>54</u>, <u>72</u>, <u>84</u>, <u>141</u>, <u>163</u>, <u>164</u>, <u>184</u>,<u>185</u>, <u>193</u>, <u>194</u>, <u>216</u>, <u>224</u>, <u>236</u>, <u>240</u>, <u>241</u>, <u>264</u>, <u>266</u>-
268, <u>270</u>, <u>275</u>, <u>279</u>, <u>287</u>, <u>291</u>,<u>293</u>, <u>300</u>, <u>303</u>-307, <u>309</u>, <u>310</u>, <u>313</u>-315, <u>321</u>, <u>323</u>-
325, <u>329</u>, <u>337</u>, <u>339</u>, <u>345</u>, <u>361</u>, <u>371</u>,<u>377</u>, <u>379</u>, <u>380</u>, <u>383</u>, <u>399</u>, <u>400</u>-
403, <u>405</u>, <u>411</u>, <u>413</u>, <u>422</u>, <u>427</u>, <u>435</u>-439, <u>441</u>, <u>442</u>, <u>448</u>,<u>450</u>, <u>481</u>, <u>482</u>, <u>483</u>, <u>487</u>;
ii. <u>60</u>, <u>62</u>, <u>68</u>, <u>116</u>, <u>154</u>, <u>209</u>, <u>290</u>, <u>300</u>, <u>333</u>, <u>378</u>; iii. <u>45</u>,<u>278</u>, *et passim*.
- Ithaca, isl. and t. (*Thiaki* or *Ithaco*), i. <u>33</u>, <u>42</u>, <u>53</u>, <u>93</u>, <u>161</u>, <u>187</u>, <u>460</u>;
ii. <u>5</u>, <u>25</u>, <u>26</u>, <u>50</u>,<u>161</u>-167; iii. <u>8</u>.
- Ithacans, i. <u>33</u>; ii. <u>173</u>.
- *Ithaco. See* Ithaca.
- Ithōme, mtn and t. of Messenia, i. <u>426</u>; ii. <u>35</u>, <u>38</u>, <u>141</u>.
- ——, t. of Thessaly, ii. <u>141</u>.
- Itium, t. and port of Gaul, i. <u>297</u>.
- Itōnus, c. of Thessaly (*Armyrus*), ii. <u>135</u>, <u>138</u>.
- Itumon, ii. <u>26</u>.
- Iturii or Ituræans, a people of Syria, iii. <u>166</u>, <u>170</u>, <u>171</u>.
- Ityca, c. of the Carthaginians, iii. <u>284</u>, <u>285</u>.
-
- Juba, king of Numidia, i. <u>440</u>; iii. <u>280</u>, <u>282</u>-284, <u>297</u>.
- Judæa, part of Syria, iii. <u>160</u>, <u>171</u>-185, <u>189</u>, <u>209</u>, <u>241</u>, <u>266</u>.
- *Judicello. See* Amenanus.
- Jugurtha, king of Numidia, iii. <u>284</u>.
- *Iviça. See* Ebusus.
- Julia, i. <u>213</u>.
- ——, Ioza, t. of Bætica, i. <u>210</u>.
- Juliopolis, t. of Phrygia, ii. <u>330</u>.
- Iulis, c. of Ceos, ii. <u>210</u>.
- Julius. *See* Cæsar.
- [Pg 361]Iulus, son of Æneas, ii. <u>356</u>.
- Junc Plain, in Spain, i. <u>240</u>.
- Juno (Hēra), i. <u>5</u>, <u>41</u>, <u>393</u>; ii. <u>29</u>, <u>39</u>, <u>341</u>; iii. <u>11</u>.
- ——, Argive, temple of, i. <u>375</u>; ii. <u>110</u>, <u>127</u>.
- ——, named Cupra, by the Tyrrheni, i. <u>357</u>.
- ——, Pharygæa, ii. <u>127</u>.
- ——, island of, i. <u>253</u>, <u>255</u>.

- Ixiŏn, king of the Lapithæ, i. 507.
-
- *Kaisaruh. See* Mazaca.
- *Kaki-Scala. See* Taphiassus.
- *Kandili. See* Alyzia.
- *Kankri. See* Cenchrea.
- *Kapurna. See* Chæroneia.
- *Karabogher. See* Priapus.
- *Karaburun. See* Mimas.
- *Karadje-Burun. See* Criumetopon.
- *Kara-dagh. See* Masias.
- *Kara-Gedik. See* Andriclus.
- *Kara-Hissar. See* Tyana.
- *Karasi. See* Mysia.
- *Karasu. See* Melas.
- *Kara-sui. See* Cydnus.
- *Karlas. See* Bœbeis.
- *Karmelis. See* Gaugamela.
- *Kas, el. See* Casium.
- *Kas-Owa. See* Dazimonitis.
- *Kastri. See* Delphi.
- *Kelikdni. See* Calycadnus.
- *Kelisman. See* Clazomenæ.
- Keltæ, Kelti. *See Kelts.* Keltiberia, i. 222, 243-245.
- Keltiberians, i. 52, 214, 222, 228, 229, 238, 242-244, 246, 250.
- Keltic nations, i. 247, 291, 442, 443, 454, 481, 482.
- ——, zone, i. 147.
- ——, isthmus, i. 206.
- Keltica (*France*), i. 13, 99, 101, 111, 113-
116, 141, 174, 184, 192, 193, 206, 223, 226,240, 243, 267, 279, 289, 296, 298, 309, 323-325, 357, 442, 443, 447.
- ——, Transalpine, i. 264, 266, 296, 300, 309, 325.
- ——, Cisalpine, i. 303, 315, 336, 337.
- ——, Citerior, i. 324.
- ——, Gallia Cispadana, i. 325.
- [Pg 362]Keltici, people of Spain, i. 227, 230.
- Kelto-ligyes (*Ligurians*), i. 302.
- Keltoscythians, i. 52; ii. 240.
- Kelts,
i. 18, 52, 116, 208, 232, 238, 241, 246, 264, 265, 277, 282, 296, 297, 299, 308,316, 317, 438, 439, 443, 449, 463, 466.
- ——, Transalpine, i. 302, 316.

- Kemmenus, (*the Cevennes*), i. 219. *See* Cemmenus.
- Kenæum (*Kabo Lithari*), i. 94. *See* Cenæum.
- Kent, i. 99, 288, 296.
- Kentrones. *See* Centrones.
- *Kerasun. See* Paryadres.
- *Kerempi-Burun. See* Carambis.
- Kerkina, isl. (*Kerkeni*), i. 185.
- *Kerman. See* Carmania.
- Kerne, isle of, i. 75.
- Kerretani, people of the Iberians, i. 243.
- *Kertsch. See* Bosporus; Panticapæeon.
- *Kharput. See* Carcathiocerta.
- Khersobleptes, i. 516.
- *Khosistan. See* Susiana.
- *Kidros. See* Cytorum.
- *Kilandria. See* Celenderis.
- Kimbrians. *See* Cimbri.
- Kimmerians. *See* Cimmerians.
- *Kinoli. See* Cinolis.
- *Kisamos. See* Cisamus.
- *Kiutahia. See* Cotiaeium.
- *Kizil-Ermak. See* Halys.
- *Kodscha. See* Sirbis.
- *Koft. See* Coptus.
- *Koluri. See* Salamis.
- *Konia. See* Iconium.
- *Kopru-su. See* Eurymedon.
- *Kormakiti. See* Crommyum.
- *Kosseir. See* Philotera.
- *Krio, Cape. See* Criumetopon.
- *Krisso. See* Crissa.
- *Kulat-el-Mudik. See* Apameia.
- *Kulp. See* Colapis.
- *Kur. See* Cyrus.
- *Kurds. See* Gordyæi.
- *Kyno. See* Cynus.
-
- *La Punta. See* Actium.
- *La Riccia. See* Aricia.
- Labanæ, baths in Italy, i. 354.
- Labicum, i. 341.

- Lampeis, t. of Crete, ii. 194.
- *Lampeni. See* Amphissa.
- Lampesis, Lamptreis, t. of Attica, ii. 89.
- Lampōnia, t. of the Troad, ii. 382.
- Lampsacēnē, ii. 350.
- Lampsacēni, ii. 347, 349.
- Lampsacus, city of Mysia (*Lampsaki*), i. 518; ii. 340, 347, 349, 350, 352; iii. 6.
- Lamus, r. and t. of Cilicia, iii. 54, 55.
- Landi, people of Germany, i. 447.
- Langobardi, nation of Germany, i. 445.
- Lanuvium (*Civita Lavinia* or *Città della Vigna*), i. 344, 355.
- Laodicēa, city of Lycaonia, iii. 43.
- ——, c. of Cœle-Syria (*Iouschiah*), iii. 170.
- ——, c. of Syria (*Ladikiyeh*), iii. 161, 162, 167.
- ——, c. of Media, ii. 264.
- ——, c. of Phrygia (*Urumluk*), ii. 332, 334, 336, 408, 409; iii. 43.
- Laodiceia, mother of Seleucus, iii. 161.
- Laodiceians, ii. 334, 336.
- Laomedon, ii. 359.
- Laōthoë, ii. 395.
- Lapathus, t. of Cyprus (*Lapito*), iii. 69.
- Lapē, t. of Lesbos, ii. 127.
- Lapersæ, ii. 42.
- Lapithæ, people of Thessaly, i. 15, 507; ii. 144-148.
- Lapithēs, *same as* Mopsus.
- *Lapito. See* Lapathus.
- Laranda, t. of Lycaonia (*Caraman*), ii. 322.
- Larisa, daughter of Piasus, ii. 397.
- ——, Cremaste, city of Pelasgiotis, i. 94; ii. 138, 144, 373, 374, 395-397.
- ——, city of Phthiotis, ii. 145.
- ——, c. of Thessaly, ii. 77, 272.
- ——, c. of Attica, ii. 145.
- ——, c. of Crete, ii. 144.
- ——, c. on the confines of Elis and Achæa, ii. 145.
- ——, Phriconis in Asia, ii. 145.
- ——, c. of Syria, ii. 145, 165.
- ——, c. of Pontus, ii. 145.
- ——, c. of the Troad, i. 329; ii. 145, 374, 395.
- ——, Ephĕsia, ii. 145.
- ——, Phricōnis, c. of Æolis, ii. 145, 397.
- ——, citadel of the Argives, ii. 51, 144.

- Larisæan rocks, at Lesbos, ii. 145.
- Larisian plain, ii. 144.
- ——, Jupiter, ii. 145.
- Larisus, r. of Achæa, ii. 74, 145.
- Larius (*Lake of Como*), i. 287, 304, 312, 317.
- Laroloni, i. 337.
- Lartolæētæ, people of Spain, i. 239.
- Larymna, t. of Bœotia, ii. 98, 100.
- ——, Upper, t. of Locris, ii. 100.
- Lās, t. of Laconia, ii. 42.
- Lasion sea-coast, ii. 9.
- Lathōn, Lēthæus, r. of Cyrenaica, iii. 21, 291.
- Latin towns, i. 278.
- ——, coast, i. 344.
- ——, cities, i. 356.
- Latina, Via, i. 351, 352, 353, 356, 370.
- Latine. *See* Latium.
- Latini, Latins, i. 227, 325, 326, 340, 343-346, 349, 438.
- Latinus, i. 339.
- Latium, i. 325, 338, 339, 344, 345, 348, 351, 352, 360, 371, 378.
- Latmic Gulf, Ionia, iii. 6.
- Latmus, mtn of Caria, iii. 6.
- [Pg 364]Latmus, t. of Caria. iii. 6.
- Latomiæ, islands in the Arabian Gulf, iii. 194.
- Latona, ii. 208; iii. 11, 29.
- ——, temple of, ii. 24, 207, 239.
- Latopolis, city of Egypt, iii. 257, 263.
- Latopolītæ, iii. 257, 263.
- Latus, iii. 263.
- Laurentum, t. of Latium, i. 339, 346.
- Laüs, city of Lucania, i. 376, 377.
- ——, r. of Lucania (*Lao*), i. 376, 379-381.
- ——, gulf of Lucania, i. 376.
- Laviansene, ii. 278, 285, 310.
- Lavicana, Via, i. 352.
- Lavicum, t. of Latium, i. 352.
- Lavinia, daughter of Latinus, i. 339.
- Lavinium, city of Latium, i. 343, 345, 398; ii. 378.
- Leap, the, ii. 162.
- Lebadeia, city of Bœotia, ii. 111, 122.
- Lebedos, city of Ionia (*Lebedigh*), iii. 2, 16.

- Lĕbēn, t. and port of Crete, ii. 199.
- Lebenii, ii. 199.
- *Lebida. See* Leptis.
- Lebinthus, island (*Levita*), ii. 212.
- *Lebrixa. See* Nebrissa.
- Lecanomanteis, iii. 180.
- Lechæum, port of Corinth (*Pelagio*), i. 88; ii. 62, 63.
- Lectum, prom. of the Troad (*Baba Kalessi*), ii. 339-342, 372-376, 388, 390.
- Lĕda, wife of Tyndareus, and daughter of Thestius, ii. 173.
- Lĕĕtani, people of Spain, i. 239.
- *Lefka; see* Leuctra.
- Legæ, or Leges (*Legi*), ii. 235.
- Leimōn, iii. 26.
- Leimōne, same as Elēnē, ii. 145.
- Lĕlantum, plain of, i. 90; ii. 154, 178.
- Leleges, i. 493, 494; ii. 93, 327, 328, 343, 374-376, 381, 383, 394, 395; iii. 2, 4, 11,40, 63, 65.
- Lelegia, i. 493.
- Lĕmenna, lake (*Lake Leman, the Lake of Geneva*), i. 277, 303, 310.
- Lĕmnos, island (*Stalimene*), i. 43, 66, 71-73, 187, 329, 512, 513, 515; ii. 21, 158, 168,180, 190, 298, 394.
- Lemovices, people of Gaul (*the Limousins*), i. 284.
- Leōcorium, ii. 86, 87.
- Leon, rocks on the Ethiopian shore, iii. 201.
- Leōnidas, i. 17; ii. 130, 181.
- Leōnides, stoic, iii. 33.
- Leonnatus, friend of Alexander, ii. 136.
- Leonnorius, leader of the Galatæ, ii. 319.
- Leontes, ii. 350.
- Leontini, i. 31, 412, 414.
- Leontopolis, c. of Egypt, iii. 171, 240.
- Leontopolītæ, iii. 240, 257.
- Leontopolite nome, in Egypt, iii. 240.
- Leōsthenes, ii. 136.
- *Lepanto; see* Naupactus.
- ——, *Gulf of; see* Corinth, Gulf of.
- Lepidum. *See* Rhegium.
- Lepidus, Marcus, i. 323.
- Lĕpontii, Alpine race, i. 304, 306.
- Lepreātæ, ii. 18, 31, 45.
- Lepreātis, ii. 18.

- Lepreum, city of Triphylia, ii. 15, 17, 19, 21, 22, 31.
- Leptis, city of Africa (*Lebida*), iii. 289.
- Leria. *See* Lerus.
- Lerians, ii. 212.
- *Lerida, see* Ilerda.
- Lerna, r. of Argolis, ii. 48.
- ——, lake of Argolis, ii. 48, 52.
- Lērō (*Ile Ste Marguérite*), i. 276; ii. 212, 214; iii. 5. *See Leros.*
- Lesbia. *See* Lesbos.
- Lesbians, ii. 365.
- Lesboclēs, ii. 392.
- [Pg 365]Lesbos, island (*Mytilini*), i. 71, 93, 187, 329, 518; ii. 32, 213, 303, 339-345, 351, 353,375, 384, 388, 390-394, 398; iii. 8, 19, 36, 250.
- *Lesina*, isl. *See* Pharos.
- Lēthæus, r. of Crete (*Maloniti* or *Messara*), ii. 199.
- ——, r. of Magnesia, ii. 303; iii. 21.
- ——, r. of Thessaly, iii. 21.
- ——, r. of the Cyrenaic. *See* Lathōn.
- Lēthē, r. of Lusitania, i. 229, 230. *See* Limæa.
- Lethus, ii. 395.
- Lēto, temple of. *See* Latona.
- Letopolite nome, in Egypt, iii. 247.
- Letoum, iii. 47.
- Leuca, t. of Calabria, i. 429.
- ——, mtn of Crete (*Aspra-vuna* or *Sfakia*), ii. 194.
- ——, Leuce, island of (*Ilan-Adassi*), i. 188, 470; ii. 41.
- ——, Leuce-Come, t. of the Nabatæi, iii. 211.
- ——, Leuce-Acte, prom. of Eubœa, ii. 90.
- ——, prom. of Libya, ii. 213; iii. 235.
- ——, in Thrace, i. 518.
- ——, Capo di. *See* Iapygia.
- Leucadian Sea, i. 505.
- Leucadians, i. 494; ii. 162.
- Leucadius, brother of Penelope, ii. 162.
- Leucæ, t. of Ionia (*Leokaes*), iii. 20, 21.
- Leucani, i. 315, 339, 373-380, 392, 397, 427, 431.
- Leucania, i. 374-376, 380.
- Leucas, isl. and t. (*Sta Maura*), i. 91, 159; ii. 159, 161, 163, 171, 174.
- Leucaspis, iii. 236.
- Leucatas, prom. of Leucas, ii. 161, 167, 173.
- Leuci, people of Gaul, i. 288.

- Leucimmē, prom. of Corcyra (*C. Bianco*), i. 497.
- Leucippus, i. 399.
- Leuco, i. 463.
- Leucocomas, ii. 199.
- Leucolla, port of Cyprus, iii. 69.
- Leucōn, king of Bosporus, i. 476-478.
- Leuconotus (name of a wind), i. 45; iii. 292.
- Leucopĕtra, prom. of Italy, i. 315, 388.
- Leucophryēne, Artemis, iii. 22.
- Leucophrys, same as Tenedos, ii. 373.
- Leucōsia, island, i. 185, 375, 387.
- Leucosyri, ii. 288.
- Leucothea, temple of, ii. 228.
- Leuctra (*Lefkà*), c. of Bœotia, ii. 110, 111, 152.
- ——, battle of, ii. 68, 70.
- Leuctri, ii. 36.
- Leuctrum, t. of Laconia, ii. 36, 38, 39.
- ——, vill. of Achæa, ii. 73.
- *Levita. See* Lebinthus.
- Leuternian coast of Calabria, i. 429.
- Leuternians, giants of Phlegra, i. 429.
- Lexovii, i. 281, 290.
- Libanus, mtn, iii. 149, 169-171.
- Libēs, priest of the Chatti, i. 447.
- Libēthra, Leibēthrum, city of Pieria, i. 508; ii. 105, 187.
- Libēthriades, ii. 105, 187.
- Libophœnices, people of Libya, iii. 289.
- Libs (S. W. wind), i. 45; ii. 303.
- Liburni, i. 407, 487.
- Liburnia, i. 484.
- Liburnian islands, i. 186, 484, 487.
- Libya,

i. 2, 8, 15, 25, 41, 50, 51, 55, 56, 60, 63, 64, 76, 78, 88, 103, 148, 150, 154,155, 159, 160-164, 174, 180-187, 191, 197, 200, 206, 216-218, 226, 236, 255, 334,400, 416, 422, 439, 440, 458, 459, 504; ii. 169, 303; iii. 219, 226, 247, 253, 266,270.

- Libyan Sea, i. 185, 403, 496; ii. 5, 6, 35, 36; iii. 68.
- Libyans, i. 17, 256; iii. 271.
- [Pg 366]Libyans, Hesperītæ, iii. 22.
- Libyrnē, *same as* Scardon, i. 484.
- Libyrnides. *See* Liburnian Islands.
- Licattii, people of the Vindelici, i. 307.

295

- Licha, iii. 199.
- Lichades Islands (*Litada*), i. 94; ii. 126.
- Lichas, companion of Hercules, ii. 126, 155; iii. 201.
- Licymna, citadel of Tirynthes, ii. 54.
- Licymnius, iii. 31.
- Licyrna, ii. 171.
- Liger, r. (*Loire*), i. 265, 281-284, 286, 288, 289, 291, 292, 295, 297.
- Liguria, (*Genoa*), i. 193, 247, 265, 279, 302, 308, 311, 313-316, 324-326, 330, 439.
- Ligurian headlands, i. 139.
- Ligurisci, i. 454.
- Ligyes, Ligurians, i. 193, 267, 269, 274, 275, 300-303, 322, 323, 331, 462.
- ——, Oxybian, i. 276, 301, 314.
- Ligystica. *See* Liguria.
- ——, coast, i. 184, 185.
- ——, nations, i. 193, 194.
- ——, Sea of, (*Gulf of Genoa*), i. 160, 185, 193.
- Lilæa, c. of Phocis, i. 25; ii. 101, 123, 124, 128.
- Lilybæum, prom. of Sicily (*Capo Boeo*), i. 400-403, 411, 421; ii. 378.
- ——, c. of Sicily (*Marsalla*), i. 411; iii. 287.
- Limæa, r. of Lusitania (*Lima*), i. 229, 230.
- Limena, Limenera, ii. 48.
- Limenia, t. of Cyprus, iii. 70.
- Limnæ, t. of Messenia, i. 385; ii. 39, 40.
- ——, t. of the Thracian Chersonese, i. 517; iii. 5.
- ——, suburb of Sparta, ii. 40.
- Limnæum, suburb of Sparta, ii. 41.
- *Limousins. See* Lemovices.
- Limyra, t. of Lycia, iii. 47.
- Limyrus, r. of Lycia, iii. 47.
- Lincasii. *See* Lingones.
- Lindii, Lindians, iii. 33.
- Lindus, c. of Rhodes (*Lindo*), ii. 374; iii. 29, 33, 55.
- Lingones, Lincasii, people of Gaul, i. 278, 288, 310.
- Līnum, t. of Mysia, ii. 349.
- Linx, c. of Mauritania. *See* Lynx.
- Lipari Isles, i. 31, 84, 89, 185, 369, 383, 386, 415, 417-421.
- Liris, r. of Latium (*Garigliano*), i. 347, 352, 353.
- *Lisbon. See* Ulyssea.
- *Lissa. See* Issa.
- Lissus, t. of Dalmatia (*Alesso*), i. 485.
- *Litada. See* Lichades Islands.

- Liternum, t. of Campania (*Torre di Patria*), i. 361.
- Liternus, r. of Campania, i. 361.
- *Lithada. See* Cenæum.
- Lithrus, mtn of Pontus, ii. 306.
- *Livadhia. See* Lebadeia.
- Livia, piazza of, i. 351.
- Lixus, t. of Mauritania, iii. 279, 281.
- ——, r. of Mauritania (*Lucos*), i. 150.
- Lochias, prom. of Egypt, iii. 226, 230.
- Locri, people of Greece, i. 389-392, 494; ii. 85, 113-115, 124-130, 134, 135, 159, 365.
- ——, Epizephyrii, in Italy, i. 168, 381, 383, 388; iii. 289.
- ——, Epicnemidii, ii. 78, 113, 114, 124, 125, 128, 132.
- ——, Ozolæ or Hesperii, ii. 2, 114-116, 125-128, 158.
- ——, Opuntii, i. 389; ii. 114, 124-126.
- Locria, ftn in Locris, i. 389.
- Locris, in Greece, ii. 6, 42, 114, 124-132, 137, 151, 171, 340.
- ——, in Italy, i. 186, 388, 390.
- Locrus, i. 494.
- *Loire. See* Liger.
- Lopadūssa, island (*Lampidusa*), iii. 288.
- [Pg 367]Lŏryma, mtn and shore of Caria, iii. 34.
- Lŏtophagi, i. 37, 236, 237; iii. 281.
- Lŏtophagitis, name of the Lesser Syrtis, iii. 288.
- *Loubadi. See* Apolloniatis.
- Lucas, i. 494.
- Lucca, t. of the Ligyri, i. 323.
- Lucĕria, t. of the Daunii (*Lucera*), i. 398, 433.
- Lucius Tarquiuius Priscus. *See* Tarquinius.
- ——, Mummius, ii. 64, 65.
- *Lucos*, r. *See* Lixus.
- Lūcotŏcia, city of the Parisii, i. 290.
- Lucrine Lake and Gulf (*Lago Lucrino*), i. 362, 364.
- Lucullus, ii. 278. 292. 294. 307. 320.
- ——, Marcus, i. 490; ii. 65.
- Lūcŭmo, son of Demaratus, i. 326.
- Lūdias, r. and lake of Macedonia, i. 508, 509.
- Luerion, i. 302.
- Luerius, i. 285.
- Lugdūnum, t. of Aquitania (*Lyons*), i. 286.
- ——, c. of the Segosiani, i. 265, 277, 284, 288, 289, 309, 310.

- Lūgeum, i. 482.
- Lūji, people of Germany, i. 444.
- Lūna, i. 323, 329, 330.
- Lūpiæ, t. of Calabria, i. 430.
- Lūpias, r. of Germany (*Lippe*), i. 445.
- Lusitania, i. 181, 228-230, 250.
- Lusitanians, i. 209, 221, 228, 229, 231, 245, 250.
- Lūsōnes, people of Spain, i. 243.
- Lux Dubia, i. 211.
- Lycabēttus, mtn of Attica, ii. 90, 164.
- Lycæum, mtn of Arcadia (*Myntha*), i. 311; ii. 22, 75, 76.
- Lycaōn, i. 329.
- ——, son of Priam, i. 66; ii. 344, 346.
- Lycaonia, i. 202; ii. 276, 281, 284, 319, 321, 322, 332; iii. 44, 65.
- Lycaonians, i. 195; ii. 277, 304, 322; iii. 64.
- Lycastus, ii. 200.
- Lyceum, at Athens, ii. 87, 88, 90.
- Lychnidus, t. of Epirus (*Lago d' Ochrida*), i. 495, 500.
- Lycia, i. 8, 32, 38, 189, 195, 201; ii. 54, 259, 313, 317, 328, 329, 409; iii. 27, 28, 44-48, 54, 59, 68, 73, *et passim*.
- ——, in the Troad, ii. 317, 328, 329.
- Lycii, Lycians, ii. 277, 304, 327-329, 344, 346, 353, 360, 362, 405, 410; iii. 1, 41, 49,63.
- ——, in the Troad, ii. 162, 327, 344, 346, 360, 362.
- ——, Carian, ii. 327, 329.
- Lycomēdes, priest of the Comani, ii. 308.
- ——, king of Scyrus, ii. 140.
- ——, son of Pharnaces, ii. 311.
- Lycopolis, c. of Egypt, iii. 240, 257, 258.
- Lycopolītæ, iii. 257.
- Lycōreia, t. of Phocis, ii. 116.
- Lycormas, ii. 160.
- Lyctii, ii. 194.
- Lyctus, c. of Crete (*Lytto*), ii. 196, 200.
- Lycurgus, Lacedæmonian, ii. 43, 44, 203, 204; iii. 179.
- ——, king of the Edoni, ii. 187; iii. 76.
- ——, orator, ii. 368.
- Lycus, ii. 334.
- ——, son of Pandiones, ii. 81; iii. 49.
- ——, r. of Assyria (*Erbil*), i. 123; iii. 143, 144.
- ——, r. of Syria (*Nahr-el-Kelb*), iii. 170.

- ——, r. of Phrygia, ii. 334.
- ——, r. of Armenia and Pontus, ii. 270, 295, 306.
- Lydia, i. 91, 96, 326; ii. 68, 102, 185, 298, 327, 333, 351, 407, 410; iii. 22, 60.
- [Pg 368]Lydian temples, ii. 185.
- ——, gates, ii. 386.
- Lydians, i. 41, 328, 397, 453;

ii. 277, 317, 326, 329, 332, 346, 384, 386, 396, 402,403, 406, 407; iii. 24, 38, 63-65, 140, 141.

- Lydus, son of Atys, i. 326, 467.
- Lygæus, ii. 173.
- Lygdamis, leader of the Cimmerii, i. 96.
- Lyncēstæ, people of Macedonia, i. 495, 500, 501.
- Lyncestis, i. 500.
- Lynx, iii. 277, 278, 281, 282.
- *Lyonnaise, the,* i. 285-290.
- *Lyons. See* Lugdunum.
- *Lyons, Gulf of. See* Galatic Gulf.
- Lyrceium, mtn of Argolis, ii. 51, 58, 124.
- ——, vill. of Argolis, i. 410.
- Lyrnēssis, in the Troad, ii. 345.
- Lyrnēssus, t. of the Troad, ii. 343, 345, 377, 384.
- ——, t. of Pamphylia (*Ernatia*), iii. 49, 61.
- Lysias, fortress of Judæa, iii. 181.
- ——, t. of Syria, iii. 166.
- ——, t. of Phrygia, ii. 332.
- Lysimachia, city of the Thracian Chersonese (*Eksemil*), i. 202, 517.
- ——, t. and lake of Ætolia, ii. 172.
- Lysimachus, son of Agathocles, i. 464, 469, 490, 517;

ii. 315, 355, 361, 371, 377, 399,400; iii. 11, 20, 22.

- Lysiœdi, iii. 23.
- Lysippus, sculptor, i. 424; ii. 171, 350.
- Lysis, iii. 23.
- *Lytto. See* Lyctus.
-
- Mā, temple of Enyus, ii. 279.
- Macæ, people of Arabia Felix, iii. 186, 187.
- Macar, ii. 32, 346.
- Macaria, part of Messenia, ii. 38.
- Macaria, ftn of Attica, ii. 59.
- Macaros-polis, same as Lesbos.
- Macedonia, i. 42, 187, 194, 425, 432, 466, 481, 493, 495, 496, 499, 500, 501, 504-

516, 519; ii. 1-3, 10, 64, 92, 94, 129, 132, 140, 141, 147-154, 157; iii. 220, 297, *et passim.*

- ——, Upper, i. 500, 506.

- Magarsa, t. of Cilicia, iii. 60.
- Magi, Persian priests, i. 35, 149, 152; iii. 116, 136, 137, 140, 141, 180.
- Magnēsia, Magnētis, distr. of Thessaly, i. 506, 510; ii. 132-135, 139-141, 146, 148.
- ———, t. of Caria, ii. 299, 303, 335; iii. 6, 7, 21-24, 43.
- ———, t. of Lydia (*Manisa*), ii. 326, 335, 397.
- Magnesian Sea, ii. 135.
- Magnētes, of Thessaly, i. 43; ii. 131, 140, 146-148; iii. 7.
- ———, of Caria, ii. 333; iii. 22, 23.
- Magnētis, i. 507.
- Magnopolis, c. of Pontus, ii. 306.
- Magōdi, same as Lysiōdi, iii. 23.
- *Majorca and Minorca. See* Gymnasiæ.
- *Makro Teichos. See* Cnossus.
- Malaca, c. of Bætica (*Malaga*), i. 235, 238, 241, 245.
- Malaus, ii. 340.
- Maleæ, prom. of Laconia (*Cape-Malio* or *St. Angelo*), i. 38, 140, 163, 164; ii. 40, 41,47-49, 60, 77, 195.
- Maleōs, i. 335.
- Malia, prom. of Lesbos (*Sta. Maria*), ii. 390, 391.
- Maliac Gulf (*G. of Zeitun*), i. 17, 512; ii. 4, 96, 110, 126, 130-138.
- ———, war, ii. 153.
- Malians, Malienses, i. 43; ii. 2, 5, 135, 136, 147, 151.
- Malii, people of Mesopotamia, iii. 158.
- Malli, people of India, iii. 94.
- Mallus, city of Cilicia, ii. 283; iii. 59, 60.
- Malŏthas, c. of Arabia, iii. 212.
- *Malta. See* Melite.
- Malūs, in the Troad, ii. 371.
- Mamaus, r. of Triphylia, ii. 16.
- Mamertīni, in the c. of Messana, i. 404, 405.
- Mamertium, t. of the Bruttii, i. 391.
- Mana. *See* Larisus.
- *Mandani. See* Milania.
- Mandanis, iii. 113, 114, 116.
- Mandūbii, people of Gaul, i. 285.
- Manes, Phrygian name, ii. 126?.
- ———, Paphlagonian name, ii. 302.
- ———, r. of Locris, *see* Boagrius, ii. 126.
- *Mangalia. See* Callatis.
- *Manijas. See* Miletopolitis.
- Manius Aquillius, iii. 21.

- Mantiane, lake of Armenia, ii. 270. *See* Matiana.
- Mantineia, c. of Arcadia, ii. 8, 75; iii. 378.
- Manto, daughter of Tiresias, ii. 148; iii. 15, 59.
- Mantua, c. of Cisalpine Gaul, i. 317.
- Maracanda, ii. 254.
- Maranītæ, people of Arabia, iii. 204.
- Marathēsium, t. of Ionia (*Scala Nova*), iii. 10.
- Marathon, vill. of Attica, ii. 57, 59, 67, 86, 90.
- ——, Tetrapolis of, ii. 153.
- ——, field of Spain, i. 240.
- Marathus, t. of Phocis, ii. 122.
- ——, t. of Phœnicia, iii. 167.
- Marcellus, founder of Corduba, i. 212.
- ——, Marcus, i. 244.
- [Pg 370]Marcellus, son of Octavia, iii. 59.
- Marcia, wife of Cato, ii. 250.
- Marcian water, i. 356.
- Marcina, c. of Campania (*Vietri*), i. 374.
- Marcomanni, people of Germany, i. 444.
- Mardi, people of Persia and Armenia, ii. 240, 264.
- Mardonius, ii. 108.
- *Mare Morto. See* Acherusia.
- Mareōtis, Mareia, lake of Egypt, iii. 223, 228, 230, 236, 241, 247.
- Margala, Margalæ, t. of Triphylia, ii. 23, 24.
- Margiana, distr. of Asia, i. 112, 113.
- Margiani, ii. 245, 251, 252.
- Margus, r. of Margiana, ii. 252.
- ——, r. of Illyria, i. 488.
- Mariaba, city of the Sabæans, iii. 190, 207.
- Mariandyni, people of Paphlagonia, i. 453; ii. 18, 286-288, 290, 314; iii. 63.
- Mariandynus, ii. 287.
- Marinum, i. 337.
- Marisus, r. of Dacia (*Maros*), i. 468.
- *Maritza. See* Hebrus.
- Marius, i. 274, 354.
- Marmaridæ, people of Africa, i. 198; iii. 275, 294.
- Marmarium, t. of Eubœa, ii. 153.
- Marmōlītis, distr. of Paphlagonia, ii. 314.
- *Marmora, Sea of. See* Propontis.
- Marobodus, i. 444.
- Marōnia, t. of Thrace, i. 515, 516.

- *Maros. See* Marisus.
- Marrucina, i. 358, 359.
- Marrucini, people of Italy, i. 358, 359.
- Mars, i. 232, 277, 340, 357, 372; ii. 328, 362, 395, 409.
- *Marsa-al-Halal. See* Naustathmus.
- *Marsalla. See* Lilybæum.
- *Marseilles. See* Massalia.
- *Marseillese, the. See* Massilians.
- Marsi, people of Italy, i. 326, 349, 351, 353, 356, 358.
- Marsi, people of Germany, i. 443.
- Marsiaba, city of Arabia, iii. 212.
- Marsian or Marsic war, i. 353, 358, 388.
- Marsyas, ii. 186, 334, 390.
- ——, r. of Phrygia, ii. 303, 333, 334.
- Martius Campus, i. 350, 371.
- *Martos. See* Tukkis, i. 213.
- Marucini. *See* Marrucini.
- Maruvium, city of Italy, i. 359.
- Masæsylii, Masæsyli, people of Numidia, i. 198; iii. 279-282, 287, 289, 291.
- Masanasses, king of Numidia iii. 282, 285, 286.
- Masēs, t. of Argolis, ii. 54, 58.
- Masiani, people of India, iii. 90.
- Masius, mtn of Armenia (*Kara-Dagh*), ii. 238, 261, 268; iii. 157.
- Massabatica, distr. of Media, ii. 264; iii. 154.
- Massaga, t. of India, iii. 90.
- Massagĕtæ, Scythian race, ii. 240, 245, 247, 248; iii. 75.
- Massalia, city of Gaul (*Marseilles*),

i. 100, 110, 111, 114, 116, 117, 150, 160, 161,173, 174, 184, 202, 217, 221, 238-240, 247, 265-285, 301, 310, 376, 452; ii. 331,368; iii. 21, 30, 32, 279.

- ——, Gulf of, i. 271.
- Massilians, Massilienses (*the Marseillese*),

i. 194, 267, 269, 270, 271, 276, 282, 283,504.

- Massyas, distr. of Syria, iii. 166, 170, 171.
- Mastaura (*Mastauro*), t. of Lydia, iii. 26.
- Masthles, iii. 41.
- *Mastico. See* Phanæ.
- Masylies, people of Numidia, i. 198; iii. 282, 284, 285.
- Matalum, port of Gortyna, ii. 200.
- *Matapan, Cape. See* Tænarum.
- Mataurus, t. of Sicily, i. 416.
- Mater Isodroma, temple of, ii. 145.
- Matiana, distr. of Media, i. 78, 112; ii. 242, 262, 264, 270.

- [Pg 371]Matiani, Matiēnī, inhabitants of Matiana, ii. 249, 273.
- Matrīnum, port of Adria, i. 358.
- Matrīnus, r. of Picenum (*Piomba*), i. 358.
- *Matzua. See* Tabaïtic mouth of the Nile.
- Mauretania. *See* Maurusia.
- Mauri, same as Maurusii, iii. 276.
- *Maurolimne. See* Molycreia.
- Maurusia (*Algiers* and *Fez*), i. 3, 150, 151, 197, 201, 210, 215, 226, 236, 439; iii. 275-284.
- Maurusians, i. 7, 198, 206; ii. 36; iii. 276.
- Mausōleium, of Halicarnassus, iii. 34.
- ——, (*Tomb of Augustus*), at Rome, i. 351.
- Mausōlus, king of Caria, ii. 383; iii. 34, 35.
- Maximus Æmilianus. *See* Æmilianus.
- Mazaca (*Kaisarieh*), ii. 282, 283; iii. 44.
- Mazacēni, ii. 283.
- Mazæi, people of Pannonia, i. 483.
- Mazēnēs, king of the island Doracta, iii. 188.
- Mazūsia, prom. of the Thracian Chersonese, i. 517.
- Mecestus, r. of Phrygia (*Simau-Su*), ii. 332.
- Mecōne, same as Sicyon.
- Mecyberna, port of Olynthus, i. 511.
- Medea, i. 31, 72, 73, 321, 332, 484; ii. 266, 273.
- Mĕdeōn, t. of Phocis, ii. 106, 122.
- ——, t. of Bœotia, ii. 106, 122.
- Medes, i. 41, 196; ii. 125, 216, 230, 239, 249, 264-270; iii. 239.
- Mēdia, i. 72, 76, 112, 115, 123; ii. 235, 238, 240, 242, 250, 259-271, 273; iii. 109,124, 125, 129-134, 153, 154, 158.
- ——, Atropatian, ii. 260, 264, 267.
- ——, the Greater, ii. 260, 264, 267.
- Mediolanium (*Saintes*), i. 283.
- Mediolanum (*Milan*), i. 317.
- Mediomatrici, people of Gaul, i. 288, 289.
- Mediterranean Sea, i. 8, 56, 60, 62, 75, 78, 81, 82, 85, 87, 88, 105, 120, 128, 173,174, 183, 184, 189, 190, 192, 206, 210, 216, 234, 241, 244, 245, 253, 264, 266, *et passim*.
- Medius, historian, ii. 272.
- Medma, t. of Magna Grecia, ii. 383, 384.
- Medoaci, people of Italy, i. 321.
- Medoacus, r. of Italy, i. 318.
- ——, port of Patavia, i. 318.
- Mĕdon, ii. 134.

- Medus, son of Medea, ii. 266; iii. 132.
- Medusa, ii. 62.
- Mĕdylli, people of Gaul, i. 276, 303.
- Megabari, people of Ethiopia, iii. 203, 219, 266.
- Megabates, leader of the Persians, ii. 96.
- Megabyzi, priests of the Ephesians, iii. 13.
- Megalagyrus, ii. 391.
- Megalocephali, i. 458.
- Megalopolis, city of Arcadia, ii. 37, 71, 75.
- ———, city of Pontus, ii. 306, 310.
- Megalopolītæ, iii. 145.
- Megalopolītis, distr. of Arcadia, ii. 5, 15, 72.
- ———, distr. of Pontus, ii. 306, 309.
- Megara, city of Greece, ii. 3, 48, 57, 80-84, 108, 122; iii. 30.
- ———, city of Sicily, i. 403, 404, 406, 407.
- ———, city of Syria, iii. 165.
- Megaræans, Megareans, Megarians, i. 404, 412, 490, 494; ii. 2, 4, 63, 70, 81, 82, 84,315.
- Megarenses, Hyblæi, ii. 73.
- Megarici, sect of philosophers, ii. 82.
- Megaris, i. 256, 506; ii. 4, 6, 62, 63, 78-81, 84, 91, 99.
- ———, mountains of, ii. 99.
- [Pg 372]Megasthenes, of Chalcis, founder of Cumæ, i. 361.
- ———, historian, i. 107-109, 117, 120; iii. 75, 79, 80, 84, 96, 97, 101, 103, 110, 116.
- Megēs, son of Phyleus, ii. 9, 143, 167, 170.
- Megillus, iii. 83.
- Megiste, island, iii. 47.
- *Mekran. See* Gedrosia.
- Melæna, prom. of Ionia, iii. 18, 19.
- ———, prom. of Chios, iii. 18, 19.
- Melænæ, vill. of the Troad, ii. 371.
- ———, Melania, city of Cilicia (*Mandane*), ii. 371; iii. 52, 177.
- Melamphyllus, same as Samos, ii. 168; iii. 8.
- Melampus, ii. 20.
- Melanchus, tyrant of Lesbos, ii. 391. *See* Megalagyrus.
- Melanēis, same as Erĕtria, ii. 155.
- Melania, same as Melænæ.
- Melanippe, mother of Bœotus, i. 399.
- Melanippus, ii. 344.
- Melanthus, father of Codrus, ii. 35, 81, 82.
- Melantian rocks, in the Ægean (*Stapodia*), iii. 7.

- Mēn Ascæus, temple of, at Pisidia, ii. 307, 333.
- ——, Carus, temple at Antioch, ii. 307, 336.
- ——, Pharnaci, temple in Pontus, ii. 306, 307.
- Menander, comic poet, i. 455; ii. 162, 210; iii. 8, 9.
- ——, king of the Bactrians, ii. 252, 253.
- [Pg 373]Menapii, people of Germany, i. 289, 290, 297, 298.
- Mĕnas, founder of Mesembria, i. 490.
- *Menavyat-su. See* Melas.
- Mende, t. of Macedonia, i. 511.
- *Mender-Tschai. See* Mæander.
- Mendes, c. of Egypt, iii. 240.
- Mendesian mouth of the Nile, iii. 239.
- ——, nome, vill. of Egypt, iii. 240.
- Mendesians, iii. 257.
- Menĕbria, same as Mesēmbria.
- Menĕcles, iii. 34, 40.
- Mĕnēcrates, disciple of Xenocrates, ii. 299, 300, 326, 396.
- ——, disciple of Aristarchus, iii. 26.
- Menedēmus, an Eretrian philosopher, ii. 82, 156.
- Menelaïte nome, iii. 239.
- Menelaüs, son of Atreus, i. 3, 15, 18, 47, 59-64, 67, 68, 72, 76, 225, 333; ii. 35, 43,47, 53, 173, 186, *passim.*
- ——, brother of Ptolemy I., iii. 238, 239.
- ——, t. in the Delta of Egypt, iii. 242.
- ——, t. and port of Cyrenæa, i. 64; iii. 238, 239, 294.
- Menestheus, i. 392; ii. 83, 398.
- ——, port of (*Puerto Santa Maria*), i. 211.
- ——, son of Spercheus, ii. 136.
- ——, oracle of, i. 211.
- Meninx, island (*Zerbi*), i. 37, 185, 237; iii. 288.
- Menippus, of Gadara, iii. 175.
- ——, of Stratonice, iii. 40.
- Mennæus, iii. 166.
- Mēnŏdōrus, iii. 24.
- Mēnŏdŏtus, ii. 401.
- Menœtius, father of Patrocles, ii. 126.
- Mĕnōn, companion of Alexander, ii. 271.
- Mentes, king of the Taphii, ii. 166, 171.
- Mēŏnes, inhabitants of Lydia. *See* Mæones.
- Mēŏnia. *See* Mæonia.
- Mercury, i. 158; ii. 16; iii. 263.

307

- *Merida. See* Augusta Emerita.
- *Merim, Al. See* Moro.
- Mermadalis, r. in the land of the Amazons, ii. 235.
- Meroë, sister of Cambyses, iii. 225.
- ——, island, i. 50; iii. 195, 217-220, 270.
- ——, metropolis of Ethiopia,

i. 50, 98, 99, 106, 107, 110, 112, 119, 120, 144, 171,172, 174, 199, 200, 203; iii. 84, 217-220, 270-272, 275.

- Meropidæ, ii. 345.
- Meropis, name of the island of Cos, i. 459.
- Mĕrops, i. 52.
- ——, of Percotè, ii. 345, 346.
- *Mersivan. See* Neapolis.
- Mērus, mtn of India, iii. 76.
- *Mesarlyk-Tschai. See* Cydnus.
- Mesēmbria, t. of Thrace (*Missemvria*), i. 490.
- Mesēmbriani, i. 490.
- Mĕsēne, distr. of Babylon, i. 129.
- Mesēni, Arabians, iii. 146.
- Mĕsōgis, mtn of Lydia, ii. 145, 407, 408; iii. 7, 8, 24-26.
- ——, wine of, iii. 26.
- Mesola, ii. 37.
- Mesopotamia, i. 66, 122, 124, 127, 134, 137, 196, 416; ii. 260, 261, 267, 271, 284;

iii.109, 142, 150, 154, 156, 157, 161, 166, 185.

- Messa, t. of Laconia, ii. 41.
- Messala, i. 305.
- Messapia, same as Iapygia, i. 422, 423, 430; ii. 98.
- Messapian language, i. 431.
- Messapii, i. 427, 428.
- Messapius, mtn (*Ktypa-vuna*), ii. 98.
- Messapus, ii. 98.
- Messēis, ftn of Thessaly, ii. 134.
- Messēne, c. of Messenia, i. 186, 236, 425-427; ii. 5, 8, 14-

18, 22, 24, 25, 27, 28, 31,35, 37-40, 42, 45-47, 54, 77, 81.

- [Pg 374]Messēne, Messana, Messenia, in Sicily (*Messina*), i. 402-404.
- Messenian Gulf (*Gulf of Coron*), ii. 6, 35, 36, 40.
- ——, war, i. 385, 424-427; ii. 39.
- Messenians, i. 385, 404, 405, 425, 426; ii. 5, 7, 11, 31, 35, 36, 38, 39, 82; iii. 2.
- ——, in Sicily, i. 376.
- Messina, Strait of, i. 37, 39, 69, 85, 86, 105, 110, 140, 158-

160, 163, 173, 179, 180,184, 193, 256, 313-315, 360, 379, 383, 384, 386, 401-404; ii. 60.

- Messŏa, part of Sparta, ii. 41.

- Messōla, t. of Messenia (*Messthles*), ii. 403.
- Metabum, same as Metapontium, i. 399.
- Metabus, i. 399.
- Metagonians, i. 255, 256.
- Metagōnium, prom. of Mauritania (*Ras-el-Harsbak*), iii. 279, 282.
- Metapontium, city of Magna Græcia (*Torre di Mare*),
i. 330, 379, 380, 398, 399, 422,423.
- Metapontus, i. 399.
- Metaurus, r. of Umbria, i. 337.
- ———, r. of the Bruttii (*Metauro*), i. 383, 384.
- *Meteline. See* Lesbos.
- Metellus (Q. Metellus Pius), i. 244.
- ———, surnamed Balearicus. *See* Balearicus.
- Methana, Methōne, t. and penins. near Trœzene, ii. 56.
- Methōnē, city of Messenia, i. 92; ii. 36, 37.
- ———, c. of Macedonia, i. 508, 509.
- ———, c. of Magnesia, ii. 140.
- ———, of Thrace, ii. 140.
- Methydrium, t. of Arcadia, ii. 75.
- Mēthymna, city of Lesbos (*Molyvo*), ii. 127, 145, 390, 391, 393.
- Mēthymnæeans, ii. 351, 382.
- *Metochi d' Hagia. See* Mychus.
- Mētrodōrus, of Scepsis, ii. 235, 380; iii. 202.
- Mētrodōrus, disciple of Epicurus, ii. 350.
- Mētropolis, t. of Thessaly, ii. 141.
- ———, t. of Magna Phrygia (*Ballyk*), ii. 332 ; iii. 43.
- ———, t. of Lydia, iii. 1.
- Mētropolītæ, ii. 141; iii. 8.
- Mētrōum, temple of the mother of the gods, iii. 20.
- Mētūlum, t. of the Iapodes, i. 309, 483.
- Mēvania, city of Umbria (*Bevagna*), i. 337.
- Micipsa, son of Masinissa, iii. 282, 285, 286.
- Micythus, i. 376.
- Midaeium, c. of Phrygia Epictetus, ii. 332.
- Midas, king of Magna Phrygia, i. 96; ii. 321, 326; iii. 66.
- ———, Phrygian name, i. 467.
- Midea, t. of Bœotia, ii. 54.
- ———, t. of Argolis, ii. 54.
- Mideia, t. of Bœotia, i. 92; ii. 110.
- *Midjeh. See* Salmydessus.
- *Milan. See* Mediolanum.

- Minyeians, ii. 111.
- Minyeius, Minyeïus, r. of Triphylia, ii. 10, 20, 27, 55.
- Misēnum, prom. and port of Campania (*Punta di Miseno*), i. 93, 360, 362, 364, 368.
- Misēnus, companion of Ulysses, i. 39, 364.
- Mithracina, ii. 271.
- Mithras, god of the Persians, the Sun, iii. 136.
- Mithridates, of Pergamus, ii. 401.
- ——, Ctistes, king of Pontus, ii. 314.
- ——, Euergetes, king of Pontus, ii. 197, 198.
- ——, Eupator, king of Pontus, i. 22, 114, 440, 471, 472, 475-479; ii. 198, 209, 285,286, 290, 292, 294, 304-307, 313, 314, 356, 357, 380, 387, 401, 405; iii. 13, 232.
- Mithridatic war, ii. 89, 305-307, 312, 356, 386; iii. 25.
- Mithridatium, citadel of the Galatæ, ii. 320.
- Mithropastes, iii. 188.
- Mitylenæans, i. 5, 16; ii. 366, 374, 376, 392.
- Mitylene, c. of the island of Lesbos, ii. 145, 220, 365, 366, 391, 392.
- Mnasalces, ii. 108.
- Mnasyrium, iii. 33.
- Mneyis, god of the Egyptians, iii. 241, 245.
- Mnōans, ii. 287.
- Moagetes, tyrant of Cibyra, ii. 410.
- Moaphernes, ii. 228, 307.
- Moasada, fortress of Judæa, iii. 183.
- Mōchus, philosopher, iii. 173.
- *Modena. See* Mutina.
- Modra, t. of Phrygia, ii. 289.
- Mœris, lake in Egypt, i. 79; iii. 223, 253, 255, 257, 258.
- Mœsi, people of Thrace, i. 453, 454, 466 ; ii. 287.
- Mōlō, iii. 34, 40.
- Molochath, r. of Mauritania (*Muluwi*), iii. 279, 281.
- Molossi, Molotti, nation of Epirus, i. 427, 493, 495, 496, 499, 500, 502, 504; ii. 131,137, 356.
- Molycreia, t. of Ætolia (*Xerolimne* or *Maurolimne*), ii. 128, 160, 171, 172.
- Molycrium, Rhium, same as Antirrhium, ii. 6.
- *Molyvo. See* Methymna.
- Mōmemphis, city of Egypt, iii. 241, 242.
- Mōmemphitæ, iii. 241.
- Momemphite nome, iii. 241.
- Monarites wine, ii. 278.
- *Mondego. See* Mundas.
- Monētium, t. of the Iapodes, i. 309, 483.

- [Pg 376]Monœci Portus, in Liguria (*Port Monaco*), i. 300, 301.
- Monœcus, Hercules, temple of, i. 301.
- Monommati, i. 68, 458; iii. 108.
- *Monte Nero. See* Ænus.
- Moon, temple to the, ii. 234.
- Mopsium, t. of Thessaly, ii. 146, 148.
- Mopsŏpia, ancient name of Attica, ii. 87, 148.
- Mopsopus, ii. 87, 148.
- Mopsuhestia, t. of Cilicia, iii. 61.
- Mopsus, one of the Lapithæ, ii. 148.
- ——, son of Mantus, ii. 148; iii. 15, 50, 59, 60.
- *Morea, the. See* Apian land.
- Mŏrēna, distr. of Mysia, ii. 330.
- Morgantium, t. of Sicily, i. 386, 408.
- Morgētes, the, emigrate to Sicily, i. 385, 386, 407, 408.
- Morimēnē, distr. of Cappadocia, ii. 278, 281, 284, 321.
- Morimēni, ii. 321.
- Morini, people of Gaul, i. 289, 290, 297, 298.
- Mormolyca, i. 29.
- Moro, t. of Lusitania (*Al-Merim*), i. 228.
- Morys, ii. 317.
- Morzeus, king of Paphlagonia, ii. 314.
- Moschi, ii. 225.
- Moschic mountains, i. 96; ii. 219, 226, 260, 267, 296.
- Moschice, distr. of Colchis, ii. 228, 229.
- Mōsēs, iii. 177, 178.
- Mosynœci, people of Pontus, ii. 269, 297.
- *Mualitsch-Tschai. See* Rhyndacus.
- *Mudania. See* Myrleani.
- Mūgilŏnes, people of Germany, i. 445.
- Mulius, ii. 10.
- *Muluwi. See* Molochath.
- Mūnda, metropolis of Turdetania (*Mondo*), i. 213, 241.
- Mundas, r. of Lusitania (*Mondego*), i. 229.
- Mūnychia, port of the Athenians, ii. 85.
- Murēna, ii. 410; iii. 53.
- *Murviedro. See* Saguntum.
- Mūsæus, ii. 187; iii. 180.
- Muses, the, i. 66; ii. 11, 23, 24, 183, 187.
- Mūseum of Alexandria, iii. 229.
- Mūsicanus, distr. in India, iii. 85, 86, 95, 106.

- Mūtina, t. of Italy (*Modena*), i. 305, 322, 324.
- Mycale, prom. of Ionia (*Samsun Dagh*), i. 8; ii. 376; iii. 2, 7, 10.
- Mycalessus, vill. of Bœotia, ii. 96, 99, 106.
- Mycēnæ, c. of Argolis, i. 329; ii. 48, 53, 54, 59, 60.
- Mycēnæa, ii. 53.
- Mycenæans, ii. 53, 55.
- Mychus (*Bay of Metochi d' Hagia*), ii. 104, 122.
- Mycŏnus, island (*Myconi*), ii. 208, 211.
- Mygdones, i. 453, 506, 514; ii. 316, 330.
- ——, people of Mesopotamia, ii. 268; iii. 142, 157.
- Mygdonia, Mygdonis, part of Mysia, i. 515; ii. 299, 332, 348.
- ——, part of Mesopotamia, iii. 157.
- Mylæ, t. of Sicily (*Milazzo*), i. 401, 412.
- Mylasa, c. of Caria, iii. 37-40.
- Mylasians, iii. 38.
- Myndia, iii. 37.
- Myndii, Myndians, ii. 383; iii. 36.
- Myndus, c. of Caria, ii. 383; iii. 37.
- Mynēs, king of Lyrnessus, ii. 343, 384, 394.
- Myonnesus, island, ii. 138, 393.
- ——, t. of Ionia, iii. 16, 17.
- Myra, t. of Lycia, iii. 45, 47.
- Myrcinus, t. of Macedonia, i. 512.
- Myriandrus, t. of Syria, iii. 61.
- Myrina, Amazon, ii. 298, 328, 399.
- ——, c. of Æolis, ii. 237, 298, 397, 399.
- Myrinæans, ii. 397.
- [Pg 377]Myrleani (*Mudania*), ii. 330.
- Myrleātis, ii. 299.
- Myrleia, c. of Bithynia (*Mudania*), ii. 299, 315, 318.
- Myrmēcium, c. of the Tauri (*Yeni-kaleh*), i. 477; ii. 222.
- Myrmidons, ii. 50, 57, 132, 136.
- ——, tragedy of Æschylus, ii. 390.
- Myrŏn, statuary, iii. 7.
- Myrrinūs, vill. of Attica, ii. 90.
- Myrsilus, historian, i. 93; ii. 382.
- ——, tyrant of Mitylene, ii. 391.
- Myrsinus, t. of Elis, ii. 12, 13.
- Myrtŏan Sea, i. 187, 496, 518, 519; ii. 6, 49, 57.
- Myrtūntium, t. of Elis, ii. 13.
- ——, estuary near Leucada, ii. 171.

- Myscellus, Achæan, founder of Crotona, i. 394, 406, 407; ii. 73.
- Mysi, or Mysians, i. 9, 195, 453, 454, 457, 460, 461, 466, 468, 488;
ii. 277, 287, 302,316-319, 389, 402, 407; iii. 27, 38, 63, 67, *passim*.
- ——, around Olympus, ii. 319, 326, 330.
- ——, tragedy of Sophocles, ii. 32.
- Mysia (*Karasi*), i. 17, 202; ii. 32, 287, 298, 299, 314, 316, 317, 326-
332, 386, 401,403, 404, 407; iii. 67.
- ——, Abrettēnē, *see* Abrettēnē, ii. 330, 332.
- ——, Olympii, ii. 326.
- Mysian Bosporus, ii. 318.
- ——, Olympus, iii. 30.
- Mysius, r. of Mysia, ii. 390.
- Mysos, the beech tree, ii. 326, 327.
- *Mytilene. See* Lesbos.
- Myūs, c. of Caria (*Derekoi*), ii. 335; iii. 2, 6.
- ——, Hormus, port of Egypt on the Arabian Gulf (*Suffange-el-Bahri*), i. 178 ;
iii. 193,211, 213, 260, 261.
-
- Nabatæa, distr. of Arabia, iii. 204.
- Nabatæans, iii. 177, 189, 204, 209-211, 214.
- Nabiani, people inhabiting the Caucasus, ii. 239.
- Nabocodrōsor, iii. 75.
- Nabrissa, city of Bætica. *See* Nebrissa.
- Nabūrianus, mathematician, iii. 146.
- Nacoleia, t. of Phrygia Epictetus, ii. 332.
- Nagidus, t. of Cilicia, iii. 52, 69.
- *Nahr-Damur. See* Tamyras.
- *Nahr-el-Asy. See* Orontes.
- *Nahr-el-Kelb. See* Lycus.
- *Nahr-Ibrahim. See* Adonis.
- Naïs, ii. 375.
- Namnetæ, people of Aquitania (capital Nantes), i. 283.
- *Nanfio. See* Anaphe.
- Nanno, poem of Mimnermus, iii. 2, 3.
- Nantuātæ, an Alpine race, i. 303.
- Napata, city of Ethiopia, iii. 268.
- Napē, t. near Methymna, ii. 127.
- Napitinus, Gulf of, i. 379.
- *Naples. See* Neapolis.
- ——, Bay of. *See* Crater.
- Nar, r. of Umbria (*Nera*), i. 337, 349.

- Narbōn, the Narbonnaise (*Narbonne*), c. of Gaul, i. 159, 160, 184, 265-286, 302, 310.
- Narcissus, ii. 96.
- Narnia, r. of Dalmatia (*Narni*), i. 337.
- Narōn, c. of Umbria (*Narenta*), i. 484, 487.
- Narthacium, t. of Thessaly, ii. 136.
- Narthēcis, island, iii. 7.
- Narycus, t. of the Opuntian Locrians, ii. 126.
- Nasamōnes, race of Africa, i. 198; iii. 291, 294.
- Nasica, Cornelius Scipio, i. 484.
- Natison, r. near Aquileia, i. 319.
- Nauclus, son of Codrus, iii. 2.
- Naucratis, c. of Egypt, iii. 239, 242, 250, 252.
- Naulochus, t. of Mœsia, i. 490; ii. 145.
- Naupactus, c. of the Ozolean Locrians (*Lepanto*), ii. 127, 159.
- [Pg 378]Nauplia, c. of Argolis, ii. 48, 54, 55.
- Nauplius, son of Neptune, ii. 48.
- Nauportus, c. of the Taurisci. *See* Pamportus, i. 482.
- Naustathmus, ii. 360, 363, 364, 385.
- ——, port of the Cyrenaic, iii. 294.
- Naxos, island (*Naxia*), ii. 208, 210.
- ——, c. of Sicily, i. 403-406, 412, 414.
- Nea, vill. (*see* Enea and Ænea), in the district of Troy, ii. 371.
- Neæethus, r. of Magna Grecia (*Nieto*), i. 394.
- Neandria, t. of the Troad, ii. 373, 375.
- Neandris, ii. 189.
- Neanthēs, of Cyzicus, i. 71.
- Neapolis, c. of the Tauric Chersonese, i. 479.
- ——, c. of Macedonia (*Kavala*), i. 512, 513.
- ——, c. of the Samians on the coast of Ephesus, iii. 10.
- ——, c. of Pontus (*Mersivan*), ii. 311.
- ——, c. of the Carthaginians, iii. 288, 289.
- ——, c. of Campania (*Naples*), i. 34, 202, 365, 366, 369, 377.
- ——, same as Leptis, iii. 289.
- Neapolitans, i. 39, 368, 369.
- Neapolitis, same as the Halys, ii. 311.
- Nearchus, i. 109, 119; iii. 74, 80-85, 88, 100, 101, 114, 115, 120, 122, 127-129, 132,187, 188.
- Nebrissa (*Lebrixa*), i. 211, 215.
- Nebrodes, mtns, i. 415.
- Necropolis, suburb of Alexandria, iii. 230, 231, 236.

- Necyomanteis, iii. 180.
- Nĕda, r. of Peloponnesus, ii. 17, 22, 24, 26, 37.
- Nĕdōn, r. of Laconia, ii. 28, 37.
- ——, vill. of Messenia, razed by Teleclus, ii. 37.
- Nedūsian Minerva, temple of, ii. 37.
- Negra, iii. 218.
- Negrana, c. of Arabia, iii. 212.
- Negrani, iii. 212.
- *Negropont. See* Eubœa.
- Neis, ii. 394.
- Nĕïum, in the isl. of Ithaca, ii. 164.
- Neleïdæ, i. 398; ii. 35.
- Nēleus, father of Nestor, ii. 7, 17, 26, 27.
- ——, founder of Miletus, iii. 2, 4.
- ——, son of Coriscus, ii. 378, 379.
- ——, r. of Eubœa, ii. 157.
- Nēlia, t. of Magnesia, ii. 139.
- Nemausus, city of Gaul (*Nîmes*), i. 267, 268, 278, 279, 302.
- Nĕmĕa, r. near Corinth, ii. 66.
- ——, t. of Argolis, ii. 60.
- Nemean games, ii. 60.
- ——, lion, ii. 60.
- Nemesis, statue of, at Rhamnusia, ii. 87, 90.
- ——, Adrastea, temple of, ii. 348.
- Nemōssus, head of the Arverni, i. 284.
- Nĕmus, temple of Diana near Aricia, i. 355.
- Nemydia, cognomen of Diana, ii. 14.
- *Neochori. See* Argos, Amphilochian.
- Neoclēs, father of Epicurus, iii. 9.
- Neocōmītæ, i. 317.
- Neōn, t. in the vicinity of Parnassus, ii. 143.
- Neon-teichos, ii. 396.
- Neoptolemus, son of Achilles, ii. 119, 120, 140, 343.
- ——, general of Mithridates, i. 472.
- ——, writer of glosses, ii. 350.
- ——, tower of, i. 469.
- Nepheris, fortress of the Carthaginians, iii. 287.
- Nĕpita, t. of Etruria (*Nepi*), i. 335.
- Neptune, i. 32, 53, 69, 342; ii. 6, 17, 48, 70, 72, 211, 335, 367.
- ——, temples of, ii. 16, 40, 55, 69, 71, 109, 154, 213; iii. 7, 230.
- ——, Asphalian, i. 90.

317

- *Noja. See* Netium.
- Nola, i. 367, 370.
- Nomades, Numidæ (Wanderers), i. 51, 198, 441, 461; ii. 231, 233, 240, 299, 302, 352; iii. 166, 194, 197, 198, 205, 212.
- Nomentana Via, i. 339, 354.
- Nomentum, i. 338.
- Nora, ii. 281.
- Noreia, c. of Cisalpine Gaul (*Friesach in Steiermark*), i. 319.
- Norici, Taurisci Norici, i. 306, 307, 310, 448.
- Northern Ocean, i. 451, 452.
- Nŏtium, coast and promontory of Chios, iii. 18.
- Notu-ceras, promontory of Ethiopia, iii. 200.
- Nŏtus, wind, i. 45, 97.
- Novum-comum, t. of Cisalpine Gaul, i. 317.
- Nūbæ, people inhabiting the Nile, iii. 219, 266.
- Nūcĕria, c. of Umbria (*Nocera Camellaria*), i. 337.
- ———, c. of Campania (*Nocera de' Pagani*), i. 367, 370, 374.
- Numa Pompilius, i. 338, 339.
- Numantia, c. of Spain, i. 229, 243, 244, 439.
- Numantians, the, i. 243.
- Numitor, king of Alba-longa, i. 340.
- Nycteus, father of Antiopa, ii. 97.
- ———, cave, iii. 164.
- Nymphæum, c. of the Tauric Chersonese, i. 476.
- ———, prom. of Mt Athos (*Cape St. George*), i. 512.
- ———, rock near Apollonia, i. 486.
- Nỹsa, vill. of Bœotia, ii. 99.
- Nysa or Nysaïs, distr. of Lydia, ii. 345, 408.
- ———, c. of Caria, iii. 24-27, 43.
- ———, c. and mtn of India, iii. 76.
- ———, mtn of Thrace, iii. 76.
- Nysæi or Nysæans, people of India, iii. 76, 90.
- Nysaeis, in Caria, ii. 335.
- Nysaïs, distr. of Lydia. *See* Nysa.
-
- Obelisks, iii. 245, 262.
- Obidiacēni, a Mæotic race, ii. 223.
- Obodas, king of the Nabatæi, iii. 211-213.
- Obulco (*Porcuna*), i. 213, 241.
- Ocalea, t. of Bœotia, ii. 106, 107.
- Ocean, i. 4-6, 33, 38, 39, 53, 68, 73, 74, 111, 143, *et passim*.

- Œnōnē, wife of Paris, ii. 360.
- Œnōnē, ancient name of Ægina, ii. 57.
- ——, two demi of Attica, ii. 57.
- Œnoparas, r. of Syria (*Afreen*), iii. 164.
- Œnops, ii. 375.
- Œnōtri, Œnotrians, Œnotrides, ancient inhabitants of Lucania, i. 376, 377, 379, 380,386, 399.
- Œnōtria, i. 313, 379, 400.
- Œnōtrian kings, i. 383.
- Œnōtrides, islands, i. 376, 387.
- Œta, mtn, i. 505; ii. 4, 114-116, 128-132, 135, 147, 158, 160.
- Œtæa, distr. of Thessaly, ii. 123, 132, 136, 160.
- Œtæan hellebore, ii. 116.
- Œtæi, people of Greece, ii. 114, 158, 386.
- Œtylus, t. of Laconia, ii. 35, 36.
- Œum, castle of, i. 95.
- *Ofanto. See* Aufidus.
- Ogyges, ii. 68.
- Ogygia, ancient name of Bœotia, i. 38; ii. 101.
- Ogyium, mtn, i. 459.
- Ogyris, isl. of the Red Sea, iii. 187, 188.
- Oïsci. *See* Vivisci.
- Olane, citadel of Armenia, ii. 270.
- Olba, t. of Cilicia, iii. 55.
- Olbia, city of the Massilians (*Eoube*), i. 269, 275.
- ——, c. of Pamphylia (*Tschariklar*), iii. 48, 49.
- ——, c. situated on the Borysthenes, i. 470.
- Oleastrum, t. of Spain, i. 239.
- Olĕnian rock, mtn of Elis, ii. 12-14, 27, 74.
- Olenii, ii. 71.
- Olĕnus (*Olĕnē?*), c. of Achaia, ii. 69, 71-74, 160.
- ——, c. of Ætolia, ii. 72, 160, 172.
- Olgassys, mtn of Paphlagonia, ii. 313, 314.
- Oliarus, one of the Cyclades (*Antiparos*), ii. 208.
- Oligasys, name of the Paphlagonians, ii. 302.
- Olizōn, t. of Magnesia, ii. 139, 140.
- [Pg 382]Olmeius, r. of Bœotia, ii. 101, 108.
- Olmiæ, prom. of the Bay of Corinth, ii. 63, 105.
- Oloossōn, t. of Thessaly, ii. 143, 145.
- Olophyxis, t. of Macedonia, i. 512, 513.
- Olūris, Olūra, t. of Messenia, ii. 24.

322

- Ophrynium, t. of the Troad, ii. 357.
- Opici, people of Campania, i. 360, 372; iii. 32.
- Opis, c. of Assyria, i. 124; ii. 271; iii. 146, 147.
- Opistholepria, part of Ephesus, iii. 3.
- Opitergium, t. of Cisalpine Gaul (*Oderzo*), i. 319.
- Opoeis, same as Opus, ii. 125.
- Opsicella (Ocella?), t. of Cantabria, i. 236.
- Opuntii. *See* Locri in Elis.
- Opuntian Gulf, ii. 114, 125, 126.
- Opūs, c. of the Locrians, i. 95; ii. 125, 126.
- *Orange. See* Arausio.
- Orbēlus, mtn of Macedonia (*Egrisou-dagh*), i. 505.
- Orbis, r. of Gaul (*the Orbe*), i. 272.
- Orcaorci, ii. 320, 321, 332.
- Orchēni, sect of the Chaldæan astronomers, iii. 146.
- Orchistēnē, distr. of Armenia, ii. 268.
- Orchomĕnia, ii. 93, 101.
- [Pg 383]Orchomĕnii, ii. 93, 111-113.
- Orchomenus, c. of Bœotia (*Scripa*), ii. 10, 101, 107, 111-113, 124.
- ———, Minyeian, ii. 20, 55, 111.
- ———, c. of Arcadia, ii. 10, 39, 75.
- ———, c. of Eubœa, ii. 113.
- Oreitæ, ii. 152; iii. 120, 124.
- Orestæ, people of Epirus, i. 499; ii. 137.
- ———, in Macedonia, i. 505, 508.
- Orestes, i. 499; ii. 68, 93, 279, 281, 339.
- ———, tragedy of Euripides, ii. 60.
- Orestia, part of Macedonia, i. 514.
- Orestias, distr. of Epirus, i. 499, 500.
- Orestis, distr. of Macedonia, i. 505.
- Oretani, people of Spain, i. 209, 212, 228, 234, 235, 243, 245.
- Orētania, i. 210, 228, 243.
- *Oreto. See* Oria.
- Oreus, c. of Eubœa (*Orio*), i. 94; ii. 152, 153.
- Oreus-Histiæa, ii. 152.
- Orgās, r. of Phrygia, ii. 333.
- Oria, c. of Spain (*Oreto*), i. 229.
- ———, vill. of Eubœa, ii. 152.
- Orīcum, t. of Illyria, i. 486.
- Oriōn, ii. 97, 152.
- ———, constellation, i. 5.

- Orithyia, i. 452.
- *Orleans. See* Genabum.
- Ormenium, Orminium, t. of Thessaly, ii. 134, 139, 142, 143, 146.
- Ormenus, Ormenides, ii. 142, 143.
- Orminium. *See* Orměnium.
- Orněæ, t. near Corinth, ii. 58, 59, 66, 347.
- ——, vill. of Argolis, ii. 58, 59.
- Orněates, Priapus, ii. 66.
- Ornithōpōlis, in Phœnicia, iii. 173.
- Oroatis, r. of Persia (the *Tab*), iii. 129, 132.
- Orobiæ, t. of Eubœa, ii. 98, 152.
- Orōdes, king of the Parthians, iii. 97.
- Orontēs, r. of Syria (*El-Asy*), i. 416; iii. 53, 61, 162-165, 170, 177.
- ——, king of Armenia, ii. 273.
- Orōpia, ii. 79.
- Orōpii, ii. 90.
- Orōpus, c. of Bœotia, i. 102, 103; ii. 79, 90, 92, 96.
- Orospeda, mtns of Spain (*Sierra de Toledo*), i. 241, 243, 245.
- Orpheus, Thracian, i. 508, 513; ii. 187; iii. 180.
- Orphic ceremonies, ii. 187.
- ——, arts, ii. 192.
- Orthagoras, iii. 187.
- Orthagŏria, c. of Thrace, i. 516.
- Orthanēs, god of the Athenians, ii. 348.
- Orthē, citadel of Thessaly, ii. 143, 144.
- Orthopolis, t. of Macedonia, i. 514.
- Orthōsia, c. of Syria (*Ortosa*), iii. 53, 167, 169, 171, 177.
- ——, c. of Caria, iii. 26.
- Ortilochus, father of Diocles, ii. 46.
- Ortōn, port of the Frentani (*Ortona-à-Mare*), i. 359, 360.
- *Ortona. See* Orthosia.
- Ortōnium, i. 359.
- Ortospana, c. of the Paropamisadæ (*Candahar*), ii. 249; iii. 124.
- Ortygia, nurse of Latona, iii. 11.
- ——, grove near Ephesus, iii. 11.
- ——, same as Delos, ii. 210, 211.
- ——, isl. and part of the city of Syracuse (*Island of St. Marcian*), i. 35, 92, 408, 409.
- Osca, c. of Spain (*Huesca*), i. 242.
- Osci, people of Campania, i. 346, 352, 360, 367.
- *Osimo. See* Auxumon.
- Osiris, iii. 242, 243, 247, 253, 259.

- Osismii, people of Gaul, i. 291.
- Ossa, mtn of Thessaly, i. 40, 94, 96, 311, 507; ii. 130, 131, 139, 146-148, 272.
- Ossa, mtn of Pisatis, ii. 32.
- Ossŏnŏba, c. of Spain, i. 215.
- Ostia, c. of Latium, i. 218, 325, 329, 334, 335, 339, 344, 345, 348.
- Ostimii (*al.* Ostiæi, Ostidamnii, Timii), people of Gaul, i. 99, 101, 291.
- *Osuna. See* Usor.
- [Pg 384]Othia, ii. 12.
- Othryadas, a Lacedæmonian, ii. 58.
- Othryoneus, ii. 367.
- Othrys, mtn of Thessaly (*Mt Gura*), ii. 32, 134, 135, 138.
- *Otranto. See* Hydrus.
- Otreus, ii. 318.
- *Otricoli. See* Ocricli.
- Otrœa, t. of Bithynia, ii. 318.
- Otus, Cyllenian, ii. 9, 167.
- Oxeiæ islands, ii. 169, 170.
- Oxus, r. of Bactriana (*Gihon*), i. 113; ii. 240, 243, 251-255.
- Oxyartes, ii. 254.
- Oxyathrēs, brother of Darius Codomannus, ii. 291.
- Oxybii, or Oxybian Ligurians, i. 276, 301.
- Oxybius, port in Liguria, i. 276.
- Oxylus, king of the Ætolians, ii. 3, 30, 33, 77, 176.
- Oxynia, t. of Thessaly, i. 501.
- Oxyrynchus, city of Egypt (*Bekneseh*), iii. 257.
- Ozolæ. *See* Locri.
-
- Paches, Athenian commander, ii. 366.
- Pachynus, prom. of Sicily (*Cape Passaro*), i. 160, 186, 187, 400-403, 411, 421; ii. 41.
- Pacorus, leader of the Parthians, iii. 159, 163.
- Pactōlus, r. of Lydia, ii. 303, 353, 403.
- Pactya, c. of the Thracian Chersonese, i. 517, 518.
- Pactyas, mtn, iii. 7, 21.
- *Padua. See* Patavium.
- Padus. *See* Po.
- Pæŏnia, i. 488, 489, 504, 505, 509, 512-514; ii. 131.
- ——, mtns, i. 43, 481, 496.
- ——, nations, i. 485.
- Pæŏnians, Pæones, people of Macedonia, i. 9, 489, 495, 496, 506, 508, 514, 515; ii.383, 394.
- Pæsēni, ii. 349.

325

- Pæstum, Gulf of, i. 373.
- Pæstus, c. of Lucania, i. 373.
- Pæsus, t. of the Troad, ii. 349; iii. 5.
- ——, r. of the Troad (*Beiram-dere*), ii. 349.
- Pagæ, Pēgæ, t. of Megaris (*Liba-dostani*), ii. 4, 63, 79, 82, 92, 105.
- Pagasæ, t. of Magnesia, ii. 139.
- Pagasitic Gulf, in Thessaly, ii. 140, 142.
- Pagræ, fortress of Syria (*Baghrus*), iii. 163.
- *Paitschin. See* Pedasus.
- Palacium, t. of the Tauric Chersonese, i. 479.
- Palacus, i. 471, 475.
- Palæa, t. of Mysia, ii. 387.
- ——, t. of Cyprus, iii. 69.
- Palæbyblus, c. of Phœnicia, iii. 170.
- Palæopolis, the old city where the Emporitæ dwelt, i. 240.
- Palæpaphus, c. of Cyprus, iii. 70.
- Palæpharsalus, t. of Thessaly, ii. 133; iii. 233.
- Palæphatus, ii. 299, 300.
- Palærus, t. of Acarnania (*Porto Fico*), ii. 159, 171.
- Palæscēpsis, t. of the Troad, ii. 371, 372, 375, 376.
- Palæ-tyrus, t. of Phœnicia, iii. 173.
- Palamēdēs, son of Nauplius, ii. 48.
- ——, tragedy of Euripides, ii. 186.
- Palatium, hill of Rome, i. 348, 351.
- Paleis, t. of Cephallenia, ii. 166, 167.
- *Palencia. See* Pallantia.
- *Paleocastro. See* Eretria and Thronium.
- *Palermo. See* Panormus.
- Palestine, iii. 204.
- *Palestrina. See* Præneste.
- Palibothra, Palimbothra, c. of India (*Patelputer*), i. 109; iii. 79, 80, 90, 97, 118.
- Palibothrus, iii. 97.
- Palici, in Sicily, i. 416.
- Palinthus, sepulchre of Danaus, ii. 52.
- Palinurus, prom. of Lucania, i. 376.
- [Pg 385]Paliurus, t. of the Cyrenaic, iii. 294.
- Pallantia, t. of Spain (*Palencia*), i. 243.
- Pallas, son of Pandion, ii. 81.
- Pallēne, peninsula of Macedonia, i. 510-512; ii. 154, 299.
- Palma, t. of the Baleares, i. 251.
- Palmys, son of Hippotion, ii. 317.

- Paltus, t. of Syria, iii. 130, 167.
- Palus. *See* Mæotis.
- Pambœotia, festival, ii. 108.
- *Pambuk-Kalessi. See* Hierapolis.
- Pamisus, r. of Messenia (*Pirnatza*), i. 403; ii. 28, 37, 38, 45.
- ——, r. of Laconia, ii. 38.
- ——, r. of Elis, ii. 7, 16, 38.
- *Pampeluna. See* Pompelon.
- Pamphylia (*Tekiah*), i. 190, 194; ii. 244, 325, 385; iii. 27, 28, 44-55, 59, 61, 68, 73.
- ——, cities, ii. 324.
- ——, Sea of, i. 183, 189; ii. 325; iii. 68.
- ——, Gulf of, i. 189.
- Pamphylians, i. 196; ii. 304; iii. 63.
- Pamportus, t. of the Taurisci. *See* Nauportus, i. 309.
- Pan, worshipped by the Mendesii, iii. 240.
- ——, in Meroë, iii. 271.
- Panænus, ii. 29.
- Panætius, Stoic, iii. 33, 60.
- *Panaro. See* Scultanna.
- Panchæa, i. 58, 459.
- Pandarus, king of the Lycians, ii. 317, 344, 346.
- ——, worshipped at Pinara, iii. 46, 47.
- Pandataria, island (*Vento Tiene*), i. 185, 347.
- Pandiōn, father of Lycus, ii. 328.
- ——, king of India, iii. 49, 74.
- Pandionidæ, ii. 81.
- Pandōra, mother of Deucalion, ii. 149.
- ——, name of Southern Thessaly, ii. 149.
- Pandosia, c. of the Bruttii, i. 382.
- ——, c. of Thesprotia, i. 382, 497.
- Paneium, temple of Pan, at Anaphlystus, ii. 89.
- ——, at Alexandria, iii. 231.
- Pangæum, mtn of Macedonia, i. 512, 515; iii. 66.
- Pangani, ii. 239.
- Panhellenes, ii. 50.
- Panionian festival, sacrifices, ii. 69.
- Paniōnium (*Ischanli*), iii. 10.
- Panna, t. of Samnium, i. 371.
- Pannōnia, i. 483; iii. 10.
- Pannonii, Pannonians, i. 309, 448, 482, 483, 487.
- Panŏpeis, ii. 121, 123.

- Panŏpeus, t. of Phocis, ii. 113, 122, 123, 124.
- Panŏpolis, t. of Egypt, iii. 258.
- Panormus, t. of Sicily (*Palermo*), i. 401, 411.
- ——, t. of Epirus (*Panormo*), i. 486, 497.
- ——, port of Ephesus, iii. 11.
- Pans, with wedge-shaped heads, i. 109; ii. 186.
- *Pantalaria. See* Corcyra, Cossura.
- Pantaleŏn, son of Omphalion, ii. 39.
- *Pantano, l' Osteria del. See* Gabii.
- Panticapæans, ii. 222.
- Panticapæum (*Kertsch*), i. 472, 476-478.
- ——, temple of Æsculapius at, i. 114; ii. 221.
- Panxani. *See* Pangani.
- *Papa, Cape. See* Araxus.
- Paphlagonia, i. 96, 202, 475; ii. 18, 285, 287, 289, 290, 297, 301, 314, 319, 329.
- Paphlagonian names, ii. 301.
- Paphlagonians, i. 195, 279, 291, 440, 458, 476; ii. 18, 302, 304, 310, 313, 314.
- Paphos, c. of Cyprus, i. 65; ii. 13; iii. 68, 70, 71.
- ——, Palæpaphos, iii. 70, 71.
- Parachelŏïtæ, in Thessaly, ii. 136.
- ——, in Ætolia, ii. 136.
- Parachelŏïtis, distr. of Ætolia, ii. 169.
- Parachoathras, mtn of Media, ii. 245, 249, 260, 267.
- [Pg 386]Paradeisus, t. of Cœle-Syria, iii. 170.
- Parætacæ, Parætacēni, a people of Media, ii. 261, 264; iii. 135, 142, 146, 153.
- Parætacēne, i. 123; iii. 124, 132, 152.
- Parætŏnium, port of Marmara (*El-Baretun*), i. 64; iii. 235, 253, 259, 294.
- Paralus, founder of Clazomenæ, iii. 3.
- Parapomisus, ii. 245, 248; iii. 78, 89, 124-126.
- Parapotamia, Parapotamii, t. of Phocis, ii. 101, 113, 123, 124.
- ——, distr. of Syria, iii. 166.
- Parasŏpia, distr. of Bœotia, ii. 97, 103.
- Parasŏpias, in Thessaly, ii. 136.
- Parasŏpii, in Bœotia, ii. 103.
- ——, vill. of Thessaly, ii. 66.
- Parati, people of Sardinia, i. 334.
- Pareisactus, iii. 230.
- Parianē, ii. 350.
- Pariani, ii. 347, 348, 374.
- Parii, Parians, inhabitants of the island of Paros, i. 484; ii. 210, 349.
- Paris (Alexander), i. 65, 274; ii. 360, 376.

- Parisa, ii. 235.
- Parisades, i. 476, 477.
- Parisii, people of Gaul, i. 290.
- Parisus, r. of Pannonia, i. 482.
- Parium, c. of Mysia (*Kamaraes* or *Kemer*), i. 518; ii. 340, 348-351.
- ——, in the Propontis, ii. 210.
- Parma, t. of Cisalpine Gaul, i. 322.
- Parmĕnides, i. 143, 375.
- Parmenio, ii. 272; iii. 125.
- Parmesans, i. 323.
- Parmesus, ii. 108.
- Parnassii, ii. 121.
- Parnassus, i. 40, 311, 505; ii. 2, 62, 67, 93, 105, 114-116, 121, 123, 125, 129, 143,158, 195.
- Parnēs, mtn of Attica, ii. 90.
- Parni, Aparni, ii. 241, 244, 248.
- Paropamisadæ, iii. 77, 82, 124-128.
- Paropamisus. *See* Parapomisus.
- Parōræa, distr. of Epirus, i. 498.
- Parōræi, people of Epirus, i. 499.
- Parorbelia, distr. of Macedonia, i. 514.
- Parōreatæ, people of Triphylia, ii. 19.
- Parōreius. *See* Phrygia, iii. 43.
- Paros (*Bara*), i. 332, 484; ii. 66, 208, 210, 211.
- Parrhasii, people of Arcadia, ii. 7, 75, 241.
- Parrhasius, the painter, iii. 14.
- Parsii, same as Parrhasii.
- Parthenia, same as Samos, ii. 168; iii. 8.
- Partheniæ, i. 424-426.
- Parthenias, r. of Elis, ii. 32.
- Parthĕnium, mtn of Arcadia (*Partheni*), ii. 76.
- ——, prom. of the Tauric Chersonese, i. 474.
- ——, vill. of the Cimmerian Bosporus, i. 474, 477; ii. 222.
- ——, temple of Diana in the Tauric Chersonese, i. 474.
- Parthenius, r. of Samos, *see* Imbrasus, ii. 168.
- ——, r. ii. 287-290, 351.
- ——, mtn, ii. 58, 76.
- Parthenōn, temple of Minerva, in the Acropolis, ii. 84, 86.
- Parthenopē, t. of Campania, iii. 32.
- ——, one of the Sirens, i. 34, 39.
- ——, tomb of, i. 365.

- Parthi, Parthyæi, Parthians, i. 18, 22, 195, 196, 441; ii. 216, 241-245, 250-255, 263-277; iii. 97, 124-126, 131, 135, 136, 152, 159, *et passim.*
- Parthi, Histories of, i. 178; iii. 73.
- Parthia, ii. 246, 250, 251, 262, 264-274; iii. 124, 128, 131, 141, 152, 153.
- Parthian autocrat, Labienus, iii. 39.
- Parthiene, ii. 250.
- Parthini, people of Illyria, i. 500.
- [Pg 387]Parus, one of the Liburnian islands. *See* Pharus.
- Paryadrēs, mtn of Armenia (*Kerasun*), ii. 226, 260, 267, 269, 296, 305, 306.
- Parysatis, iii. 216.
- Pasargadæ, c. of Persia (*Fesa*), iii. 116, 131-134.
- Pasiani, Scythian race, ii. 245.
- Pasitigris, iii. 131, 132.
- *Passaro, Cape. See* Pachynus.
- Patala, c. of India, iii. 95.
- Patalēne, distr. of India (*Tatta* or *Sindi*), ii. 253; iii. 80, 83, 84, 94, 95, 120.
- Patara, c. of Lycia (*Patera*), iii. 45, 47.
- Patarus, iii. 47.
- Patavini, city of the (*Padua*), i. 253.
- Patavium (*Padua*), i. 317, 324.
- Pateischŏreis, Persian nation, iii. 130.
- *Patelputer. See* Palibothra.
- Patmos, one of the Sporades (*Patmo*), ii. 212.
- Patræ, Patreis, c. of Achæa (*Patras*), ii. 6, 8, 69, 71, 73, 74.
- Patraeus, t. of the Bosporus (*Ada*), ii. 222.
- Patrŏclēs, i. 106-108, 115; ii. 242, 243, 255; iii. 79.
- Patrŏclus, leader of the Myrmidons, ii. 26, 89, 126, 136, 343, 395.
- ——, rampart of, isl., ii. 89.
- Paulus Æmilius, i. 495, 516.
- Paunitis, ii. 269.
- *Pavia. See* Ticinum.
- Pax Augusta, t. of Spain (*Beja*), i. 227.
- Pēdalium, prom. of Cyprus, iii. 69.
- Pēdasa, t. of Caria, ii. 383.
- Pēdaseis, ii. 383.
- Pēdasis, ii. 383.
- Pēdasum, t. of Caria, ii. 383.
- Pēdasus, t. of the Leleges, near Troy (*Paitschin*), i. 494; ii. 343, 375, 383, 395.
- ——, t. of Messenia, ii. 35-37.
- Pegasitic Gulf (*G. of Volo*), i. 512.
- Pēgasus, ii. 62.

- Peiræeus, port of Athens. *See* Piræus.
- Peirēne, ftn of the Acrocorinthi, ii. 62.
- Peirithous, son of Ixion, ii. 137, 144.
- Peirōssus, c. of Mysia, ii. 349.
- Peirus, r. of Achæa, ii. 14, 71.
- Peirustæ, nation of Pannonia, i. 483.
- Peisander, poet, iii. 34, 78.
- ——, son of Bellerophon, ii. 328, 409.
- Peisistratus, son of Nestor. *See* Pisistratus.
- Pĕïum, citadel of the Tolistobogii, ii. 320.
- *Pelagio. See* Lechæum.
- Pelagonia, part of Macedonia, i. 500, 508, 514, 516.
- ——, Tripolitis, i. 500, 501.
- Pelagonians, Pelagones, people of Macedonia, i. 501, 514; ii. 137.
- ——, same as Titans, i. 514.
- Pĕlana, t. of Messenia, ii. 37.
- Pelargi, same as Pelasgi, ii. 87.
- Pelasgi, Pelasgians, i. 328, 329, 335, 367, 493, 501, 502, 513; ii. 18, 87, 93, 105, 148,288, 395-397; iii. 40.
- Pelasgia, Peloponnesus, i. 329.
- Pelasgian Zeus, i. 328, 329.
- ——, Argos, i. 329; ii. 49, 133.
- ——, temple, ii. 93.
- ——, plain, ii. 139, 148.
- Pelasgicum, part of Athens, ii. 49.
- Pelasgiōtæ, same as Hellenes, i. 329; ii. 52, 132, 146.
- Pelasgiōtis, part of Thessaly, i. 503, 504; ii. 132, 139.
- Pelasgus, i. 329.
- Pĕlĕgōn, father of Asteropæus, i. 514.
- Pĕlethrŏnium, i. 460.
- Pēleus, ii. 115, 126, 135-137, 142.
- Pĕlias, i. 72, 74; ii. 139.
- ——, daughters of, i. 72.
- Pĕligni, people of Italy, i. 326, 344, 353, 358-360.
- Pelinæum, mtn, iii. 19.
- [Pg 388]Pelinnæum, t. of Histiæotis, ii. 141, 142.
- Pēliŏn, mtn of Thessaly, i. 33, 40, 311, 460, 507; ii. 130, 131, 139, 142, 144-148, 157.
- Pella, c. of Macedonia, i. 495, 508, 509, 516.
- ——, c. of Syria, iii. 165.
- Pellæan country (*Pelagonia*), i. 508.

- Pellana, t. of Laconia, ii. 72.
- Pellene, t. of Achæa, ii. 59, 71, 72.
- ——, vill. of Achæa, ii. 72.
- Pēlōdes, lake, in Epirus, i. 497.
- Pelopidæ, ii. 51, 53, 59; iii. 66.
- Peloponnesian war, ii. 366; iii. 32.
- Peloponnesians, ii. 50, 175.
- Peloponnesus,

i. 40, 105, 140, 158, 159, 186, 201, 256, 257, 329, 330, 385, 400, 404,408, 478, 492, 496; ii. 2-11, 33, 34, 38, 43, 49-51, 57, 59, 60, 68, 71, 77, 78, 80,128, 140, 142, *et passim*.

- ——, figure of, i. 128; ii. 5.
- ——, islands, ii. 192.
- ——, promontories of, i. 139; iii. 291.
- Pelops, i. 492; ii. 31, 36, 39, 43, 56, 326.
- Pelorias, i. 400-404.
- Pelorus, monument of, i. 17.
- ——, tower, i. 256.
- ——, Cape (*Cape Faro in Sicily*), i. 34, 384.
- Peltæ, t. of Phrygia, ii. 332.
- Peltinian plain, in Phrygia, ii. 407.
- Pelūsiac mouth of the Nile. *See* Nile.
- Pelūsium, c. of Egypt (*Tineh*), i. 55, 62, 79, 91, 129, 134, 135; ii. 217; iii. 171, 175-177, 222, 226, 233, 241, 243.
- Pēneius, r. of Peloponnesus, ii. 8, 9, 11.
- ——, r. of Thessaly (*Salampria*), i. 9, 328, 501, 505-507, 513; ii. 4, 67, 77, 131, 134,142, 144-148, 272, 397.
- Penelope, Penelopeia, i. 328; ii. 50, 162, 173, 300.
- Penestæ, slaves of the Thessalians, ii. 287.
- *Peniscola. See* Cherronesus.
- Penta Dactylon. *See* Taÿgetum.
- Pentelic marble, ii. 90.
- Pentheus, ii. 103.
- Penthilus, son of Orestes, ii. 94, 154, 339, 340.
- *Pentima. See* Corfinium.
- Peparēthus, isl. (*Scopelo*), i. 187; ii. 140.
- Perasia. *See* Diana.
- Percope, ii. 351.
- Percōtē, t. of Mysia (*Bergas*), ii. 344-346, 350, 351.
- Perdiccas, ii. 394; iii. 229.
- Pergamēnē, ii. 326, 332.
- Pergamum, c. of Mysia, ii. 387, 389-402; iii. 66.
- Pergamus, ii. 379; iii. 46.

- Pergē, t. of Pamphylia (*Murtana*), iii. 49.
- Periander, tyrant of Corinth, ii. 366.
- Perias, t. of Eubœa, ii. 152.
- Pericles, ii. 84, 152; iii. 9.
- *Périgord*, inhabitants of. *See* Petrocorii.
- Perinthus, c. of Thrace, i. 515, 518.
- Perisadyes, people of Epirus, i. 500.
- Periscii, i. 146.
- Permēssus, r. of Bœotia, ii. 101, 108.
- Perperēna, t. of Mysia, ii. 376.
- Perperna, iii. 21.
- Perrhæbi, people of Thessaly, i. 96, 410, 507; ii. 10, 137, 141, 143-147, 153, 158.
- Perrhæbia, ii. 144-147.
- Perrhæbic cities, ii. 145.
- Perrhæbis, ii. 144.
- Persepolis, i. 122, 123; iii. 130-133.
- Perseus, i. 202, 439, 495, 509, 516; ii. 41, 59, 108, 211, 315, 400; iii. 208, 239, 259.
- Persia, i. 117, 122-126, 131, 132, 201; ii. 240, 254-274, 293; iii. 34, 109, 113, 120,124, 125, 128-142, 188, 208, 213.
- [Pg 389]Persian Sea, ii. 219, 257; iii. 146, 149, 186, 188.
- ——, gates, iii. 132.
- ——, palaces, i. 331.
- ——, war, i. 518; ii. 7, 57.
- ——, Gulf, i. 68, 121, 123, 124, 129, 183, 196, 261; ii. 266, 267, 270, 271; iii. 88,120, 125-129, 132, 146, 185, 186, 188, 215.
- Persians, i. 17, 41, 96, 196, 463; ii. 84, 87, 94, 96, 108, 130, 155, 181; iii. 35.
- Perūsia, c. of Etruria (*Perugia*), i. 335, 336.
- *Pescara. See* Aternum.
- Pessinuntis, cognomen of Rhea, ii. 184.
- ——, (*Possene*), ii. 184.
- Pessinūs, c. of Phrygia (*Bala Hissar*), ii. 320, 332.
- *Pesti. See* Posidonia.
- Petalia, isl., ii. 151.
- Peteon, vill. of Bœotia, ii. 106.
- Peteus, ii. 83.
- Petilia, t. of the Lucani, i. 378.
- Petnēlissus, t. of Pisidia (*Kislidscha-koi*), iii. 49.
- Pĕtra Nabatæōn, c. of Arabia, iii. 189, 204, 209, 211. *See* Tilphossium.
- Pĕtrēius, legate of Pompey, i. 242.
- Pĕtrocorii, people of Aquitania (*inhabitants of Périgord*), i. 284.
- Petronius, prefect of Egypt, iii. 222, 267-269.

333

- Peucĕ, isl. of the Danube (*Piczina*), i. 463, 464, 470.
- Peucĕtii, people of Italy, i. 315, 422, 423, 428, 431, 432, 436.
- Peucini, people of the Bastarni, i. 469, 470.
- Peucolaïtis, c. of India, iii. 90.
- Phabda, t. of Pontus, ii. 296.
- Phabra, isl. near Attica, ii. 89.
- Phabrateria, t. of Latium. *See* Fabrateria.
- Phaccŭssa, c. of Egypt, iii. 245.
- Phæaces, Phæeces, Phæacians, i. 39; ii. 122, 190.
- Phædimus, i. 65.
- Phædon, ii. 82.
- Phædrus, leader of the Athenians, ii. 153.
- ——, dialogue of Plato, i. 452; ii. 91.
- Phæstus, c. of Crete (*Hodyitra*), ii. 196, *n.*, 200.
- Phaetŏn, son of the Sun, i. 320.
- ——, tragedy of Euripides, i. 52.
- Phagres, t. of Macedonia, i. 512.
- Phagroriopolis, c. of Egypt, iii. 245.
- Phagroriopolite nome, iii. 245.
- Phalacrum, prom. of Corcyra, i. 497.
- Phalanna, t. of Thessaly, ii. 144.
- Phalannæi, ii. 144.
- Phalanthus, i. 424, 425, 430.
- Phalara, t. of Thessaly (*Stillida*), i. 94; ii. 137, 138.
- Phalasarna, t. of Crete, ii. 193, 200.
- Phalces, ii. 77.
- Phalēreis, demus of Attica, ii. 89.
- Phalericum, ii. 91.
- Phalerii, people of Etruria, i. 335.
- Phalērus, iii. 70.
- Phalisci, people and city of Etruria, i. 335.
- Phaliscum, c. of Etruria, i. 335.
- Phanæ, port of the island of Chios (*Porto Mustico*), iii. 18.
- Phanagŏria, Phanagoreia, Phanagoreium, c. of the Bosporani, i. 472, 477; ii. 223.
- Phanarœa, distr. of Pontus, i. 113; ii. 295, 305, 309, 311.
- Phanias, ii. 210, 392.
- Phanŏteis, ii. 101.
- Phanŏteus, c. of Phocis, ii. 122, 123.
- Phaon, ii. 162.
- Phara, c. of Achæa, ii. 71, 74.
- ——, c. of Messenia, *see* Phēræ, ii. 74.

- ——, t. of the Carthaginians, iii. 285.
- Pharæ, village near Tanagra, ii. 99.
- ——, c. of Thessaly. *See* Pheræ.
- Pharātæ, in Messenia, ii. 74.
- Pharbetite nome, iii. 240.
- [Pg 390]Pharcadōn, c. of Thessaly, ii. 142.
- Phareis, Pharieis (Pharæeis?), in Achæa, ii. 73.
- Pharenses, inhabitants of Phara, ii. 74.
- Pharis, c. of Laconia, ii. 40.
- Pharmacussæ, islands near Salamis, ii. 85.
- Pharnaces, king of the Bosporani, ii. 224, 239, 292, 294, 306, 311, 401.
- Pharnacia, c. of Pontus, i. 190, 491; ii. 294, 296, 297, 304, 305; iii. 61.
- Pharos, isl. of Egypt, i. 46, 47, 58, 59, 88, 91, 211; iii. 226, 227, 238, 240.
- ——, tower or lighthouse on the island, ii. 280; ii. 230.
- ——, one of the Liburnian islands (*Lesina*), i. 186, 484.
- Pharsalia, ii. 132; iii. 233.
- Pharsalii, ii. 134, 135.
- Pharsalus, mtn, ii. 32.
- ——, c. of Thessaly, anciently Palæpharsalus, now Satalda, ii. 133-136; iii. 233.
- ——, New, on the Enipeus, ii. 133-136, 155.
- Pharusii, people of Libya, i. 198; iii. 277, 280.
- Pharygæ, c. of Locris, ii. 127.
- ——, c. of Argolis, ii. 127.
- Pharygæa, Juno, ii. 127.
- Pharygium, prom. of Phocis, ii. 122.
- Pharziris, same as Parysatis, iii. 216.
- Phasēlis, t. of Lycia (*Tirikowa*), iii. 48, 49, 55.
- Phāsis, c. of Colchis, i. 440; ii. 225, 227, 230.
- ——, r. of Colchis (*Rion*), i. 71, 72, 82, 138, 457; ii. 225, 227, 230, 270, 296.
- Phatnitic mouth of the Nile, iii. 239, 240.
- Phauēne (? Phasiane), distr. of Armenia, ii. 268.
- Phaunītis, distr. of Armenia, ii. 269.
- Phayllus, ii. 119.
- Phazēmon, c. of Pontus, ii. 311.
- Phazēmonitæ, ii. 311.
- Phazēmonitis, ii. 310, 311.
- Phea, Pheæ, Pheia, c. of Pisatis, ii. 16, 22, 25, 26.
- ——, prom. of Pisatis, ii. 15.
- Pheidippus, iii. 31.
- Pheidōn, ii. 34, 58.
- Phellōn, r. of Triphylia, ii. 16.

- Phellos, stronghold of Lycia, iii. 47.
- Phĕmŏnoĕ, ii. 117.
- Phĕnĕus, t. of Arcadia, ii. 75, 76.
- Phĕræ, Phĕra, c. of Messenia. *See* Phara, ii. 35-37, 46, 74.
- Phĕræ, c. of Thessaly (*Velestina*), ii. 139, 148, 235, 272.
- Phĕræa (Heræa ?), c. of Arcadia, ii. 32, 42.
- Phĕræi, ii. 143.
- Phĕrĕcydes, Syrian, i. 28, 254; ii. 167, 190, 211; iii. 2, 15.
- ——, Athenian, ii. 211.
- Phĕsti or Festi, t. of Latium, i. 341.
- Phidĕnæ or Fidenæ, t. of Latium, i. 335, 341.
- Phidias, of Athens, ii. 29, 53, 86, 87.
- Phigalia, t. of Arcadia, ii. 22.
- Philadelpheia, t. of Lydia. (*Ala Schehr*), ii. 335, 406.
- Philadelphia, t. of Judæa, iii. 177, 181.
- Philadelphus. *See* Ptolemy.
- Philæ, isl. and c. of Upper Egypt, i. 64; iii. 243, 265, 267.
- Philæni, altars of the, i. 256, 257; iii. 290, 291.
- Philalēthēs, ii. 336.
- Philĕmōn, comic poet, iii. 55.
- Philĕtærus, ii. 198, 307.
- ——, founder of the Attali, ii. 289, 399, 400.
- ——, son of Attalus, ii. 400.
- Philĕtes, poet, ii. 42; iii. 36.
- Philip, son of Amyntas, i. 463, 492, 508, 509, 512, 513; ii. 38, 39, 56, 64, 111;

iii.165, *et passim.*

- ——, city of, i. 512.
- [Pg 391]Philip, father of Perseus, i. 439; ii. 38, 146, 315.
- ——, tyrant of the Areitæ, ii. 152.
- ——, Pseudo, ii. 401.
- ——, isl. of, in the Arabian Gulf, iii. 199.
- Philipopolis, c. of Macedonia, i. 514.
- Philippi, c. of Macedonia, i. 515.
- ——, battle of, iii. 58.
- Philippus, historian, iii. 41.
- Philisteides, tyrant of Eubœa, ii. 152.
- Philo, historian, i. 119.
- ——, architect, ii. 85.
- Philochorus, i. 502; ii. 39, 81, 88, 97.
- Philoctētēs, i. 378, 411; ii. 134, 140, 148.
- Philodēmus, of Gadara, Epicurean, iii. 175.

336

337

339

- Pinarus, r. of Cilicia, iii. 60.
- Pindar, poet, i. 232, 256, 369, 405, 409, 493, 502, 519; ii. 97, 107, 109, 118, 132,184, 290, 404; iii. 16, 19, 33, 108, 240.
- [Pg 393]Pindus, mtn of Thessaly, i. 410, 501, 505, 507; ii. 128-135, 137, 141-147, 158.
- ———, t. of Locris, ii. 128, 137.
- ———, r. of Locris, ii. 128.
- *Pineta di Castel Volturno. See* Gallinarian wood.
- *Piomba. See* Matrinus.
- Pionia, t. of the Leleges in Mysia, ii. 381.
- *Pira. See* Pyrrha.
- Piræeus, same as Amisus, ii. 294.
- Piræus, i. 91; ii. 79, 85, 87, 89, 91; iii. 32.
- Pirithous, i. 76, 507.
- Pisa, tract of country, ii. 32.
- ———, c. of Elis, ii. 31, 32.
- ———, ftn, ii. 31, 32.
- ———, c. of Etruria, i. 315, 323, 329, 330, 334.
- Pisātæ, in Elis, i. 330; ii. 9, 15, 28, 30, 31, 33, 39.
- ———, in Etruria, i. 331, 334.
- Pisātis, distr. of Elis, i. 330; ii. 8, 11, 13-16, 28, 30-34, 45, 53, 56, 59.
- ———, territory of Pisa, i. 315, 330.
- Pisidia, land of Asia, i. 32, 54; ii. 307, 322-326, 332, 383, 409; iii. 48, 54, 63.
- ———, mtns, iii. 47, 48.
- ———, cities, ii. 324-326.
- ———, Taurus, i. 195; ii. 319.
- Pisidians, i. 195; ii. 216, 304, 322-324, 407, 409.
- Pisilis, t. of Caria, iii. 28.
- Pisistratus, son of Nestor, ii. 25, 46.
- ———, tyrant of Athens, ii. 83, 88.
- Piso, Cnæus, præfect of Libya, i. 197.
- ———, Carbo, i. 319.
- ———, Ahenobarbus, i. 277.
- ———, son of Pompey, i. 213.
- Pissūri, ii. 245.
- Pitanæi, ii. 383.
- Pitanātæ, in Samnium, i. 372.
- Pitane, t. of Mysia (*Tschandarlik*), ii. 339, 376, 387, 389, 398.
- Pithecūssa, Pithecūssæ, isl. (*Ischia*), i. 84, 89, 93, 185, 368, 369, 386, 387, 404.
- Pithecussæans, i. 365.
- Pitnisus, t. of Lycaonia, ii. 321.

- Pittacus, ii. 366, 391, 392.
- Pittheus, son of Pelops, ii. 56.
- Pitya, Pityeia, t. of Mysia, ii. 317, 346, 349, 371.
- Pityassus, t. of Pisidia, ii. 324.
- Pityocamptēs, ii. 80.
- Pityūs, vill. of the Troad, ii. 349.
- ——, the Great, part of the coast of Colchis, ii. 225, 226.
- Pityūssa, ancient name of Lampsacus, Salamis, and Chios, q. v. ii. 83.
- Pityūssæ, islands, i. 251.
- Pixōdarus, king of Caria, iii. 35.
- Placentia, t. of Cisalpine Gaul, i. 322-325.
- ——, (*Piacenza*), i. 322, 323, 325.
- Placus, mtn, ii. 343, 386.
- Planasia (*Isle St. Honorat*), i. 185, 239, 276.
- Planctæ, or Wandering Rocks, i. 32, 224, 256.
- Platææ, c. of Bœotia, ii. 66, 94, 100, 104, 107, 108, 111.
- ——, vill. of Sicyon, ii. 108.
- Platæans, ii. 100.
- Platamōdes, prom. of Messenia, ii. 22.
- *Platana. See* Hermonassa.
- Platanistus, shore of Cilicia, iii. 52.
- ——, same as Macistus, ii. 18.
- Plato, i. 154, 390, 452, 462, 464; ii. 91, 183, 188, 197, 353, 354, 382; iii. 34, 110,179, 222, 240.
- Plax, ii. 386.
- Pleias, Plēïas, Plēïades, constellation, iii. 59, 82, 83, 126, 274.
- Pleistus, r. of Phocis, ii. 116.
- Plēmyrium, c. of India, iii. 89.
- [Pg 394]Pleræi, people of Dalmatia, i. 484, 485.
- Pleurōn, c. of Ætolia, Old and New, ii. 72, 159, 160, 171, 172, 175, 178, 179.
- Pleurōnia, ii. 160, 178, 179.
- Pleurōnii, ii. 160, 173, 176.
- Pleutauri, people of Spain, i. 233.
- Plinthinē, c. of Egypt, iii. 236.
- Plumbaria, isl. (*S. Pola*), i. 239.
- Plūtiadēs, iii. 59.
- Plūto, i. 220; iii. 25.
- Plutonium, i. 363; ii. 408; iii. 25.
- Plutus, i. 220, 221.
- Plynos, port of Marmara, iii. 294.
- Pneuentia, t. of the Piceni (Pollentia ?), i. 357.

- Pnigeus, t. of Marmara, iii. 235.
- Po, r. of Italy, i. 287, 303, 312, 316, 317, 320, 322, 323, 360, 438, 439; ii. 71.
- Podalirius, heroum or shrine of, i. 434.
- Podarcēs, brother of Protesilaus, ii. 134.
- Pœaessa, t. of Laconia, ii. 37.
- Pœcile, iii. 54.
- Pœdicli, same as Peucĕtii, i. 423, 431.
- Pœeïssa, t. of the isl. of Ceos, ii. 210.
- Pœmandris, same as Tanagra, ii. 96.
- Pœum. *See* Bœum, i. 505.
- Pœus, mtn of Thessaly, i. 501.
- Pŏgōn, port of Trœzen, ii. 55.
- Pŏla, t. of Istria, i. 73, 313, 321, 483.
- Pŏlĕmōn, son of Pharnaces, king of Pontus, ii. 220, 223, 224, 305, 322.
- ——, of Laodicea, son of Zeno, ii. 334.
- ——, philosopher, ii. 387.
- ——, Periēgētēs, i. 23; ii. 86.
- Polentia, t. of the largest of the Balearic Islands (*Pollença*), i. 251.
- *Policandro. See* Pholegandrus.
- Polichna, ii. 84, 371, 376.
- Pŏlieum, t. of Lucania, i. 397.
- *Polina. See* Apollonia.
- *Polino. See* Prepesinthus.
- Pŏlisma, Polium, t. of the Troad, ii. 368.
- Polites, companion of Ulysses, i. 380.
- ——, son of Priam, ii. 364.
- Polium, ii. 368.
- Pollentia. *See* Pelentia.
- Poltyŏbria, t. of Thrace, i. 490.
- Polyanus, mtn of Epirus, i. 501.
- Polybius, historian, i. 1, 23, 31, 35, 36, 38, 39, 145, 147, 148, 156-164, 209, 222, 226,244, 256, 258, 259, 274, 283, 301, 309, 310, 315, 319, 330, 360, 393, 418, 4 35,438, 481, 487, 495, 518; ii. 1, 51, 64, 7, 120, 251; iii. 234, *et passim.*
- Polybōtēs, ii. 213.
- Polybus, ii. 64.
- Pŏlycasta, mother of Penelope, ii. 173.
- Polycles, ii. 288.
- Polycletus, historian, ii. 243, 244; iii. 130, 139, 150.
- ——, statuary, ii. 53.
- Polycrates, tyrant of Samos, iii. 8, 9.
- Polydamas, Trojan, ii. 364.

- Posidonius, the Stoic,

i. 1, 6, 8, 23, 44, 66, 84, 86, 90, 143, 144, 146, 148, 152, 154-156, 158, 203, 207, 208, 216, 220, 222, 229, 230, 235, 244, 245, 247, 256, 258,259, 261, 273, 280, 281, 294, 319, 325, 400-402, 406, 413, 420, 450, 453, 454, 456, 461, 475, 486; ii. 1, 217, 251, 387; iii. 33, 151, 166, 170, 173, 208, 216, 225,244, 278, 282, *et passim.*

- [Pg 396]*Possene. See* Pessinuntis.
- Potamia, distr. of Paphlagonia, ii. 314.
- Potamii, ii. 89.
- Potamō, of Mitylene, ii. 392.
- Potamus, vill. of Attica, ii. 89.
- Potentia, c. of Picenum, i. 357.
- Potidæa, c. of Macedonia, i. 511.
- Potniæ, c. of Bœotia, ii. 103, 109.
- *Pozzuoli. See* Puteoli.
- *Pozzuolo. See* Dicæarchia.
- Practius, r. of Mysia, ii. 340, 344, 346, 350, 351.
- Prænestina, Via, i. 352, 353.
- Præneste, c. of Latium, i. 353, 354, 370.
- Pramnæ, iii. 117, 118.
- Prasia, vill. of Attica (*Raphti*), ii. 89.
- Prasiæ, c. of Argolis, ii. 48, 55.
- Prasii, people of India, iii. 97.
- ——, inhabitants of the city of Prasus, ii. 189, 199.
- Prasus, c. of Crete, ii. 195, 199.
- Prausi, Gallic nation, of whom Brennus was king, i. 280.
- Praxander, iii. 69.
- Praxiphanes, iii. 33.
- Praxitĕlēs, ii. 105; iii. 13.
- Preferni, i. 344.
- Prĕmnis, t. of Ethiopia, iii. 268.
- Prepesinthus, one of the Cyclades (*Polino*), ii. 208.
- Priam, ii. 169, 301, 342, 344, 345, 360, 367, 377.
- Priamidæ, ii. 378.
- Priamōn (? Prōmōn), c. of the Dalmatæ, i. 484.
- Priapeia, songs of Euphronius, ii. 66.
- Priapēnĕ, distr. of Mysia, ii. 347.
- Priapēni, ii. 347.
- Priăpus, son of Bacchus, temple of, ii. 66, 348.
- Priăpus, t. of Mysia, i. 518; ii. 317, 332, 340, 347-349.
- Priēnē, c. of Ionia (*Samsun*), ii. 69, 299, 333, 335; iii. 2, 7.
- Priēnians, ii. 69; iii. 10.
- Priōn, mtn near Ephesus, iii. 3.

346

- Psillis, r. of Bithynia, ii. 288.
- *Psiloriti. See* Ida.
- Psygmus, on the coast of Ethiopia, iii. 200.
- Psylli or Psyllians, people of Libya, i. 198; ii. 348; iii. 260, 294.
- Psyra (*Psyrà*), isl. near Chios, iii. 19.
- Psyttalia, isl. near Salamis, ii. 85.
- Pteleasimum, distr. of Triphylia, ii. 24.
- Ptĕlĕŏs, ii. 357.
- Ptĕlĕum, mtn of Epirus, i. 505.
- ——, c. of Triphylia, ii. 23, 24.
- ——, c. of Thessaly, ii. 24, 135, 139.
- Pterelas, ii. 162.
- Ptolemaïs, t. of Phœnicia, i. 201.
- ——, t. of Pamphylia (*Alara*), iii. 50.
- ——, t. of the Cyrenaic, iii. 292.
- ——, t. of the Troglodytic, i. 200; iii. 191, 194, 204.
- ——, t. of the Thebais, iii. 258.
- Ptolemies, i. 178.
- Ptolemy Aulētes, ii. 308; iii. 231, 232, 234.
- ——, Ceraunus, ii. 400.
- ——, Cocce's son, iii. 230.
- ——, Epiphanes, ii. 231.
- Ptolemy Euergētes, iii. 231.
- ——, Euergētes II. or Physcon, i. 149, 152, 156; ii. 124, 172; iii. 231, 234.
- ——, Lathūrus, iii. 231.
- ——, Philadelphus, ii. 120, 172; iii. 47, 193, 194, 224, 231, 260.
- ——, Philomētor, iii. 164, 231.
- ——, Philopator, ii. 199; iii. 176, 231.
- ——, Sōter, iii. 290.
- ——, king of Cyprus, iii. 71.
- ——, son of Aulētes, iii. 234.
- ——, son of Juba, iii. 281, 283, 297.
- ——, son of Lagus, i. 463; iii. 123, 229, 231, 239.
- ——, son of Mennæus, iii. 166.
- Ptōum, mtn of Bœotia, ii. 109.
- Publius Crassus. *See* Crassus.
- ——, Claudius Pulcher. *See* Pulcher.
- ——, Servilius, ii. 322.
- Pulcher, Publius Claudius, iii. 71.
- Punic War, Second, i. 239.
- *Purali. See* Arbis.

348

349

- Raphia (*Refah*), t. of Judæa, iii. 176.
- Rapti, modern name of Prasia.
- *Ras-el-Kasaroun. See* Casium.
- *Ras-el-Razat. See* Phycus.
- *Ratoüs. See* Arathus.
- Rauraris, i. 272.
- *Ravee. See* Hyarotis.
- [Pg 399]Ravenna, i. 314, 318, 319, 322, 323, 326, 337.
- Reatĕ, c. of the Sabines (*Rieti*), i. 338, 339.
- Red Sea, *see* Arabian Gulf and Erythræan Sea, iii. 244, 254, 260.
- *Refah. See* Raphia.
- Reggio, i. 315, 317.
- ——, in Modena. *See* Rhegium-Lepidum.
- Regis-Villa, c. of Etruria, i. 335.
- Rĕmi, people of Gaul, i. 289, 290.
- Rĕmus, brother of Romulus, i. 340, 343.
- Reneia, isl. (*Rhena*), ii. 209.
- Rhacŏtis, part of Alexandria, iii. 227.
- Rhadamanthus, Cretan lawgiver, ii. 196, 204.
- ——, brother of Minos, i. 3, 225; ii. 122, 196, 328.
- ——, tragedy of Euripides, ii. 32.
- Rhadinē, song of Stesichorus, ii. 21.
- Rhæci, i. 343.
- Rhæti, i. 287, 304, 306, 307, 311, 317, 447, 448, 482.
- Rhætian wine, i. 306.
- Rhætica, Rhætia, i. 482.
- Rhaga, ii. 264.
- Rhagæ, Rhages, c. of Media, i. 94; ii. 250, 264.
- Rhamanītæ, people of Arabia, iii. 212.
- Rhambæi, people of Syria, iii. 166.
- Rhamis, i. 446.
- Rhamnūs, t. of Attica, ii. 90.
- Rhathĕnus, ii. 296.
- Rhatŏtes, name of the Paphlagonians, ii. 302.
- Rhea, mother of the gods, ii. 22, 183-189.
- ——, (Agdistis, Idæa, Dindymēnē, Sipylēnē, Pessinūntis, Cybĕlē, Cybēbē), ii. 184-186.
- ——, Silvia, daughter of Numitor, i. 340.
- Rhecas, ii. 224.
- Rhegians, i. 391.
- Rhegīni, i. 385, 386.

350

- Rhegium (*Reggio*), i. 94, 186, 256.
- ———, c. of the Bruttii i. 77, 384-386, 388-390, 404, 431.
- ———, t. of Gaul, beyond the Po (*Reggio*), i. 317.
- ———, Lepidum, t. of Gaul, this side the Po (*Reggio in Modena*), i. 322.
- Rhēgma, at the mouths of the Cydnus, iii. 56.
- Rhenus, r. (*Rhine*), i. 99, 192, 193, 264, 265, 285-290, 292, 296, 297, 304, 306, 308,310, 317, 442-447, 451, 480; iii. 296.
- ———, sources of the, i. 265, 289, 304, 317.
- ———, mouths of the, i. 99, 265, 288, 289, 296, 447, 451.
- Rhesus, king of Thrace, i. 514; ii. 351.
- ———, r. of the Troad, ii. 304, 341, 351, 357, 371.
- Rhetia, mother of the Corybantes, ii. 190.
- Rhiginia, r. of Thrace, i. 516.
- *Rhine. See* Rhenus.
- Rhinocolūra, Rhincocorura, t. of Phœnicia (*El-Arish*), iii. 149, 176, 211.
- Rhipē, t. of Arcadia, ii. 75.
- Rhium, prom. of Achaia (*Drepano*), ii. 6, 73, 79.
- ———, t. of Messenia, ii. 37, 38.
- Rhizæi, Bay of the, i. 485.
- Rhizōn, t. on the coast of Illyria (*Risano*), i. 485.
- Rhizonic Gulf (*Gulf of Cataro*), i. 483, 485.
- Rhizophagi, people of Ethiopia, iii. 95.
- Rhizūs, t. of Thessaly, ii. 139, 148.
- Rhoa, i. 269.
- Rhodanus, *Rhone*, r. of Gaul, i. 249, 266, 267, 269, 271, 272, 274-288, 302, 303, 310.
- Rhodaspes, son of Phraates, iii. 160.
- Rhodes, i. 38, 105, 109, 123, 131, 133, 160, 172-175, 179, 180, 184, 187, 189, 201,202, 332, 423, 486; ii. 164, 188, 189, 212, 213, 216, 217, 328, 331, 374, 409; iii.33, 34, 39, 40, 230.
- [Pg 400]Rhodians, Rhodii, i. 90, 240, 398; ii. 85, 194; iii. 33, 34.
- Rhodius, r. of the Troad, ii. 304, 357, 371.
- Rhodōpē (Rhodos, Rhode ?), t. of Spain, i. 240; iii. 32.
- ———, mtn of Thrace (*Despoto-dagh*), i. 311, 481, 489, 506, 514.
- Rhodōpis, iii. 250.
- Rhodos. *See* Rhodes.
- Rhoduntia, citadel near Thermopylæ, ii. 129.
- Rhoeitēs, r. of the Troad, ii. 371.
- Rhœtaces, r. of Albania, ii. 230.
- Rhœtium, t. of the Troad, ii. 357, 358, 361, 368.
- Rhombites, the Greater, Bay of Mæotis, ii. 221.
- ———, the Lesser, ii. 221.

- Samaria, same as Sebaste, c. of Judæa, iii. 177.
- Samarianē, c. of Hyrcania, ii. 242.
- Samē, t. of Cephallenia, ii. 163, 166, 167.
- Sami, heights, ii. 169.
- [Pg 402]Samia, Samos, ii. 212, 213; iii. 10.
- Samian strait, iii. 10.
- Samians, inhabitants of Samos, i. 518; ii. 168, 212; iii. 9, 10.
- Samicum, citadel of Triphylia, ii. 16, 17, 19, 21, 26.
- Samicus, plain, ii. 21.
- Samnītæ, Samnites, Saunītæ, i. 339, 344, 346, 357, 360, 367, 371-374, 377, 378, 380,387, 399, 431, 438.
- Samnites, c. of the, i. 353.
- ——, women of the, in an island of Gaul, i. 295.
- Samnitic mtns, i. 326.
- Samnium, i. 360.
- Samōnium, prom. of Crete. *See* Salmonium.
- ——, in the Neandris, ii. 189.
- ——, Alexandrian, ii. 189.
- Samos, isl. in the Icarian Sea (*Samo*), i. 93, 187; ii. 163, 168, 169, 303; iii. 2, 3, 7-11.
- ——, Thracian, ii. 10.
- ——, Ionian, ii. 10, 21.
- ——, c. of the island of, iii. 3.
- ——, c. of Triphylia, ii. 19, 21, 25.
- ——, and Samē, same as Cephallenia, and c. of this isl., ii. 163, 166, 167.
- ——, Threïcian, same as Samothracē, i. 516; ii. 168.
- Samosata, c. of Syria, iii. 44, 161.
- Samothracē, Samothracia, isl. (*Samothraki*), i. 43, 187, 296, 516; ii. 168, 189, 190.
- Samothracians, i. 516; ii. 180.
- Sampsiceramus, prince of the Emiseni, iii. 166, 167.
- *Samsun*. *See* Priene.
- *Samsun Dagh*. *See* Mycale.
- *San Dimitri*. *See* Antandros.
- *San Giarno*. *See* Corydallus.
- Sanā, t. of Pallene, i. 511.
- Sanaus, c. of Phrygia, ii. 332.
- Sandalium, citadel of Pisidia, ii. 323.
- Sandaracurgium, mtn of Pontus, ii. 313.
- Sandobanes, r. of Albania, ii. 230.
- Sandōn, father of Athenodorus, iii. 58.
- Sandrocottus, king of the Prasii, i. 109; iii. 97, 105, 107, 125.

- Sandyx, ii. 271.
- Sangarius, r. of the Troad (*Sakaria*), ii. 289, 314, 321, 351; iii. 66.
- Sangias, vill. of Phrygia, ii. 288.
- Sanisēnē, distr. of Paphlagonia, ii. 314.
- Sanni, people of Pontus, ii. 296.
- *Santa Maria. See* Malia and Trogilium.
- *Santa Maura. See* Leucas.
- *Santo, Mount. See* Athos.
- Santoni (*inhabitants of Saintonge*), i. 283, 284, 310.
- *Santorino. See* Thera.
- Saōcondarius, ii. 321.
- *Saone, r. See* Arar.
- Saos, r. *See Save*. Sapæ, Sapæi, people of Thrace, i. 515, 516; ii. 169, 298, 305.
- Saperdes, ii. 393.
- Saphnioeis. *See* Satnioeis.
- Sapis, r. of Cisalpine Gaul (*Savio*), i. 322.
- Sappho, i. 65; ii. 162, 388, 391, 393; iii. 250.
- Sapra limnē (*or* Putrid Lake), at the Tauric Chersonese, i. 473, 474.
- ——, in the Troad, ii. 387.
- Saraastus, king of India, ii. 253.
- *Sarabat. See* Hermus.
- *Saragossa. See* Cæsar Augusta.
- *Sarakoi. See* Zeleia.
- Saramēnē, ii. 294.
- Sarapana, fortress of Colchis (*Choropani*), ii. 227, 230.
- Saraparæ, people dwelling beyond Armenia, ii. 273.
- Sarapis, god of the Egyptians, iii. 242, 248.
- Sarapium, temple of Sarapis, iii. 230, 248.
- Saravēnē, prefecture of Cappadocia, ii. 278.
- Sardanapalus, king of Assyria, iii. 55, 143.
- [Pg 403]Sardinia, i. 78, 160, 177, 185, 216, 330-334; iii. 32, 240, 297.
- ——, Sea of, i. 78, 84, 159, 185, 216, 325.
- Sardinian Gulf, i. 216.
- Sardis, Sardeis, c. of Lydia (*Sart*), i. 96; ii. 336, 400, 402-406; iii. 23.
- Sardō, Sardōn, isl., i. 219, same as Sardinia.
- Sareisa, c. of the Gordyenes, iii. 157.
- Sargarausēnē, prefecture of Cappadocia, ii. 278, 281.
- Sarmatians, Sauromatæ, i. 453, 468, 470, 480; ii. 219, 226, 227, 230, 240, 302.
- Sarnius, r. of Hyrcania, ii. 245.
- Sarnus, r. of Campania (*Sarno*), i. 367.
- Sarōnic Gulf, ii. 6, 49, 56, 63.

- ——, Sea (*Gulf of Engia*), i. 187.
- *Saros, Bay of. See* Melas.
- Sarpēdōn, prom. of Cilicia, ii. 405; iii. 53, 69.
- ——, prom. of Thrace, i. 516.
- ——, brother of Minos, founder of Miletus, ii. 328, 347; iii. 49.
- ——, leader of the Syrians, iii. 174.
- Sarsina, t. of Umbria, i. 337.
- *Sart. See* Sardis.
- Sarus, r. of Cappadocia and Cilicia, ii. 279.
- *Sasamo. See* Segesama.
- Sasō, isl. (*Saseno*), i. 429.
- Satalca, t. of Mesopotamia, iii. 157.
- *Satalda. See* Pharsalus.
- *Satal-dere. See* Æsepus.
- Satnioeis, or Saphnioeis, r. of the Troad, i. 494; ii. 375, 379, 394.
- Satnius, ii. 375, 394.
- Saturn, i. 494; ii. 39, 183, 184, 189, 378.
- ——, temple of, i. 254.
- Satyr, painted by Protogenes, iii. 29, 30.
- Satyri, ii. 180, 184, 186.
- Satyrium, near Tarentum, i. 425.
- Satyrus, king of Bosporus, i. 462, 476.
- ——, monument of, ii. 222.
- ——, founder of the city of Philotera, iii. 193.
- Saunitæ, *see* Samnites, i. 372.
- Sauromatæ, i. 172, 194, 195, 452, 464; ii. 240, 302. *See* Sarmatians.
- Saus, Sauus, r. of Hungary. *See* Save.
- *Save, r.* i. 309, 482.
- *Savio. See* Sapis.
- Scæan gates, in the city of Troy, ii. 351, 363.
- ——, wall, ii. 351.
- Scæi, ii. 351.
- Scæus, r. in the Troad, ii. 351.
- Scamander, r. i. 90; ii. 358, 360, 361, 363, 369, 370, 378.
- ——, plain of, ii. 361, 362.
- Scamandrius, son of Hector, ii. 377, 378; iii. 66.
- Scandaria, Scandarium, prom. of the island of Cos, iii. 36.
- Scardon, a Liburnian city (*Scardona*), i. 484.
- ——, in Elis, ii. 347.
- Scardus, mtn of Macedonia (*Schar-dagh*), i. 505.
- Scarphē, c. of Bœotia, i. 95; ii. 103.

- Scarpheia, Scarphē, c. of the Epicnemidian Locrians (*Bondoniza*), i. 94; ii. 126.
- Scaurus, M. Æmilius, i. 323.
- Scēnæ, c. of Mesopotamia, iii. 159.
- Scēnītæ, Scenites, in Arabia, Mesopotamia, Syria, etc., i. 63, 196, 441; ii. 219, 252; iii.160, 166, 185, 190, 204.
- Scepsia, ii. 361, 375.
- Scēpsian territory, ii. 190.
- Scēpsians, ii. 361, 377.
- Scēpsis, ii. 300, 360, 369, 375-381; iii. 5.
- Scēpsius, Demetrius. *See* Demetrius.
- Scēptūchiæ, Scēptūchi, ii. 225.
- Schĕdia, t. of Egypt, iii. 237-241.
- [Pg 404]Schĕdieium, ii. 124.
- Schedius, ii. 124.
- *Schelidan Adassi. See* Chelidonian Isles.
- Scheria, same as Corcyra, i. 459.
- Schœnŭs, c. of Bœotia (*Morikios*), ii. 58, 103.
- Schœnūs, r. of Bœotia, ii. 103.
- ——, port of Corinth, ii. 49, 63, 79.
- ——, clans of, i. 16.
- *Schuss. See* Susa.
- Sciathus (*Sciathos*), isl. near Magnesia, ii. 140.
- Scillūs, t. of Triphylia, ii. 16.
- *Scilly Islands. See* Cassiterides.
- Scilūrus, king of the Scythians, i. 471, 475, 479.
- Scingomagum, t. in the Alps, i. 268.
- Sciōnē, c. of Pallene, i. 511.
- Scipio, Metellus, iii. 281, 284, 285.
- ——, Æmilianus, i. 283; iii. 51, 286.
- ——, Africanus, i. 361.
- ——, Caius, i. 317.
- ——, Nasica, i. 484.
- Scira, vill. of Attica, ii. 82.
- Sciras, same as Salamis, ii. 82.
- ——, Athene, ii. 82.
- Sciron, ii. 80, 81.
- Scirones (N.W. wind), i. 43.
- Scironides rocks, ii. 80, 82.
- Scirophoriōn, ii. 82.
- Sciros, rocks of, i. 43.
- Scollis, mtn of Elis, ii. 11, 13, 14, 74.

- Scŏlus, c. of Bœotia (*Kalyvi*), i. 16; ii. 58, 103, 104.
- Scombraria, isl. near Spain (*Islote*), i. 239.
- Scombrus, iii. 25.
- Scŏpas, the sculptor, ii. 373; iii. 11.
- *Scopelo. See* Peparethus.
- Scordistæ, Scordisci, i. 450, 454, 482, 483, 485, 488, 489.
- ———, Great, i. 488.
- ———, Little, i. 488.
- Scotūssa, c. of Pelasgiotis, i. 503, 504, 514; ii. 146.
- *Scripu. See* Orchomenus.
- Scūltanna, r. of Cisalpine Gaul (*Panara*), i. 324.
- Scydisēs, mtn of Armenia (*Aggi Dagh*), ii. 226, 267, 296.
- Scylacium. *See* Scylletium.
- Scylax, ii. 318, 340; iii. 37.
- ———, r. of Pontus (*Tschoterlek Irmak*), ii. 295.
- Scylla, i. 31-33, 36, 37, 39.
- ———, daughter of Nisus, ii. 55.
- Scyllæum, prom. and port of Italy, i. 35-37, 384.
- ———, prom. of Argolis (*Skylli*), ii. 47, 55, 207.
- Scyllēticus Sinus, Gulf of Scylletium (*Golfo di Squillace*), i. 380, 392.
- Scyllētium, Scylacium, t. of the Bruttii (*Squillace*), i. 392.
- Scȳrus, isl. (*Skyro*), i. 187; ii. 124, 140.
- Scythia (*Tartary*), i. 13, 52, 99; ii. 216, 352.
- ———, desert of, i. 79, 82.
- ———, Little (or Tauric), i. 478, 489; ii. 279.
- Scythian nations, i. 247, 480, 481; ii. 235.
- ———, bow, i. 188.
- ———, history, i. 32.
- ———, zone, i. 147; ii. 247.
- ———, custom, i. 299.
- Scythians,

i. 23, 51, 52, 106, 115, 172, 179, 180, 194, 195, 458, 461, 462, 464, 467,468, 475, 480; ii. 218, 219, 221, 230, 240, 244, 245, 248, 273, 288, 302, *et passim.*

- Scythians of the East, i. 172.
- Scythopolis, c. of Galilee, iii. 181.
- *Sebaket-Bardoil. See* Sirbonis.
- Sebastē, c. of Pontus, ii. 300.
- ———, same as Samaria, iii. 177.
- Sebennytic nome, iii. 240.
- ———, mouth of the Nile, iii. 239, 240.
- Sebennytice, c. of Egypt, iii. 239.

- Secinus (*Selinda*), iii. 52.
- Segeda. *See* Segida.
- Segesama, t. of the Vacciæ (*Sasamo*), i. 244.
- Segesta (*Sisseck*). *See* Segestica.
- [Pg 405]Segestes, father of Segimuntus, i. 446.
- Segestica (*Sissech*), c. of Hungary, i. 309, 482, 483, 488.
- Segida, t. of the Aruaci, i. 243.
- Segimūntus, prince of the Cherusci, i. 446.
- *Segni. See* Signia.
- Segobriga, t. of the Celtiberi, i. 243.
- Segusii, or Segusiani, people of Gaul, i. 277, 286.
- *Seide. See* Sidon.
- Seiris, r. of Lucania, i. 397. *See* Siris.
- *Selefkeh. See* Seleuceia.
- Sĕlēnē, or the Moon, goddess worshipped by the Albani, ii. 234.
- ——, by the people of Memphis, iii. 248.
- ——, cognomen of Cleopatra, iii. 161.
- ——, Greek name for Luna, c. and port of Etruria, i. 330.
- Seleuceia, c. of Susiana, iii. 154.
- ——, c. of Assyria, on the Tigris, ii. 262, 271; iii. 145, 146, 152, 156, 162.
- ——, Pierian, c. of Syria (*Suveidijeh*), i. 486; iii. 61, 161-164, 167.
- ——, fortress of Mesopotamia, iii. 161.
- ——, c. of Cilicia, iii. 53, 54.
- Seleucis, part of Syria, iii. 160, 161, 167, 171.
- Seleucus, the Babylonian, i. 8, 261.
- ——, Nicator, king of Syria, ii. 334, 400; iii. 51, 74, 125, 145, 146, 161, 162, 165.
- ——, Callinicus, king of Syria, ii. 248; iii. 162, 168.
- Selgē, c. of Pisidia (*Surk*), ii. 324.
- Selgeis, ii. 323-325.
- Selgessus, same as Sagalassus, ii. 323.
- Selgic mountains, ii. 325.
- *Selidromi. See* Icus.
- *Selindi. See* Selinus.
- Sĕlinūntia, hot springs in Sicily (*I Bagni di Sciacca*), i. 415.
- Sĕlinūntius Apollo, worshipped by the Orobii, ii. 152.
- Sĕlinūs, c. of Sicily, i. 412.
- ——, c. of Cilicia, iii. 52.
- ——, r. of Sicily, ii. 73; iii. 68.
- ——, r. near Ephesus, ii. 73.
- ——, r. of Elis, ii. 73.
- ——, r. of Achæa, ii. 73.

- ——, strait of, i. 517.
- Setabis (*Xativa*), i. 241.
- Sethroite nome, iii. 243.
- Sētia, t. of Latium (*Sezza*), i. 344, 347, 352.
- ——, wine of, i. 347.
- Sētium, prom. of Gallia Narbonensis (*Cape de Cette*), i. 271.
- Seusamora, c. of Iberia Caucasia, ii. 231.
- Seuthēs, king of the Odrysæ, i. 5, 6.
- Seven Brothers, monuments of the, iii. 278.
- *Severino, S. See* Septempeda.
- Sextiæ, hot-baths near Marseilles, i. 267, 270.
- Sextius, i. 270.
- Sextus Pompeius, i. 213, 362, 386, 388, 404, 408, 411.
- *Sezza. See* Setia.
- *Shirban. See* Artemita.
- Sibæ, people of India, iii. 77, 94.
- Sibini, people of Germany, i. 445.
- Sibyl, Erythræan, ii. 321; iii. 18, 258, 259.
- Sicambri, i. 289, 444-446, 451.
- Sicani, people of Sicily, i. 407.
- Sicenus (*Sikino*), ii. 207.
- Sicilians, i. 9, 336, 385, 407; ii. 118.
- Sicily, i. 33-78, 84, 89, 93, 128, 164, 184-186, 194, 213, 224, 334, 361, 362, 369,376-378, 383-386, 388, 389, 392, 400-404, 407-409, 411, 412, 414, 417-422, 425,430, 437, 438, 459; ii. 4, 35, 41, 71, 92, 116, 154, 158, 378, 404; iii. 32, 59.
- ——, Sea of, i. 85, 185-187, 315, 346, 379, 380, 400, 495; ii. 5, 16, 287, 288, 297.
- ——, Strait of. *See* Messina.
- Sicyōn, c. of Peloponnesus (*Basilico*), i. 410; ii. 5, 10, 53, 58, 59, 65, 66, 71, 77, 107,108, 116, 124.
- Sicyōnia, ii. 5, 62, 66, 103.
- Sicyōnii, Sicyonians, ii. 64, 66.
- Sidē, c. of Pontus, ii. 295.
- ——, c. of Pamphylia (*Eski Adalia*), ii. 323; iii. 45, 50, 68.
- Sidēnē, distr. of Pontus (*Sidin* or *Valisa*), i. 82, 190; ii. 294, 295, 296, 305.
- ——, c. and distr. of Mysia, ii. 347, 368.
- Sidētani, in Spain, i. 245.
- Sidicīni, people of Italy, i. 352, 436.
- Sidicīnum. *See* Teanum.
- *Sidin. See* Sidene.
- Sidōn, c. of Phœnicia (*Seide*), i. 15, 59, 64, 90, 201; iii. 167, 169-174.
- Sidŏnes, people of the Bastarnæ, i. 470.

- Sidōnia (Pēdōnia?), isl. on the coast of Egypt, iii. 235.
- Sidonian women, i. 65.
- Sidōnii, Sidonians, i. 2, 41, 60, 65, 66, 68, 458; iii. 173, 174, 215, 216.
- Siga, c. of the Masæsylii (*Tafna*), iii. 282.
- Sigeia, prom. in the Troad, ii. 358, 372.
- Sigēlus, monument of Narcissus, ii. 96.
- Sigertis, king of India, ii. 253.
- Sigeum, t. of the Troad (*Ienischer*), i. 517, 518; ii. 358-363, 366, 368.
- [Pg 407]Sigia, ii. 373.
- Sigimērus, Segimerus, prince of the Cherusci, i. 446.
- Siginni, people inhabiting the Caspian, ii. 258.
- *Sigistan. See* Dranga.
- Signia, t. of Latium (*Segni*), i. 352.
- Signium (wine), i. 352.
- Sigriana, distr. of Media, ii. 265.
- Sigrium, prom. of Lesbos (*Sigri*), i. 518 ; ii. 390-393.
- *Sihon. See* Iaxartes.
- *Sikino. See* Sicenus.
- Sila, forest of the Bruttii, i. 391.
- ——, r. of India, iii. 98.
- ——, t., i. 435.
- Silacēni, people of Assyria, iii. 154.
- Silanus, i. 258.
- Silaris, r. of Campania (*Silaro*), i. 374, 375, 380.
- Silēni, servants of Bacchus, i. 286, 288, 290, 291, 297.
- Silenus, ii. 186, 318.
- Silli, people of Ethiopia, iii. 16.
- Silphium (*Lucerne*), ii. 265.
- Silta, in Thrace, i. 518.
- Silvium, t. of the Peucetii, i. 432.
- *Simau-Gol. See* Ancyra.
- *Simau-Su. See* Macestus.
- Simi, iii. 197.
- Simmias, Rhodian, ii. 42; iii. 34.
- Simodia, iii. 23.
- Simoeis, r. of the Troad, ii. 358. 361. 362. 368.
- ——, r. of Sicily, ii. 378.
- ——, plain of, in the Troad, ii. 361.
- Simōnides, lyric poet, ii. 146, 210, 394; iii. 108.
- ——, Amorginus, ii. 212; iii. 130.
- Simuntis, cognomen of Troy, ii. 74.

- Simus, physician, iii. 36.
- ——, lyric poet, iii. 23.
- Simyra, c. of Syria (*Sumrah*), iii. 167.
- Sinda, c. of Pisidia (*Dekoi*), ii. 324, 409.
- Sindi, Mæotic race, ii. 223.
- Sindic harbour, ii. 225.
- Sindica, distr. by the Cimmerian Bosporus, i. 478; ii. 219, 224, 305.
- ——, Sea, ii. 219.
- Sindomana, c. of India, iii. 95.
- Singitic Gulf, Bay of Macedonia (*G. of Monte Santo*), i. 511, 512.
- Singus, c. of Macedonia, i. 511.
- *Sinigaglia. See* Sena.
- Sinna, citadel of, iii. 170.
- *Sinno. See* Siris.
- Sinōpe, c. of Pontus, colony of the Milesians, i. 72, 106, 113, 114, 202, 216, 491; ii.198, 225, 227, 284, 291-294, 302, 310; iii. 44, 61-63.
- Sinopenses, ii. 291.
- Sinōpis, Sinopītis, Sinopic district, ii. 313.
- Sinŏria, fortress of Armenia, ii. 305.
- Sinōtium, t. of the Dalmatians, i. 484.
- Sinti, Sinties, or Saii, people of Thrace, i. 514, 515; ii. 169, 298.
- Sinuessa, Sinoessa, t. of Latium (*Monte Dragone*), i. 325, 347, 351, 360, 361, 431.
- *Siphanto. See* Siphnus.
- Siphnian bone, ii. 207.
- Siphnus, isl. (*Siphanto*), ii. 207, 208.
- *Sipuli. See* Sipylene.
- Sipūs, c. of Apulia (*Siponto*), i. 433, 434.
- Sipylēnē, cognomen of Rhea, ii. 184.
- ——, (*Sipuli Dagh*), ii. 184.
- Sipylus, mtn of Lydia, i. 91; ii. 326, 335, 337; iii. 66.
- ——, c. of Lydia, i. 91; ii. 326.
- Siraces, Siraci, inhabitants of the Caucasus, ii. 219, 238, 239.
- Siracēnē, ii. 236.
- Sirbis, r. of Lycia (*Kodscha*), iii. 47.
- Sirbōnis, Sirbōnitis, lake in Egypt (*Sebaket-Bardoil*), i. 79; iii. 176, 177, 182, 253.
- Sirens, i. 34, 35, 375, 387.
- [Pg 408]Sirenusæ, Sirenussæ, prom. (*Punta della Campanella*), i. 34, 35, 39, 368, 374, 375.
- Siris, c., i. 397-399.
- ——, r. Sinno, i. 397.
- Siritis, i. 380.

- Sirmium, c. of Pannonia, i. 483.
- Sisapō, c. of Spain, i. 214.
- Siscia, t. of Pannonia, i. 483.
- Sisimythres, stronghold of, in Bactriana, ii. 254.
- Sisinus, treasure-hold of, ii. 281.
- Sisis, ii. 304.
- Sisypheium, in the Acrocorinthus, ii. 62.
- Sisyrba, an Amazon, iii. 3.
- Sisyrbītæ, iii. 3.
- Sitacēnē, distr. of Babylon, of Apollōniātis (*Descura*), ii. 264; iii. 135, 146, 152.
- Sitacēni, ii. 223.
- Sithōnes, people of Macedonia, i. 506.
- *Sitia. See* Dicte.
- *Sizeboli. See* Apollonia.
- *Skilli. See* Scyllæum.
- *Skio, isl. See* Chios.
- Smintheas. *See* Apollo.
- Sminthia, ii. 374.
- Sminthium, temple of Apollo, near Hamaxitus, ii. 190, 374.
- Smyrna, c. of Ionia, ii. 237, 298, 303, 336 ; iii. 1, 4, 8, 20, 43.
- ——, part of Ephesus, iii. 3.
- ——, Bay of, iii. 20, 21.
- ——, an Amazon, iii. 3.
- Smyrnæans, iii. 3, 20.
- Soandus, t. of Cappadocia, iii. 44.
- Soanes, people bordering on the Caucasus, ii. 225, 229.
- Soatra, t. of Lycaonia, ii. 321.
- Sōcrates, i. 452; ii. 95; iii. 114.
- Sodom, c. of Judæa, iii. 183.
- Sogdiana, i. 113; ii. 245, 248, 253-255; iii. 125, 126.
- Sogdiani, Sogdii, Sogdians, i. 112, 195; ii. 245, 248, 253.
- *Solfaterra, la. See* Forum Vulcani.
- Soli, c. of Cilicia (*Mesetlii*), ii. 74, 347, 382; iii. 45, 46, 50, 53-55, 59-61.
- Soli, c. of Cyprus, iii. 70.
- Solii, iii. 70.
- Solmissus, mtn near Ephesus, iii. 11.
- Solōcē, iii. 154.
- Sōlōn, i. 154; ii. 83.
- Solyme, mtns of Lycia, i. 53; iii. 48.
- Solymi, people of Lycia, i. 8, 32, 54; ii. 328, 409, 410; iii. 48, 49, 63, 65.
- Solymus, mtn of Pisidia, iii. 409.

- Somnus, ii. 341.
- Sōpeithēs, king of the Indians, iii. 92, 93.
- Sōphēnē, distr. of Armenia (*Dzophok*), ii. 260, 261, 268, 278, 304; iii. 44.
- ——, prince of, ii. 278.
- Sōphēni, ii. 272, 273, 278.
- Sophoclēs, i. 410; ii. 32, 42, 51, 81, 90, 135, 170, 186, 191, 377; iii. 9, 15, 59, 60, 76.
- Sōra, t. of Latium, i. 353.
- Soracte, mtn of Latium (*Monte di S. Silvestro*), i. 336.
- *Sorgue. See* Sulgas.
- *Sorrento. See* Surrentum.
- Sosicrates, ii. 193.
- Sōsipŏlis, Jupiter, worshipped at Magnesia, iii. 23.
- Sōssinati, people of Sardinia, i. 334.
- Sōstratus, tomb of, ii. 74.
- ——, grammarian, iii. 26.
- ——, of Cnidus, iii. 227.
- Sōtades, ii. 19.
- ——, poet, iii. 23.
- Sōteira, harbour in the Arabian Gulf, iii. 194.
- Sōtēres, or Saviours, cognomen of the Dioscuri, i. 345.
- *Sour. See* Tyre.
- Southern Sea, i. 183.
- Spadines, ii. 239.
- Spain,
i. 3, 9, 13, 33, 43, 54, 72, 100, 101, 128, 141, 151, 152, 157, 160, 161, 163,165, 175, 180, 184, 192, 205, 206, 208, 210, 213, 215-219, 222-226, 228, 229, 233,234, 236, 240-242, 244, 245, 249, 251, 252, 255, 263, 264, 267, 269, 279, 296,302, 310, 325, 439, 442; iii. 32, 108, 117, 283, 286, *et passim.*
- [Pg 409]Spain, Citerior, i. 249, 250.
- ——, Ulterior, i. 240, 245; iii. 297.
- Sparta, i. 274; ii. 15, 25, 36, 40-44, 47, 49, 59, 153, 203. *See* Lacedæmon.
- Spartans, i. 385; ii. 42, 202, 203.
- Spartarium, plain of, i. 241.
- Spauta (Capauta ?), marsh in Media (*Urmiah*), ii. 262.
- Spedon, ii. 113.
- Spercheius, r. of Thessaly (*Agriomela* or *Ellada*), i. 95; ii. 55, 129, 130, 136, 137, 148.
- Spermophagi, people of Ethiopia, iii. 195.
- Sphagia, Sphactēria, island, ii. 22, 36.
- Sphēttus, t. of Attica, ii. 88.
- Spina, c. of Cisalpine Gaul (*Spinazino*), i. 318.
- Spinītæ, i. 318; ii. 119.

- Spitamenes, ii. 248, 255.
- Spŏlētium, c. of Umbria (*Spoleto*), i. 338.
- Sporades, islands, i. 187; ii. 192, 207, 211-213; iii. 33.
- Stadia, ancient name of Rhodes, iii. 31.
- Stagirus, Stagira, t. of Macedonia, i. 512, 513.
- *Stalimene, isl. See* Lemnos.
- *Standia. See* Dia.
- *Stanko, isl. See* Cos.
- Staphylus, ii. 195.
- *Stapodia. See* Melantian rocks.
- Stasanōr, iii. 70.
- Statanian wine, i. 347, 361.
- Statōnia, t. of Etruria, i. 335.
- Steganopodes, i. 68, 458.
- Steiria, vill. of Attica, ii. 89.
- Stēlæ. *See* Pillars.
- Stenyclarus, c. of Messenia, ii. 38.
- Sternophthalmi, i. 68, 458.
- Stĕrŏpa, ii. 198.
- Stēsichorus, i. 67; ii. 21, 32.
- Stēsimbrŏtus, ii. 189.
- Sthĕnĕlus, king of the Mycenæ, ii. 59.
- Sthĕnis, ii. 293.
- *Stillida. See* Phalara.
- Stiphane (*Ladik-Gol*), ii. 311.
- Stoa Pœcile, ii. 87.
- Stŏbi, t. of Macedonia, i. 504; ii. 77.
- Stœchades, islands, i. 276.
- Stoics, i. 24, 156.
- Stŏmalimnē, a salt lake, i. 275; ii. 358, 361.
- ——, village, iii. 36.
- Stŏni (*Sténéco*), i. 304.
- Stony Plain, the, i. 273.
- Stŏras, r. of Latium, i. 346.
- Strabo, geographer, i. 505, 507, 516, 517;
- ——, his country, ii. 195, 197, 311;
- ——, his ancestry, ii. 307;
- ——, his masters, iii. 26, 27, 53, 173;
- ——, his friends, i. 178, 209, 262;
- ——, his age, 439;
- ——, his travels, i. 91, 178, 332; ii. 61, 208, 255; iii. 102, 247, 262, 265;

- ——, t. of Spain, i. 238.
- Sudinus, iii. 146.
- Suessa, c. of the Volsci, i. 344, 352.
- Suessiones, people of Gaul, i. 289, 293.
- Suessūla, t. of Campania (*Castel di Sessola*), i. 370.
- Suevi, i. 289, 308, 444, 445, 448, 452.
- *Suez*, Isthmus of, i. 62, 458. *See* Heroopolis.
- *Suffange-el-Bahri. See* Myos-hormos.
- Sūgambri, people of Germany. *See* Sicambri.
- Suidas, i. 503.
- Sulchi, t. of Sardinia, i. 333.
- *Suleimanli. See* Blaudus.
- Sūlgas, r. of Gaul (*Sorgue*), i. 277, 285.
- Sulmō, c. of the Peligni (*Sulmona*), i. 359.
- *Sultan-Dagh. See* Paroreia.
- *Sultan-Hissar. See* Nisa.
- *Sumrah. See* Simyra.
- Sun, Colossus of the, iii. 29.
- Sūnium, prom. of Attica (*Cape Colonna*), i. 140, 164, 188, 496, 506; ii. 78-80, 89-96, 150, 151, 154, 193, 208; iii. 7.
- Sūnium, demus of Attica, ii. 89.
- *Sur. See* Tyre.
- Sūrena, iii. 31.
- *Surk. See* Selge.
- Surrentum, c. of Campania (*Sorrento*), i. 34, 361, 368.
- Sūsa, c. of Susiana (*Schuss*), ii. 75, 122, 123, 132; iii. 130-134, 152.
- Sūsiana, distr. (*Khosistan*), i. 201; iii. 83, 130-135, 142, 146, 151-154.
- Sūsians, i. 196; ii. 264, 266.
- Sūsis, Sūsias, same as Susiana.
- Suspiritis, ii. 235.
- Sūtrium, c. of Etruria (*Sutri*), i. 335.
- *Suveidijeh. See* Seleucia, Pierian.
- *Swiss. See* Helvetii.
- Sybaris, c. of Lucania, i. 394-396; ii. 73.
- ——, c. of the Bruttii, a colony of the Rhodians, i. 398, 399; iii. 33.
- ——, r. of Lucania, i. 394-397; ii. 73.
- ——, ftn of Achaia, ii. 73.
- Sybaritæ, Sybarites, i. 373, 376, 399; ii. 119.
- Sybota, islands (*Syvota*), i. 187, 497.
- Sycaminopolis, c. of Judæa, iii. 175.
- Sydracæ (al. Oxydraceæ), people of India, iii. 76, 94, 95.

- Tænarum, prom. of Laconia (*Cape Matapan*), i. 187, 403; ii. 5, 40, 41, 46, 55, 393; iii. 292.
- ———, t. of Laconia, ii. 36, 37.
- *Tafna. See* Siga.
- Tagus, r. of Spain, i. 161, 208, 209, 214, 227-231, 243.
- *Takli. See* Acra.
- Talabrŏcē, t. of Hyrcania, ii. 242.
- Talæmenes, ii. 403.
- *Talanta. See* Atalanta.
- Talares, people of Epirus, ii. 137.
- *Taman. See* Corocondame.
- Tamarus, prom. of India, ii. 257.
- Tamassus, t. of Cyprus (*Borgo di Tamasso*), i. 381; iii. 71.
- Tamna, t. of Arabia Felix, iii. 190.
- Tamynæ, t. of Eubœa, ii. 155.
- Tamyracas, or Corcinitic Gulf, in the Tauric Chersonese, i. 471, 473.
- ———, promontory, i. 473.
- Tamyras, r. of Phœnicia (*Nahr-Damur*), iii. 171.
- Tanagra, c. of Bœotia, ii. 66, 95-97, 99, 104, 105.
- Tanagræa, Tanagricē, ii. 92, 95, 96, 143.
- Tanaïs, c., situated on the river of that name, ii. 239, 224.
- ———, river (*Don*), i. 102, 157, 162, 163, 190, 191, 194, 442, 443, 457, 470, 477, 480; ii. 1, 215, 216, 219-221, 224, 239, 240, 243, 244, 303; iii. 296.
- Tanis, c. of Egypt on the Delta, iii. 240.
- ———, c. of the Thebaid, iii. 258.
- Tanitic nome, iii. 240.
- ———, mouth of the Nile, iii. 239, 240.
- Tantalus, i. 91; ii. 326, 337; iii. 66.
- Taŏcē, c. of Persia (*Taug*), iii. 131.
- *Taormina. See* Tauromenium.
- Tapē, c. of Hyrcania, ii. 242.
- Taphiassus, mtn of Ætolia (*Kaki-Scala*), ii. 127, 160, 171, 172.
- Taphii, Taphians, ii. 166, 170, 173.
- [Pg 412]Taphītis, prom. on the Carthaginian coast (*Cape Aclibia*), iii. 288.
- Taphos, Taphiūs, island near Acarnania, ii. 166, 167, 170, 173.
- Taphrii, in the Tauric Chersonesus, i. 473.
- Tapŏseiris, c. of Egypt, iii. 236.
- ———, the Less, c. of Egypt, iii. 236, 238.
- Taprŏbanē, isl. (*Ceylon*), i. 99, 111, 114, 180, 196, 200; iii. 81.
- Tapyri, people of Asia, ii. 248, 250, 258, 263.
- Taracōn, c. of Spain. *See* Tarraco.

- Taranto, Gulf of. *See* Tarentum.
- Taras. *See* Tarentum.
- Tarbassus, c. of Pisidia, ii. 324.
- Tarbelli, people of Gaul, i. 283.
- Tarcon (Tarquin), governor of Tarquinia, i. 326.
- Tarcondimŏtus, king of Cilicia, iii. 60.
- Tarentini, Tarentines, i. 372, 389, 397-399, 427, 430, 438.
- Tarentum, i. 347, 377, 379, 393, 399, 400, 423, 425, 427-434, 497.
- ——, Gulf of, i. 313, 315, 377, 378, 393, 423, 429.
- Taricheæ, c. of Judæa, iii. 183.
- Taricheiæ, islands near Carthage, iii. 288.
- Tarnē, t. of Bœotia, ii. 110.
- ——, t. of Lydia, ii. 110.
- Taronitis, ii. 268, 269.
- Tarpētĕs, ii. 223.
- Tarphē, c. of the Locrians, ii. 110, 127.
- Tarquin. *See* Tarquinius.
- Tarquinia, c. of Etruria, i. 326.
- Tarquinii, the, i. 327.
- Tarquinius Priscus, Lucius, i. 327, 344.
- ——, Superbus, i. 327, 344, 438.
- Tarracina, t. of Latium, i. 344, 346, 347.
- Tarraco (*Tarragona*), i. 239, 241, 242, 251.
- Tarsius, r. of Mysia (*Karadere*), ii. 347.
- Tarsus, c. of Cilicia (*Tarsous*), i. 190; iii. 45, 50, 52, 55-59, 162.
- Tartarus, i. 223, 224.
- *Tartary. See* Scythia.
- Tartēssians, i. 51.
- Tartēssis, i. 223.
- Tartēssus, c. of Spain, i. 224, 226.
- ——, r. of Spain (*Guadalquiver*), i. 222, 223.
- Tarūsco, t. of Gaul, i. 267, 268, 279.
- *Tasch Kopri. See* Pompeiopolis.
- *Tasch Owa. See* Phanarœa.
- Tasius, leader of the Roxolani, i. 471.
- Tatta, marsh in Phrygia (*Tuz-Tscholli*), ii. 321.
- Taucheira (*Tochira*), c. of the Cyrenaic, iii. 291, 292.
- *Taug. See* Taoce.
- Taulantii, people of Epirus, i. 500.
- Tauri, Scythian race, i. 476, 478.
- ——, Troglodytic mountains, iii. 194.

- Tauriana, distr. of the Bruttii, i. 379.
- Tauric Chersonesus. *See* Chersonesus.
- ——, coasts, i. 475.
- Taurīni, people of Liguria, i. 303, 311.
- Taurisci, Tauristæ, a people of Gaul, i. 307, 309, 310, 317, 450, 454, 466, 482.
- Taurisci Norici, i. 310.
- Tauroentium, t. of Gaul (*Taurenti*), i. 269, 275.
- Tauromĕnia (*Taormina*), i. 404.
- Tauromĕnītæ, i. 412.
- Tauromĕnium, c. of Sicily (*Taormina*), i. 402, 403, 405; iii. 12.
- Tauropŏlium, temple of Diana, iii. 10, 186.
- Taurus, fortress of Judæa, iii. 181.
- ——, mtn of Asia,

i. 32, 82, 105, 106, 113, 120, 125, 126, 128, 131, 136, 139, 179,184, 194, 195, 439; ii. 215, 216, 218, 226, 244, 250, 255, 256, 259-262, 267, 277-279, 281, 284, 290, 314, 321-325, 329, 333, 347, 355, 399, 400, 407, 409, 410; iii.27, 40, 54-57, 60, 61, 73, 78, 120, 142, 143, 156, 297, *et passim.*

- [Pg 413]Taurus, Pisidian, i. 195; ii. 319.
- ——, Cilician, ii. 276, 278, 319.
- ——, Anti-, ii. 259, 260, 278, 279.
- Tavium, ii. 320.
- Taxila, c. of India, iii. 82, 90, 111, 112.
- Taxiles, king of the Taxili, iii. 90, 92, 114.
- Taÿgĕtum, mtn of Laconia (*Penta Dactylon*), i. 311; ii. 37, 40, 46, 164, 194.
- Teanum, Sidicinum, c. of Campania (*Teano*), i. 352, 370.
- ——, Apulian, i. 359, 436.
- Tearco, Ethiopian, i. 96; iii. 74.
- Teatea, c. of the Marrucini (*Chieti*), i. 359.
- *Tech. See* Ilibirris.
- Tectosages, people of Gallia Narbonensis, i. 279, 280, 282.
- ——, people of Galatia, ii. 319, 320.
- Tĕgĕa, c. of Arcadia, ii. 8, 54, 58, 64, 75, 76.
- ——, territory of, ii. 76.
- Tegeatæ, ii. 60.
- Teichiūs, fortress near Thermopylæ, ii. 129.
- Teirĕsias, ii. 107, 111; iii. 15, 180.
- *Tekieh. See* Pamphylia.
- Telamōn, ii. 83; iii. 69.
- Telchīnes, inhabitants of Rhodes, ii. 180, 188; iii. 31, 32.
- Telchīnis, name of the island of Rhodes, iii. 31, 32.
- Tēlĕboæ, i. 494; ii. 166, 170-173.
- Tēlĕbŏas, i. 494.

374

- [Pg 414]Termĕssians, ii. 409.
- Termĕssus, c. of Pisidia, ii. 409, 410; iii. 48.
- Termilæ, inhabit Lycia, ii. 328; iii. 49, 63.
- *Terni. See* Interamna.
- Terpander, ii. 393.
- *Teseni. See* Themisonium.
- *Tet. See* Ruscino.
- Tetrapolis, Athenian, ii. 56, 57, 67, 88.
- ——, the Dorian, ii. 114, 115, 125, 128, 195.
- ——, of Marathon, ii. 153.
- Tetra-pyrgia, in the Cyrenaic, iii. 294.
- Teucer, son of Telamon, i. 236; iii. 55, 56, 69.
- ——, of Attica, ii. 374.
- Teucrians, inhabitants of the Troad, i. 96; ii. 373, 374.
- Teumĕssus, t. of Bœotia, ii. 104, 108.
- Teutamis, ii. 395.
- Teuthĕa, t. of Achæa, ii. 14.
- Teuthĕas, r. of Achæa, ii. 14.
- Teuthrania, distr. of Mysia, ii. 299, 326, 389.
- Teuthras (? Traeis), river of the Bruttii, i. 398.
- ——, king of Teuthrania, ii. 326-328, 346, 389.
- Teutons, i. 292.
- Teutria, isl. of, i. 434.
- *Teverone. See* Anio.
- Thala, t. of Numidia, iii. 284.
- Thalamæ, t. of Laconia, ii. 36.
- Thales, Milesian, i. 12; iii. 5.
- ——, of Crete, poet, ii. 202, 204.
- Thalĕstria, queen of the Amazons, ii. 237.
- Thamyris, the Thracian, i. 513; ii. 10, 23, 24, 187.
- Thapsacus, c. of Babylonia (*Elder*), i. 120, 122-127, 130, 131, 134-139; iii. 148, 150,156, 157, 187.
- Thapsus, t. of Africa (*Demass*), iii. 284, 288.
- Thasian pottery, i. 487.
- Thasii, i. 515.
- Thasos, island (*Thaso*), i. 43, 187, 515, 516; ii. 50, 189, 210.
- Thaumaci, c. of Phthiotis, ii. 77, 136.
- Thaumacia, c. of Magnesia, ii. 140.
- *Theaki. See* Ithaca.
- Thebaic keep, in Egypt, iii. 258.
- Thebaïs, part of Egypt, i. 67; iii. 84, 211, 221, 225, 243, 258, 260.

- Thōrax, mtn of Lydia (*Gamusch-dagh*), iii. 22.
- Thoreis, village of Attica, ii. 89.
- Thoricus, t. of Attica, ii. 88-90, 208.
- Thornax, ii. 41.
- Thrace, Thracia (*Roumelia*),

i. 42, 43, 44, 93, 110, 164, 187, 194, 311, 439, 443, 453,466, 468, 481, 505-507, 510, 512, 515, 516; ii. 140, 147, 187, 188, 197, 327, 339,340, 352, 358; iii. 66.

- Thracian Chersonesus (*Gallipoli*), i. 164, 188, 194. *See* Chersonesus.
- [Pg 416]Thracian Bosporus. *See* Bosporus.
- ——, army, ii. 67.
- ——, tribes, i. 247, 482, 483, 496.
- ——, mtns, i. 41, 488, 492, 504.
- ——, race, i. 9.
- ——, Sea, i. 42.
- ——, coast, i. 9.
- Thracians, i. 164, 453-

455, 460, 461, 463, 466, 468, 478, 481, 485, 488, 493, 496,506, 519; ii. 93, 105, 151, 187, 286, 287, 301, 316, 318.

- ——, Cabrenii, ii. 351.
- ——, Xanthii, ii. 351.
- Thrasō, sculptor, iii. 13.
- Thrasyalces, of Thasos, i. 44; iii. 225.
- Thrasybulus, Athenian, ii. 87.
- Thrax, castle of Judæa, iii. 181.
- Thriasian plain, in Attica, ii. 81, 84.
- Thrinacia, name of Sicily, i. 400.
- Throni, prom. and t. of Cyprus (*Cape Grego*), iii. 69.
- Thronia, i. 67.
- Thronium (*Paleocastro*), i. 95; ii. 126, 127.
- Thryum, Thryoessa, t. of Triphylia, ii. 23, 24, 27, 28.
- Thucydides, i. 499; ii. 2, 36, 50, 56, 58, 122, 174, 366; iii. 41.
- Thule (*Iceland*), i. 99, 100, 157, 173, 299.
- Thumæum, same as Ithome or Thome, ii. 141.
- Thūmelicus, son of Arminius, i. 446.
- Thūnatæ, nation of Illyria, i. 485.
- Thuria, t. of Messenia, ii. 36-38.
- ——, ftn, i. 396.
- ——, c. of the Thurii, i. 427.
- Thurian wine, i. 397.
- Thuriatic Gulf, ii. 37.
- Thurii or Thurians, t. of Lucania, i. 379, 380, 390, 396-398, 427; iii. 35.
- Thurius, same as Herodotus, iii. 35.

378

- Timia, i. 349.
- Timon, iii. 230.
- Timōnītis, distr. of Paphlagonia, ii. 313.
- Timōnium, iii. 230.
- Timosthenes, i. 44, 139, 141, 142, 210; ii. 120; iii. 279.
- Timŏthĕus, Patriŏn, ii. 293.
- *Tine. See* Tenos.
- *Tineh. See* Pelusium.
- Tingis, or Tiga, c. of Mauritania (Tiga), i. 210; iii. 278.
- *Tino. See* Telos.
- Tirizis, prom. and citadel of Hæmus, i. 490.
- Tiryns, c. of Argolis, ii. 49, 54, 58; iii. 31.
- Tisamenus, son of Orestes, ii. 68, 77.
- Tisiæūs, c. of Numidia, iii. 284.
- Titanes, same as Pelagones, i. 514; ii. 188.
- Titanus, t. of Thessaly, ii. 143.
- Titarēsius, r. of Thessaly, i. 507; ii. 145, 146.
- Titarius, mtn of Thessaly, i. 507; ii. 146.
- Tithōnus, father of Memnon, ii. 347; iii. 130.
- Titius, iii. 160.
- Titus Quintius, ii. 146.
- ——, Flaminius, i. 421.
- ——, Tatius, i. 338, 342, 343, 348.
- Tityri, servants of Bacchus, ii. 180, 183, 186.
- Tityrus, mtn of Crete, ii. 200.
- Tityus, ii. 121-123.
- *Tivoli. See* Tibura.
- Tlepolemus, son of Hercules, ii. 9; iii. 31, 32.
- Tlōs, c. of Lycia (*Duvar*), iii. 45.
- Tmarus. *See* Tomarus.
- Tmōlus, mtn of Lydia (*Bouz Dagh*), ii. 102, 185, 303, 353, 381, 396, 402, 403, 407; iii. 8, 26.
- Tochari, Scythians beyond the Iaxartes, ii. 245.
- *Todi. See* Tuder.
- Togati, i. 227, 250.
- Tolistobōgii, people of Galatia, i. 279; ii. 294, 319, 320.
- Tolōssa, Toulouse, i. 280, 281.
- Tomaruri, same as Tomuri, i. 503.
- Tomarus, Tmarus, mtn near Dodona, i. 501-503; ii. 137.
- Tŏmis, c. of Mœsia, i. 489, 490.
- Tomisa, Tomisæ, fortress of Cappadocia, ii. 278; iii. 44.

- Tomūri, i. <u>502</u>, <u>503</u>.
- Tŏpeira, c. of Thrace, i. <u>515</u>.
- Toreatæ, ii. <u>223</u>.
- *Tornese. See* Chelonatas.
- Torŏnæan, Torŏnic, Gulf, in Macedonia (*G. of Cassandra*), i. <u>511</u>, <u>512</u>.
- *Torre di Patria. See* Liturnum.
- ——, *Macarese. See* Fregena.
- *Tortona. See* Derthon.
- *Tortosa. See* Dertossa.
- *Toulouse. See* Tolŏssa.
- Tŏygeni, i. <u>274</u>, <u>450</u>.
- Tracheia, iii. <u>3</u>.
- Trachin, t. of Phocis, ii. <u>123</u>.
- ——, t. of Thessaly, i. <u>94</u>; ii. <u>123</u>, <u>129</u>, <u>132</u>, <u>135</u>, <u>136</u>.
- ——, Heracleian, ii. <u>103</u>, <u>130</u>.
- Trachina, same as Tarracina.
- Trachinia, distr. of Thessaly, ii. <u>66</u>, <u>135</u>, <u>156</u>.
- Trachinii, ii. <u>123</u>.
- Trachiŏtæ, i. <u>196</u>; iii. <u>50</u>.
- Trachiŏtis, Tracheia, *see* Cilicia, iii. <u>50</u>, <u>56</u>.
- Trachŏnes, mtns near Damascus, iii. <u>169</u>, <u>171</u>.
- Traclinia, ii. <u>4</u>.
- Tragææ, islands not far from Miletus, iii. <u>6</u>.
- Tragasæan salt-pan, ii. <u>374</u>.
- [Pg 418]Tragium, c. of Laconia, ii. <u>37</u>.
- Tragurium, isl. (*Traw*), i. <u>186</u>, 484.
- Tralles, c. of Lydia, ii. <u>145</u>, <u>305</u>; iii. <u>24</u>, <u>25</u>, <u>43</u>.
- Tralliani, ii. <u>336</u>; iii. <u>25</u>.
- Transpadana, i. <u>316</u>, <u>321</u>.
- Trapĕzŏn, hill of Syria, iii. <u>164</u>.
- Trapĕzūs, c. of Pontus (*Trebizond*), i. <u>476</u>, <u>491</u>, <u>517</u>; ii. <u>226</u>, <u>294</u>, <u>296</u>, <u>304</u>, <u>305</u>.
- ——, hill of the Tauric Chersonese, i. <u>476</u>.
- Trapontium, t. of Latium, i. <u>352</u>.
- Trarium, c. of Mysia, ii. <u>376</u>.
- Trasumennus, lake, i. <u>336</u>.
- Trebias, r. of Cisalpine Gaul, i. <u>323</u>.
- *Trebizond. See* Trapezus.
- Trebŏnius, iii. <u>20</u>.
- Trĕbūla, t. of the Sabines (*Monte Leone della Sabina*), i. <u>338</u>.
- *Tremiti, islands of. See* Diomede, isl.
- Trephea, lake in Bœotia, ii. <u>102</u>.

- Trēres, i. 93, 96; ii. 246, 301, 329, 346, 405; iii. 22.
- Trērus, r. of Latium (*Sacco*), i. 352.
- Trēta, c. of Cyprus (*Capo Bianco*), iii. 70.
- Trētum, prom. of Numidia (*Ebba-Ras*), iii. 281, 282, 284, 285.
- Treviri, people of Gaul, i. 289.
- Triballi, Thracian race, i. 463, 468, 485, 488.
- Tribocchi, people of Gaul, i. 288, 289.
- Tricca (*Tricola*), c. of Thessaly, i. 501; ii. 36, 56, 141, 142, 156; iii. 22.
- Triccæus, Æsculapius, ii. 36.
- Trichæces, cognomen of the Dorians, ii. 195.
- Trichōnium, c. of Ætolia, ii. 159.
- Triclari, in Thessaly, i. 508.
- Tricorii, people of Gaul, i. 276, 303.
- Tricorythus, Tricorynthus, t. of Attica, ii. 59, 67, 90.
- Tridentini, i. 304.
- Triērēs, t. of Syria, iii. 169.
- Trieteric dance, ii. 186.
- Trieterides, ii. 185.
- *Trikeri. See* Cicynethus.
- Trinacria, same as Sicily, i. 400.
- Trinĕmeis, vill. of Attica, ii. 91.
- Trinx (*al. Tinx*), t. of Mauritania, iii. 276.
- *Trionto*, i. 398.
- Triphylia, part of Elis, ii. 8, 11, 14-19, 21, 22, 33-35, 45, 53, 155.
- Triphyliac towns, ii. 17.
- Triphylian Sea, ii. 22, 28.
- Triphylii, ii. 8, 16, 22, 28, 31.
- Triphyllus, ii. 409.
- Tripodes, Tripodiscium, t. of Megaris, ii. 84.
- Tripolis, c. of Phœnicia, iii. 169.
- Tripolītis. *See* Pelagonia.
- Triptolemus, father of Gordyes, i. 40; iii. 57, 157, 162.
- ——, tragedy of Sophocles, i. 40, 41.
- Tritæa, c. of Achaia, ii. 14.
- Tritæenses, Tritæeis, ii. 14, 71.
- Tritōn, t. of Bœotia, ii. 101.
- Tritōnis, iii. 291.
- Trōad, i. 8, 91, 172, 187, 189, 195, 202, 453, 517; ii. 56, 189, 277, 317, 332, 338-390.
- ——, Pelasgic, i. 329.
- Trōades. *See* Trojans.

- Trōas, Alexandreia (*Eski Stamboul*, or *Old Constantinople*), ii. 339.
- Trocmi, people of Galatia, i. 279; ii. 312, 319, 320.
- Trœzen, son of Pelops, ii. 56.
- ——, (*Damala*), ii. 49, 55, 56, 58.
- Trœzenians, iii. 35.
- Trōgilius, prom. of Ionia (*Cape Santa Maria*), iii. 7.
- ——, isl. of Ionia, iii. 7.
- Trogītis, marsh, in Lycaonia, ii. 322.
- Troglodytæ, inhabiting the Arabian Gulf, i. 202, 203, 267, 489; iii. 203, 210, 215, 217,219, 266, 280.
- ——, in the Caucasus, ii. 238, 239; iii. 203.
- Troglodytic, i. 197; iii. 88, 191, 193, 210, 235.
- Trŏphōnius, brother of Agamedes, ii. 119; iii. 180.
- [Pg 419]Trophonius Zeus, oracle of, at Lebadea, ii. 111.
- Trojan war, i. 31, 61-63, 76, 224, 316, 377, 404; ii. 30, 72, 200, 201.
- ——, colony, i. 397.
- ——, forces, i. 518.
- ——, Minerva, i. 397.
- ——, territory, i. 17.
- Trojans, i. 64, 274, 394, 397, 453, 508, 516; ii. 18, 162, 163; iii. 41, 184, 299, *et passim.*
- ——, Aphneian, ii. 344, 346.
- Troy, Troja, Ilium, ancient city of the Troad, i. 25, 26, 33, 55, 64, 65, 76, 77, 91, 224,330, 394, 398, 494, 499, 508, 509, 511, 514, 519; ii. 16, 30, 74, 113, 132, 157, 174,186, 191, 317, 339, 350-363; iii. 34, *et passim.*
- ——, village of Egypt, iii. 252.
- Truentum, r. (*Tronto*), i. 357.
- ——, t. i. 358.
- Tryphōn, *see* Diodotus, iii. 51, 165.
- *Tsana, see* Psebo.
- *Tschandarlik, see* Pitane, ii. 339.
- *Tschariklar. See* Olbia.
- *Tschiraly. See* Olympus.
- *Tschol-Abad. See* Apollonias.
- *Tschorocsu. See* Glaucus.
- *Tschoterlek Irmak. See* Scylax.
- Tūbattii, people of Germany, i. 447.
- Tūder, t. of Umbria (*Todi*), i. 338.
- Tūisi, t. of Cantabria, i. 234.
- Tūkkis (*Martos*), i. 213.
- Tūllum, mtn, i. 308.
- Tunis, c. of the Carthaginians, iii. 287.

- *Turchal. See* Gaziura.
- Turdētani, people of Spain, i. 209, 221, 226, 227.
- Turdētania, i. 209, 210, 212, 214, 216-219, 224, 226, 235.
- Turdūli, same as Turdētani, i. 209, 223, 227, 230.
- Tūriva, distr. of Bactriana, ii. 253.
- *Tuscany. See* Tyrrhenia.
- Tusci, same as Tyrrheni.
- *Tuscolo. See* Tusculum.
- Tusculan mountain, i. 351.
- Tusculum, c. of Latium (*Tuscolo*), i. 351-353, 355.
- Tyana, c. of Cappadocia (*Kara-Hissar*), ii. 281, 284, 347.
- Tyanītis, prefecture of Cappadocia, ii. 278, 281.
- *Tyche. See* Fortune.
- Tychius, ii. 102, 403.
- Tychōn, ii. 348.
- Tymbrias, c. of Pisidia, ii. 324.
- Tymphæi, Tymphæans, i. 499, 501, 505.
- Tymphē, mtn of Epirus, i. 498.
- Tyndareian rocks, four islands on the coast of Marmora, iii. 235.
- Tyndaris, c. of Sicily (*S. Maria di Tindaro*), i. 401, 411.
- Tyndarus, ii. 173.
- Typanĕæ, c. of Triphylia, ii. 17.
- Typhoëus, ii. 403.
- Typhōn, i. 368, 369; ii. 336, 404, 406; iii. 163, 243.
- Typhōneia, iii. 260.
- Typhrēstus, mtn of Thessaly, ii. 136.
- Tyrambē, c. on the Cimmerian Bosporus, ii. 221.
- Tyranniōn, ii. 296, 380.
- Tyras, r. of Sarmatia (*Dniester*), i. 22, 162, 442, 468, 469, 478.
- Tyre, c. of Phœnicia (*Sur*), i. 91, 201; iii. 162, 169, 171-174.
- ——, isl. in the Persian Gulf (*Ormus*), iii. 187, 286.
- Tyregetæ, i. 177, 194, 443, 452, 470.
- Tyrians, i. 238, 255.
- Tyriæum, c. of Phrygia, iii. 43.
- Tyrō, daughter of Salmoneus, ii. 32.
- Tyrrheni, Tyrrhenians, i. 319, 322, 325-328, 331, 334, 335, 357, 360, 367, 385, 404,417, 438; ii. 197, 404.
- Tyrrhenia, Tyrrhenicē, i. 31, 35, 177, 301, 313, 323-330, 335-338, 349, 415, 502; ii.61, 387.
- Tyrrhenian cities, i. 331.
- ——, Gulf, i. 139.

385

- Valeria, i. 353.
- ——, Via, i. 351, 353, 402.
- Valerius Flaccus, consul, ii. 356.
- *Van. See* Arsene *and* Thopitis.
- Vapanes, t. of Corsica, i. 333.
- Var, r., i. 267, 275, 302, 313.
- Varagri, Alpine race, i. 303.
- *Varassova. See* Chalcis.
- Vardæi. *See* Ardiæi, i. 484.
- *Vardari, the. See* Axius.
- Varia, t. of Keltiberia, i. 243.
- ——, t. of Latium, i. 353.
- Varius Flaccus, ii. 356.
- Varus Quintilius, i. 446.
- Vascōns, people of Spain, i. 233, 242.
- Vates, i. 294.
- *Vathi. See* Aulis.
- *Vathy. See* Eretria.
- *Vedene. See* Vindalum.
- Veii, i. 335.
- *Velestina. See* Pheræ.
- Velitræ, t. of Latium (*Velletri*), i. 352.
- Vellæi, people of Gaul (*inhabitants of Vélai*), i. 284.
- Venafrum, t. of Campania (*Venafro*), i. 353, 361, 371.
- Venasii, ii. 281.
- Vendōn, t. of the Iapodes (*Windisch Grätz, or Brindjel*), i. 309, 483.
- Venĕti, people of Gaul, *see* Henĕti, i. 290, 291, 316.
- ——, in the Adriatic, i. 291.
- ——, of Paphlagonia, i. 316.
- *Venice, Gulf of. See* Adriatic and Illyrian Seas.
- Vennōnes, people of the Vindelici, i. 304, 307.
- *Venosa. See* Venusia.
- Ventidius, iii. 163, 164.
- *Vento Tiene. See* Pandataria.
- Venus. *See* Aphrodite.
- Venusia (*Venosa*), t. of the Samnites, i. 371, 379, 431.
- Vera, ii. 263.
- Verbanus (*Lago Maggiore*), i. 311.
- Vercelli, vill. of Cisalpine Gaul, i. 325.
- [Pg 421]Vercingetorix, i. 285.
- Verestis, r., i. 355.

- Veretum, i. 429, 430.
- Verona, i. 306, 317.
- Vertinæ, t. of Lucania (*Verzine*), i. 379.
- Vescini. *See* Vestini, i. 347.
- Vesta, i. 327, 340.
- Vestīni, i. 326, 338, 358, 359.
- Vesuvius, Mount, i. 39, 367.
- Vettōnes, people of Spain, i. 209, 228, 229, 243, 246.
- Vibo-Valentia, t. of the Bruttii, i. 383.
- Vicĕtia, t. of Cisalpine Gaul (*Vicenza*), i. 319.
- *Victimolo. See* Ictimuli.
- Vienne, capital of the Allobroges, i. 277, 278.
- Villa Publica, in the Campus Martius, i. 371.
- Viminal Gate, in Rome, i. 348.
- ——, Hill, i. 348.
- Vindalum, t. of Gaul (*Vedene*), i. 277.
- Vindelici, i. 287, 306, 307, 447, 448, 482.
- Viriathus, i. 238, 439.
- Visurgis (*Weser*), i. 445.
- Vitia, t. of Hyrcania, ii. 273.
- Vitii, people of Hyrcania, ii. 241, 248.
- Vivisci ([**Greek: Ioskôn]), i. 283.
- *Vlacho. See* Enipeus.
- Vocontii, people of Gaul, i. 268, 276, 279, 302, 303.
- Volaterræ, c. of Etruria (*Volterra*), i. 329.
- Volaterrani, i. 331.
- Volcæ Arecomisci, people of Gaul, i. 278, 279, 302.
- ——, Tectosages, i. 279.
- *Volo. See* Iolcos.
- Volsci, people of Italy, i. 339, 343, 344, 353.
- Volsinii, c. of the Tyrrhenians (*Bolsena*), i. 335, 336.
- *Volterra. See* Volaterræ.
- *Volturno. See* Vulturnus.
- *Vona. See* Genetes.
- *Vouga. See* Vacua.
- Vulcan, i. 65, 418; ii. 190; iii. 248.
- Vulturnus, c. and r. of Campania (*Volturno*), i. 353, 361, 370.
-
- Wain, the (constellation), i. 4, 5.
- *Weser. See* Visurgis.
- *Wesir Kopti. See* Gadilon.

- Xanthia, t. of the Ciconi, i. 515.
- Xanthii, in Lycia, iii. 47.
- ——, Thracians, ii. 351.
- ——, a tribe of the Dahæ, ii. 245, 251.
- Xanthus, c. of Lycia (*Eksenide*), i. 201; iii. 45, 47.
- ——, r. of Lycia, iii. 47.
- ——, r. of the Troad, ii. 351.
- ——, king of the Bœotians, ii. 82.
- ——, of Lydia, historian, i. 78, 80; ii. 326, 336, 406; iii. 66, 67.
- *Xativa. See* Sætabis.
- *Xelsa. See* Celsa.
- Xenarchus, iii. 53.
- Xenŏclēs, guardian of Alexander's treasure, i. 108.
- ——, orator, ii. 387; iii. 40.
- Xenocrates, Bithynian, ii. 299, 318, 382.
- Xenophanes, natural philosopher and poet, iii. 16.
- ——, tyrant, iii. 56.
- Xenophōn, ii. 73, 95.
- *Xerolimne. See Molycreia* and *Uria.* Xerxēne, district of Lesser Armenia, ii. 268.
- Xerxes, son of Darius, i. 17, 96, 516, 517; ii. 83, 84, 148, 254, 347, 352;
iii. 4, 6, 145.
- ——, canal of, i. 512, 513.
- ——, bridge of, i. 518.
- Ximēnē, distr. of Pontus, ii. 312.
- Xiphonia, prom. in Sicily, i. 403.
- Xoïs, isl. and c. of Egypt, iii. 240.
- *Xucar,* r. *See* Sucro.
- Xūthus, father of Ion, ii. 67.
- Xypěteon, ii. 374.

- *Yeni-kaleh. See* Myrmecium.
- *Yniesta. See* Egelastæ.
- [Pg 422]Ypsilo Nisi, iii. 16.

- *Zab, the Little. See* Caprus.
- *Zabache, Strait of. See* Bosporus, Cimmerian.
- Zacynthians, i. 239.
- Zacynthus, isl. (*Zante*), i. 187; ii. 5, 163, 167, 169; iii. 291.
- *Zafra. See* Zephyrium.
- *Zagaro Voreni. See* Helicon.

- Zeugma, at Thapsacus, ii. 263, 274; iii. 44, 157.
- ——, at Commagene, iii. 157.
- ——, at Samosata, ii. 274; iii. 44, 157.
- ——, Straits of, ii. 352.
- Zeus. *See* Jupiter.
- Zeuxis, physician, ii. 336.
- *Zia. See* Ceos.
- *Zigos. See* Amacynthus.
- Zincha, t. of Numidia, iii. 284.
- Zoïlus, orator, i. 410.
- Zōnas. *See* Diodorus.
- Zōstēr, prom. of Attica, ii. 89.
- Zūchis, lake and people of Libya, iii. 289.
- Zūmi, people of Germany, i. 444.
- Zygi, inhabitants of the Caucasus, i. 195; ii. 219, 224, 225; iii. 296.
- Zygopolis, city of Pontus, ii. 296.

LIST OF MODERN SPELLINGS OF THE TURKISH PLACE NAMES

As the Latin transliterations of Turkish place names were made from Ottoman Arabic-Persian alphabet they are quite different from the modern spelling of these names. The table below is an attempt to interprete those in modern spellings.

Transcriber.	
Afıum-Karahissar.	Afyon Karahisar
Ak-Sera	Aksaray?
Aiaghi-Dag	Ağrı dağı?
Ak-Gol	Ak göl
Ak-Schehr	Akşehir
Al-gol	Ala göl?
Altun-Suyi	Altun Su
Balkesi	Balıkesir?
Beiram-dere	Bayram dere
Beiram-koi,	Bayram köy
Beiramkoi	Bayram köy
Bakir-Tschai,	Bakır Çay
Bakyr-Tschai	Bakır Çay
Boli	Bolu?
Bouz-dagh	Bozdağ?
Bounar-bachi	Pınarbaşı
Bojuk Mender Tschai	Büyük Menderes Çayı
Delidsche	Delice Irmak

390

Irmak		
	Dikeli-koy	Dikili köy
	Doloman	Doloman Çayı
Ischai		
	Erekli	Ereğli
	Eski-Schehr,	Eskişehir
	Eski-Scheur	Eskişehir
	Eske-Adatia	Eski-Antalya?
	Giaur-Kalessi	Gavur kalesi
	Gedis-Tschai,	Gediz Çayı
	Godis-Tschia	Gediz Çayı
	Gok-su	Gök su
	Gok-Irmak	Gök Irmak
	Gumusch-dagh	Gümüş dağ
	Gunescth-	Güneş Dağ
Dagh		
	Jekil-Irmak,	Yeşil Irmak
	Jeschil Irmak,	Yeşil Irmak
	Ieschil Irmak	Yeşil Irmak
	Jenikoi	Yeni köy
	Ienischer	Yenişehir?
	Ilan-Adassi	Yılan adası
	Iéni-Kalé	Yeni Kale
	Isnik-gol	Iznik Gölü
	Jeralagoz-Dagh	Yaralıgöz dağı?
	Kara-Koi	Kara köy
	Kara-sui	Karasu
	Karadgeh-dagh	Karacadağ?
	Kara-Aghatsch	Karaağaç
	Karadje-Burun	Karaca Burun
	Kas-Owa	Kaz ova?
	Kasdagh	Kaz dağı
	Kas-dagh	Kaz dağı
	Karatepe-	Karatepe burnu
bournou		
	Kerasun	Giresun
	Kera-sun	Giresun
	Keschisch-	Keşiş dağı
Dagh		
	Kopru-Su	Köprü suyu
	Kodscha-	Koca Çay

391

Tschai

Koum-kale	Kum kale
Kostambul	Kastamonu
Tschai	Çayı?
Kutschuk-	Küçük
Meinder	Menderes
Kizil-Ermak,	Kızıl Irmak
Kizil-Irmak	Kızıl Irmak
Ladik-Gol	Ladik Gölü
Kiutahia	Kütahya
Menavyat-su	Manavgat suyu
Menavgat-su	Manavgat suyu
Mender-Tschai	Menderes Çayı
Mesarlyk-Tschai	Mezarlık Çayı
Pascha-Liman	Paşa Limanı
Pistatia-nut	Pistachio-nut
Roum-Kala	Rum Kale
Satal-dere	Çatal dere?
Sarikawak-Dagh	Sarıkavak dağı
Simau-Gol	Simav Gölü
Simau-Su	Simav suyu
Siwri-Hissar	Sivrihisar
Tuz-Tscholli	Tuz Gölü
Tschandarlik	Çandarlık
Tschandarlyk	Çandarlık
Tschileh	Şile?
Tasch-Kopri	Taş Köprü
Tchadir-Dagh	Çadır dag?
Olou-Degniz	Ulu Deniz?

JOHN CHILDS AND SON, PRINTERS.

www.ingramcontent.com/pod-product-compliance
Lightning Source LLC
Chambersburg PA
CBHW051209170526
45166CB00005B/1814